Fuel Additives

Fuel Additives

Chemistry and Technology

Robert D. Tack
London, UK

WILEY

This edition first published 2022

Registered Office
John Wiley & Sons, Inc., 111 River Street, Hoboken, NJ 07030, USA

Editorial Office
111 River Street, Hoboken, NJ 07030, USA

For details of our global editorial offices, customer services, and more information about Wiley products visit us at www.wiley.com.

Wiley also publishes its books in a variety of electronic formats and by print-on-demand. Some content that appears in standard print versions of this book may not be available in other formats.

A catalogue record for this book is available from the Library of Congress

Hardback ISBN: 9781119707417; ePub ISBN: 9781119707363; ePDF ISBN: 9781119707370; oBook ISBN: 9781119707394

Cover image: © Turgay Koca/Getty Images
Cover design by Wiley

Set in 9.5/12.5pt STIXTwoText by Integra Software Services Pvt. Ltd, Pondicherry, India

Contents

Acknowledgements

I owe a great deal to my friend and colleague Graham Jackson for his encouragement and for his painstaking reviewing and editing of every sentence in this book. Many thanks also to Dhanesh Goberdhan for his careful reading and many useful suggestions. Several diagrams and pictures come with the Chevron copyright; I thank Dan Foscalina and his colleagues at Chevron Media Relations for gaining my permission to use them.

Robert D. Tack

Preface

It is not only chemists who realise the ubiquitous nature of chemical technology in our modern environment, though it may be that only chemical technologists fully appreciate just how wide ranging are the products of the chemical industry. The scope of the chemical technology can be thought of, broadly, as the applications – such the dyes, pharmaceuticals, and textiles – and within each application and subdivision, there is a whole technology specific to that application. One broad application is chemicals in the oil industry, which includes a few subdivisions: those for exploration and recovery, those in refineries and those for the finished products of oils and fuels.

Additives for lubricating oils have been well covered in the available literature while fuel additives have been covered in much less detail and provide an area of expertise whose existence is largely unknown to the public. This book is concerned with the problems addressed by fuel additives, their chemistries, and scientific insights into their actions.

In this book, chapters on the individual additive types generally follow a discussion of the problem that they address, what they do to alleviate the problem, their chemistries – including their preparations – and some understanding of how they work. Many fuel additives are derived from existing areas of technology in other applications; degradation by corrosion and autoxidation, for example, are wide ranging problems that have spawned studies to understand them and additives to overcome them. Some additives started life in other parts of the petroleum industry such as fuel detergents, that are derived from lubricating oil dispersants, and demulsifiers, that are heavily used in crude oil recovery. As a result of this wide relationship with other applications, the sources drawn on for each chapter are also wide ranging in technical area and in literature type.

There are some publications that provide useful summaries of fuel additives and their use. In particular, the booklets provided by the Technical Committee of Petroleum Additive Manufacturers in Europe (ATC) provide a paragraph on each additive/fuel combination. The Automotive Fuels Reference Book provides extensive coverage of the production, distribution and use of fuels along with details of internal combustion engines, along with few chapters on gasoline and diesel fuel additives. Books on refining are plentiful, but they usually pay little or no attention to fuel additives. However, to understand fuels and the roles of and incentives for the use of fuel additives, it is necessary to know something about refineries. So, the first chapter summarises refinery operations to an extent that discussions of fuels in the following chapters is understandable. This first chapter is an attempt to cover problems other than those occurring from of the way in which fuels are transported and used, so it also includes such items as the proportions of different fuels used at different times and places – the demand barrel; it also introduces the range of fuels and additives, and the influence of taxation.

Degradation of fuels resulting from autoxidation is widespread and is the source of deposits that interfere with the uses of fuels. Direct treatment of such problems is needed from the moment a fuel is blended in the refinery, so the problem of stabilisers is dealt with first. The use of detergents is needed to treat the deposits that escape stabiliser treatment, so this follows on directly.

There is no obvious further natural progression for the remaining problems that appear to be independent of each other in the end use. They are, however, often linked by the way the refiner has to operate to meet production of the demand barrel, or to meet the conflicting needs of various legal specifications. Cold flow improvers enable fuels to flow without problem when they precipitate wax in winter; this a valuable aid to refiners meeting the demands for different fuels in the most cost-effective way. Wear through physical contact and corrosion both cause gradual loss of metal in fuel systems that can lead to failure and leakage; lubricity and corrosion inhibitors are similar compounds designed to prevent these problems. Combustion improvers are concerned with flame chemistry in the different environments of engines and burners. Finally. There are several problems that arise during the movement and storage of fuels that are dealt with in the final chapter.

Sources for this book have ranged from newspaper articles to academic journals. Specialist technical journals, such as Infineum's Insight magazine, and biannual reviews, such as BP's review of World Energy and the Infineum Worldwide Winter Diesel Fuel Quality Surveys, provide valuable, wide ranging views of the business, changing specifications and fuel qualities. Academic journals are generally specialised such those on Energy & Fuels, Tribology, and Colloids. Websites of additive suppliers, oil companies and those set up by government agencies, such as the European CONCAWE[1] and the International Energy Agency, are sources of wide-ranging relevant information.

For details of specific chemical additives, what they do and how they are made, there is no better source than patents. Details of the problem being addressed, tests used, the structures of commercial and proposed additives and how they are made, may be found in patents. A little more patent searching can also reveal a history of additive developments. While many materials may, in patents, be claimed to be effective solutions to the problem under discussion, the author has tried to keep to those that to have been or are commercial additives. Sometimes the exact structure of a commercial additive may not be easily accessible, but patent literature and company sponsored academic papers, combined with the suppliers' broad descriptions, often identify structure type sufficiently to aid an understanding of its activity. If I have deduced an incorrect assumption from these sources, I offer my apologies and ask forgiveness.

For the organic chemist, the author has used much of the pre-IUPAC nomenclature for the alkenes (olefins). While ethene, propene and butene are the names used in the academic and some industrial occupations, the oil industry is mostly familiar with the older names ethylene, propylene, and butylene. Products such as poly-ethylene and poly-iso-butylene (PIB) are too well established to be replaced by poly-ethene and poly-(2-methyl-propene). However, the word 'olefins' is not specific enough when discussing the position of a double bond so the word 'alkene' is then used in preference.

The author hopes that this book has a wider utility than in the petroleum industry alone. Chapter subjects are mostly applicable in other spheres of industry; as such, the general aspects of their topics may be of use to those looking for quick understanding of, for example, oxidative stability or corrosion or crystallization. Finally, I hope that the chemistry is described in sufficient detail to provide examples that chemists in further education might use to answer student questions along the lines of 'So, what is the use of this chemistry (or compound)?'.

Note

1 CONservation of Clean Air and Water in Europe, the European oil industry body that monitors health and safety of petroleum products.

Abbreviations

ACEA	European Automobile Manufacturers Association
AcOH	Acetic Acid
AKI	Anti-Knock Index (average of RON and MON)
APFC	Alkyl-Phenol-Formaldehyde Condensate
ARES	Atmospheric Residual fuel
ASA	Alkyl Salicylic Acid
ASTM	American Society for Testing and Materials
ATC	Additive Technical Committee
ATRP	Atom Transfer Radical Polymerization
Avgas	Aviation Gasoline
AWCD	All Weather Chassis Dynamometer
B10	Biodiesel, 10% FAME and 90% Petroleum Diesel
BBD	Broad Boiling Distillate
BCF	Burton, Cabrera and Franck
BHT	Butylated Hydroxy Toluene
BOCL	Ball-On-Cylinder Lubricity Evaluator
BS	British Standard
BTL	Biomass To Liquids
CAFE	Corporate Average Fuel Economy
cc	cubic centimetre
CCCD	Cold Climate Chassis Dynamometer
CCD	Combustion Chamber Deposits
CCT	Catalytic Chain Transfer
CEC	Confédération Européenne des Cadres
CEIC	Cision European Institutional Investor Company
CFPP	Cold Filter Plugging Point
CI	Cetane Index
CN	Cetane Number
COPD	Chronic Obstructive Pulmonary Disease

CP	Cloud Point
CPD	Cloud Point Depressant
CSPIT	Cold Spark Plug Immersion Test
CTL	Coal To Liquids
DDSA	DoDecenyl-Succinic Acid or Anhydride
DETA	Di-Ethylene-Tri-Amine
DI	Direct Injection
DIEGME	DiEthylene Glycol Monomethyl Ether
DMAPA	3-DimethylAminoPropylAmine
DPF	Diesel Particulate Filter
DRA	Drag Reducing Agent
DSC	Differential Scanning Calorimetry
DTBHQ	Di-Tert-ButylHydroquinone
DTBP	Di-Tertiary Butyl Peroxide
E30	Gasohol, gasoline with 30% ethanol
ECA	Emission Control Area
EDA	1,2-Ethylene-DiAmine
EDTA	Ethylene Diamine Tetra-acetic Acid
EGR	Exhaust Gas Recycle
EHN	2-EthylHexyl-Nitrate
EI	Energy Institute
EN	European Norms (standards)
EU	European Union
EV2-EH	Ethylene Vinyl-2-EthylHexanoate copolymer
EVA, EVAC	Ethylene Vinyl Acetate Copolymer
EVEC	Ethylene Vinyl Ester Copolymer
FAME	Fatty Acid Methyl Ester
FBC	Fuel Born Catalyst
FBP	Final Boiling Point
FCC	Fluid Catalytic Cracker
FO	Fuel Oil
FT	Fischer-Tropsch
FVAC	Fumarate Vinyl-Acetate Copolymer
GDI	Gasoline Direct Injection
GLC	Gas Liquid Chromatogram or Chromatography
GTL	Gas To Liquids
GC-MS	Gas Chromatography - Mass Spectroscopy
HAGO	Heavy Atmospheric Gas Oil
HCGO	HydroCracked Gas Oil
HFO	Heavy Fuel Oil
HFRR	High-Frequency Reciprocating Rig

HGO	Heavy Gas Oil
HGV	Heavy Goods Vehicle
HLB	Hydrophile Lipophile Balance
HSR	Heavy Straight Run
HT	Hydrogenated Tallow
HVO	Hydrogenated Vegetable Oil
IATA	International Air Transport Association
IBP	Initial Boiling Point
ICE	Internal Combustion Engine
IDI	InDirect Injection
IP	"Institute of Petroleum, merged with the Institute of Energy to form the Energy Institute"
IVD	Intake Valve Deposits
JFTOT	Jet Fuel Thermal Oxidation Tester
JP-8	Jet Propellant 8 or Jet Propulsion fuel 8
kg/m3	Kilograms per cubic metre
kJ	Kilo Joule
Km	Kilometre
KPa	Kilo Pascals
LGO	Light Gas Oil
LPG	Liquefied Petroleum Gas
LSP	Low Speed Pre-ignition
LSR	Light Straight Run
LTFT	Low Temperature Flow Test
MBT	Methylene Bis (Thiocyanate)
MD	Middle Distillate
MDA	Metal Deactivating Additive
MDFI	Middle Distillate Flow Improver
mEq	Milli-Equivalent
mg/kg	milligrams per kilogram (also ppm by weight)
MGO	Medium Gas Oil
ml	millilitre
mm	millimetre
MMT	Methylcyclopendadienyl Manganese Tricarbonyl
Mn	Number average molecular weight
MON	Motor Octane Number
MPa	Mega Pascal
MSDS	Material Safety Data Sheet
MT	Million Tonnes
MTBE	Methyl t-Butyl Ether
Mtoe	million toe

MW	Mega Watt
Mol.Wt.	Molecular Weight
NBD	Narrow Boiling Distillate
NBP	Normalised Boiling Point
nm	nanometre
NMP	Nitroxide Mediated Polymerization
NOx	Nitrogen Oxides
OMA	Olefin/Maleic Anhydride
ORI	Octane Requirement Increase
pa	per annum
PAA	Poly-Alkyl-Acrylate
PAH	Poly-Aromatic Hydrocarbon
PAM	Poly Amine (usually poly-ethylene diamine)
PAMA	Poly-Alkyl-MethAcrylate
PAO	Poly Alpha Olefin
PBO	Poly-Butylene-Oxide
PBTPA	PIB-ThioPhosphonic Acid
PDA	Para-phenylene DiAmine additives
PDMS	Poly-Di-Methyl Siloxane
PE	PolyEthylene
PEG	Poly-Ethylene-Glycol
PEO	Poly-Ethylene-Oxide
PEPEB	Poly-Ethylene/Poly-Ethylene-Butylene diblock
PEPEP	Poly-Ethylene/Poly-Ethylene-Propylene diblock
PFI	Port Fuel Injectors
PFT	Programmed Fluidity Test
PIB	Poly Iso Butylene
PIBSA	Poly Iso Butylene Succinic Anhydride (or acid)
PPD	Pour Point Depressant
ppm	parts per million
PPO	Poly-Propylene-Oxide
pS	PicoSiemens
RAFT	Reversible Addition-Fragmentation chain Transfer
RON	Research Octane Number
ROP	Ring Opening Polymerization
rpm	revolutions per minute
SEM	Scanning Electron Microscopy
TAPs	Trans-Alaska-Pipeline-systems
TCMBT	2-(ThioCyanoMethylthio)BenzoThiazole
TDC	Top Dead Centre
TEL	Tetra Ethyl Lead

TEM	Transmission Electron Microscopy
TEPA	Tetra-Ethylene-PentAmine
TETA	Tri-Ethylene-Tetra-Amine
TEU	Twenty-foot Equivalent Unit (shipping container)
toe	tonnes oil equivalent
TOFA	Tall Oil Fatty Acids
TPSA	Tetra-Propenyl Succinic Acid
UK	United Kingdom
ULSD	Ultra Low Sulphur Diesel
USA	United States of America
USAF	United States Air Force
V/V	Volume per Volume (as % V/V)
VAT	Value Added Tax
VisRES	Visbreaker Residual fuel
VRES	Vacuum Residual fuel
WAFI	Wax Anti-settling Flow Improver
WASA	Wax Anti-Settling Additive
WAXD	Waxy Distillate
WCM	Wax Crystal Modifier (or Modifying)
WDFQS	World Winter Diesel Fuel Quality Survey
WSD	Wear Scar Diameter
ZDDP	Zinc Dialkyl-Dithio-Phosphate

1

Fuels and Fuel Additives – Overview

1.1 Introduction

The fuels under consideration here are the liquid fuels obtained from crude oil by fractional distillation and other refinery processes, together with biofuels and synthetic Gas to Liquid (GTL) and related fuels. The gases methane, ethane, propane and butane are not covered as they generally need few, if any, additives other than odorants.

Coal is not considered here either. Most coal is burned in fluidised beds in electricity power stations, which provided about 38% of the world's electricity (22% in Europe) in 2017 [1]; overall, coal provided 28% of the world's energy and oil provided 34%, in 2017. Coal additives are added to reduce soot, fly-ash, bottom-ash, slag and clinker in power stations, but a European Commission research document concludes that only the use of lime or limestone effectively reduces deposits, fly-ash and sulfur oxides [2,3]. Magnesium additives are used in furnaces burning heavy fuel but are not effective in coal, so links with the additives used for petroleum derived fuels are almost non-existent.

This chapter will give a condensed explanation of the refinery operations that convert crude oil into fuels and compositions of the different fuels. Also, in this chapter are presented some of the statistics of quantities of various refinery products and global variations. The author's intention is to provide sufficient understanding of refining to support the discussions of the applications of fuel additives. Much has been written about refineries, specifications and internal combustion engines, so the reader is guided to some of the relevant literature for more detailed explanations – for an example of the engineers' perspective see the *Automotive Fuels Reference Book* [4].

Finished fuels are put together from the refinery components to provide the properties that they need to have and to meet fuel specifications. However, many of the properties that a fuel should have are difficult to achieve by blending

Fuel Additives: Chemistry and Technology, First Edition. Robert D. Tack.

refinery output streams. Fuel additives provide the additional necessary properties – usually in the most economical way and sometimes the use of additives is the only way for the different fuels to meet their specifications. This chapter introduces the reader to the range of fuel additives used in the different fuels.

When fuels reach the market and are sold, not surprisingly there is a whole tax regime, particularly for motor fuels. And, as in other areas of taxation, there are some concessions of low taxation on fuels for particular applications which carry the potential for criminal diversion of low-tax fuels into high-tax applications. So, it is here that we are first introduced a discussion of fuel additives: dyes and markers that are used to identify low-tax fuels.

While petroleum fuels and their production form a worldwide industry, most of the data referred to is from Europe and the USA because information of these markets is that which is most widely available. The oil industry started in the USA, followed closely by Europe [5] – the industry in other parts of the world followed in the footsteps of these two regions and still set specifications which are closely related to those in the USA and in Europe.

1.2 Refinery Operations and Processes

There is much literature on refineries, and useful, easy-to-follow descriptions and diagrams are provided in oil company websites [6–12]. A schematic of refinery operations helps to visualise how the different fuel components arise and how they are combined into the finished fuels (Figure 1.1) [8]; this will help in understanding the discussions of how particular fuel problems arise and the refinery limitations which lead to additives providing the best solutions.

1.2.1 Distillation

Crude oil coming into the refinery is preheated in the heat exchangers that, at the same time, cool the exiting refinery product streams. Steam is then injected into the crude which moves into a furnace (at about 500°C to 550°C) where it is heated to almost 400°C. The furnace is at the bottom of the atmospheric (pressure) distillation tower – more commonly known as the atmospheric pipestill or topping unit. Here, the more volatile fractions rapidly vaporise, and the hot vapours pass up the tower through a series of perforated metal trays – the perforations are holes with a collar and a supported cap known as a bubble cap. As the vapours cool to the lower temperatures higher in the tower, the less volatile components condense into liquids that collect on the trays and pass down the column. During this process, the bubble caps force the rising vapours to bubble through the condensing liquids, thus improving the efficiency of the fractional distillation: the cooler liquid condenses more of the less volatile

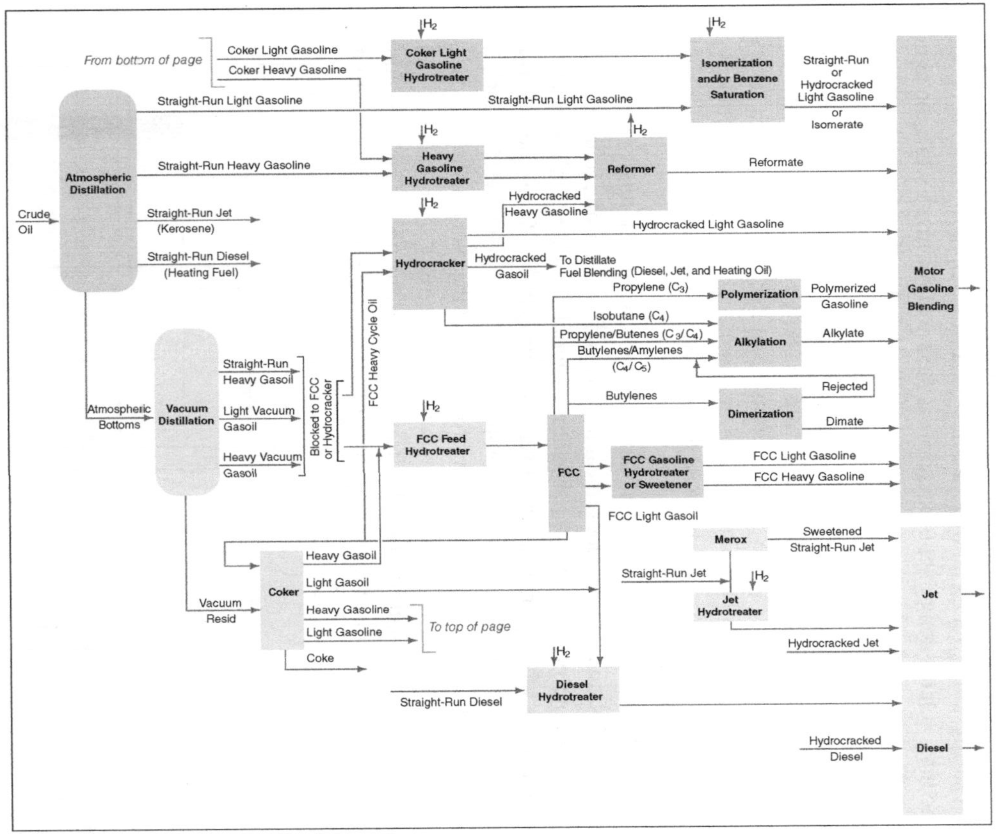

Figure 1.1 Refinery Operations.

liquids from the vapour and the vapour takes out the more volatile components from the liquid. Every few trays, the condensed liquids are continuously removed as a 'side-stream' or 'cut'. These side-streams are collected and blended for use as the distillate fuels gasoline, jet, kerosene, diesel fuel and gas oil or are transferred to various conversion processes (Figure 1.1).

A steady, falling temperature gradient is established in the distillation tower. The most volatile components pass over at the top of the tower while the non-volatile material, known as atmospheric residual fuel, collects at the bottom of the pipestill, after removal of more volatile fractions with the help of a current of steam. Boiling temperatures of the side-streams are highest at the bottom of the tower – for example heavy gas oil (up to 380°C) – and lowest at the top (below 100°C) – such as the gasoline component light naphtha (up to 70°C). Exact boiling ranges of the side-streams and their degree of overlap (efficiency of separation) depend upon the exact details of operations at each refinery but the variations are not huge.

The proportions of side-streams, hence products, vary with type of crude oil. 'Light crudes' and 'heavy crudes', meaning low and high density, contain different proportions of lower and higher boiling material. West Texas Intermediate is a light crude which contains a relatively high proportion of the valuable low-boiling gasoline fractions and a low proportion of high-boiling residual fuel; in contrast, Arabian heavy crude has low proportions of the valuable light products and a large (about 60%) of low-value high-boiling material ('other' in Figure 1.2), of which half is vacuum distillable leaving half as an undistillable vacuum residual fuel [13]. More importantly, the distributions of products possible from

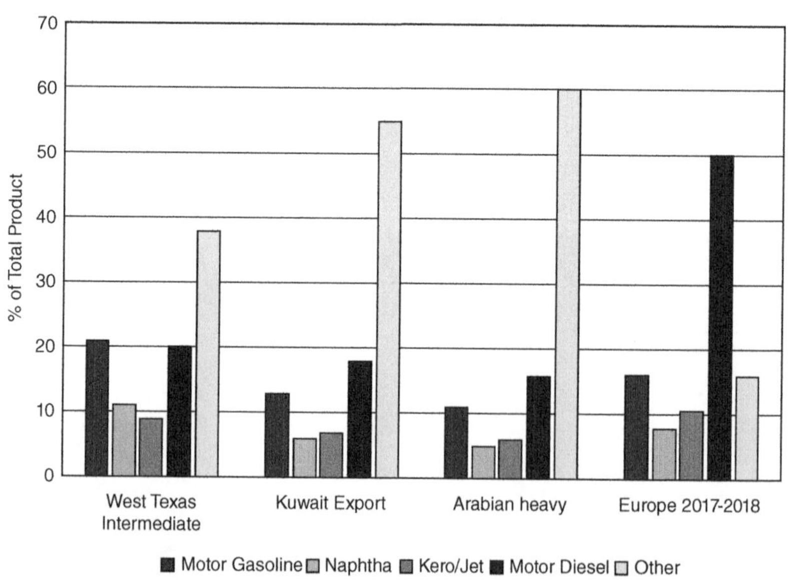

Figure 1.2 Straight-Run Yields (Wt.% of Total) of a Range of Crude Oils, Compared with European Union-28 Refinery Output 2017.

both crudes after atmospheric distillation, known as 'straight-run', and blending do not match that of the average refinery output. A region's average refinery output is designed to meet customers' needs (such as the EU-28 [14,15] in Figure 1.2), which is termed the demand barrel.

1.2.2 Balancing Production to the Demand Barrel

Distillate fuels are needed in larger proportions than are naturally present in crude oil. In Europe, this is particularly true for the middle distillates gas oil and diesel, while the USA needs more gasoline. The problem lies in the high proportions of atmospheric residual fuels, such as 32% to 54% of the crude oil ('other' in Figure 1.2), which outstrips demand for this fuel. Clearly, this excess cannot be treated as waste; the quantities are huge – the reality of refining is that all that comes into a refinery as crude oil must go out as products. For example, the Fawley refinery supplies 14% of the UK's oil products; it processed around 22 million tonnes of crude oil in 2019 (60 thousand tonnes per day[1]) [11]. If one-third of this came out as atmospheric residual fuel, that would pose an enormous problem of disposal.

Most refineries, then, have more complex operations than those providing only straight-run, atmospheric distillation. Through the second half of the twentieth century, almost all simple, straight-run only refineries (also known as 'hydroskimming' refineries) in Europe and North America have closed and the more complex refineries have expanded to meet demand. Atmospheric residual fuel is further refined first by being vacuum distilled up to a temperature equivalent[2] of 550°C. This takes place in a vacuum distillation pipestill, which has the same arrangement as for the atmospheric pipestill but with the addition of a partial vacuum (reduced pressure), provided by steam injectors. From this pipestill, a range of vacuum-distilled fractions (known as 'vacuum gas oils') are produced; some of these are converted into lubricating oils but most is used as feed for the crackers (section 1.2.3). Finally, left behind at the bottom of the vacuum pipestill is a certain proportion of a high-boiling, undistillable vacuum residue. Some of the vacuum residue is converted to bitumen or heavy lubricating oil, some is blended with low-value distillates into residual fuels and some is processed further to produce lighter products in cokers (section 1.2.5). Residual fuels are used in heavy marine diesel fuels, in refinery furnaces and in some power stations.

After the atmospheric and vacuum distillations, the refinery products are typically those shown in Table 1.1 [6] from a heavy Middle Eastern crude. For this Kuwait crude oil distillation, the proportion of vacuum residue is 26.7% while an assay of Kuwait export crude put the proportion of atmospheric residue at 54.5% (Figure 1.2) [13], i.e. 'fairly heavy'.

Refinery fuel products from complex European refineries are made up of higher proportions of non-residual, higher-value products than are possible by blending the straight-run distillation cuts of either the Kuwait or Arabian

Table 1.1 Side-streams in a typical modern refinery from Kuwait export crude.

Cut Number	Product Name	Cut Vol% of Whole	Distillation[3] End Point	Average NBP	Average Mol. Wt.
1	Gases	1.3	10	2.5	56
2	Light naphtha	7.3	70	44	71
3	Naphtha	16.6	180	132	112
4	Kerosene	10.1	240	210	160
5	Light diesel	7.8	290	264	200
6	Heavy diesel	7.0	340	314	244
7	Atmospheric gas oil	3.8	370	355	285
8	Vacuum gas oil	2.4	390	380	313
9	Vacuum distillate	18.3	550	467	435
10	Vacuum residue	26.7	–	689	1150

Table 1.2 Petroleum refinery products in Europe 2018.

Product	Refiney Output, Mtoe	Refinery Output, % of Total
Total output:	638.5	100
LPG	30.3	2.8
Naphtha	42.2	7.0
Gasoline	80.1	18.5
Kerosene + Jet	62.8	8.6
Gas oil /Diesel oil	292.6	39.4
Fuel oil	49.3	11.7
Other products	53.1	12.1
Refinery use/losses	27.7	–

crudes (Figure 1.2[4], cf. Table 1.2 [15]). The difference between total refinery output (638.5 Mtoe[5]) and the use as liquid fuel (transport plus heating, 508 Mtoe) is made up of other petroleum product applications such as heavy (marine) fuel, industrial use (many varied sectors), liquefied petroleum gases, lubricating oils, waxes and chemicals (Figure 1.3 [16]). Many refineries produce items such as solvents that contain aliphatic and aromatic hydrocarbons,

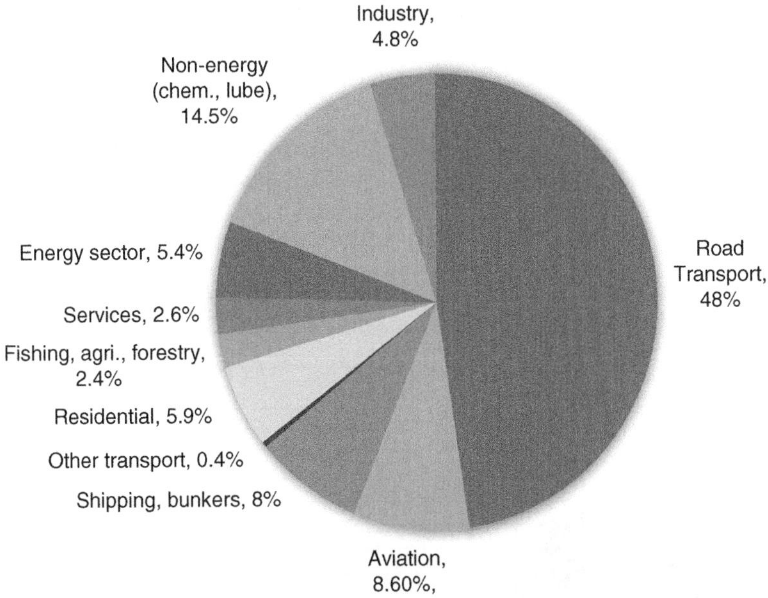

Figure 1.3 Distribution of End Uses for Petroleum Products in the EU-28, 2016.

ketones and alcohols; raw materials for the chemical and plastics industries such as ethylene, propylene and butylene, and higher alkenes, such as tetrapropylene, are other refinery products. These other products are made from both the liquid fractions and the gases. Diesel is also used for non-transportation engines such as generators. Import/export balances can confuse the quantities, and the counting of residual fuel may be short because sales are often private.

(Quantities in million tonnes oil equivalent (Mtoe) are close to actual tonnes as the factors for conversion of quantities, by weight, to tonnes oil equivalent are close to 1.0:1.01 for diesel and 1.105 for gasoline; residual fuel oil is 0.955.)

Imbalances between the supply and demand barrels vary between regions, seasonally and over years with macroeconomic changes, such as the adoption of fuel-efficient diesel in Europe and heating oil being replaced by natural gas. They are corrected, in part, by refinery conversion facilities such as crackers (section 1.2.3). Such facilities raise costs and prices but enable global refining to meet the global balance in demand. Regional imbalance is corrected by trade that reflects different regional needs. As a result, excess of a fuel in one region may be exported to a region that has a shortage of that fuel. The import/export effect is illustrated by the European and USA markets. A comparison between the two illustrates the differences between the USA and the European petroleum fuel markets. These differences in fuel consumption explain the differences in emphasis on additives that have developed in these two markets.

Transportation fuels, road, aviation and shipping, take up the major part of fuel consumption (Figure 1.3, in Europe [16]) so dominate refinery operations and trade. Europe and the USA have quite different distributions of transport fuel consumption – European transport relies heavily upon diesel while gasoline dominates in the USA [17] (Figure 1.4). In a worldwide comparison, while Europe depends upon middle distillates and the USA upon gasoline, the Asia-Pacific region depends somewhat evenly upon these two (Figure 1.5 [18]).

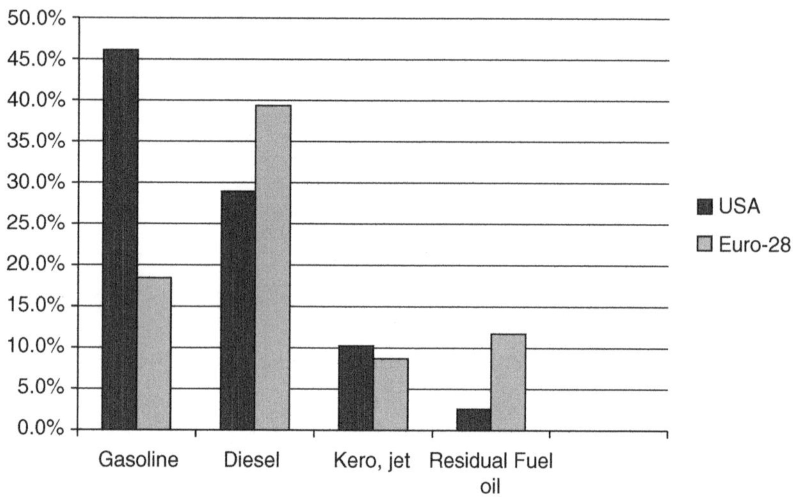

Figure 1.4 Distribution of Liquid Transport Fuels in the USA and Euro 28 Countries, 2018.

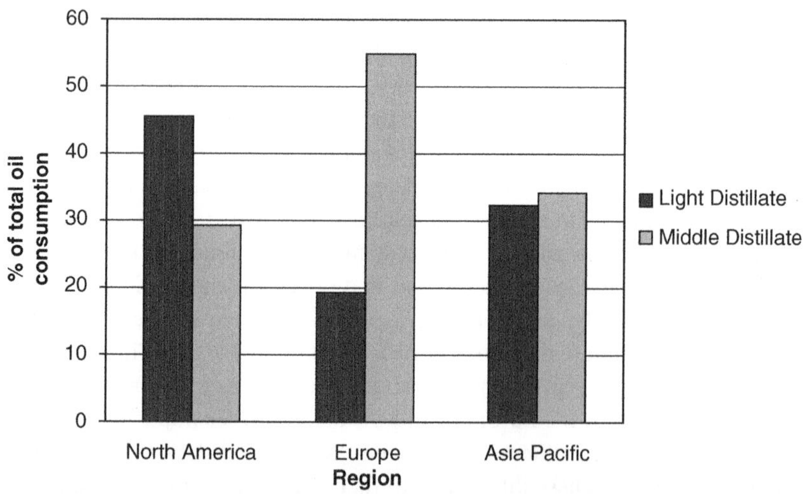

Figure 1.5 Distribution of Liquid Fuel Use by Type in Main Economic Blocs, 2018 [18][6].

As might be expected, import/export figures for these fuels show that Europe imports substantial quantities of middle distillates, while exporting similar quantities of gasoline [15] (Figure 1.6). Historically, the USA imported more gasoline than it exported, for example 25 Mtoe in 2011 [19]; however, since the recent resurgence of oil production in the USA, it has become a net exporter of both diesel (58 million tonnes ULSD in 2017) and gasoline (31 million tonnes) [17]. Europe's major supplier of diesel is now Russia, and the Middle East is its major supplier of jet fuel [15].

In the world of petroleum and petroleum products, such changes are not surprising given the background that over the ten years from 2008 to 2018, the total consumption of oil has changed little in North America (from 1105 to 1113 Mtoe) while falling in the European Union (731 to 647 Mtoe) and growing strongly in Asia Pacific (1250 to 1695 MT pa) [18].

Worldwide, the 4,474 million tonnes of petroleum products consumed in 2018 had a distribution of petroleum cuts [18] (Figure 1.7) most like that in the Asia Pacific. It is perhaps worth reminding the reader that diesel engines are used to propel buses, HGVs/trucks, ships and smaller boats, tractors, construction vehicles and back-up electricity generators as well as passenger cars and light vans. By reason of the greater fuel efficiency of diesel over gasoline engines (and the rising cost of oil), the growth in the demand for diesel was outstripping that for gasoline in the noughties, such that in 2012, 51% of new cars were diesel in the UK and 56% in Europe [20] – except in America. In fact, the USA had been dubbed 'Refiner to the World' because the high demand for diesel outside the USA was providing USA refineries with profitable production and sales of diesel for export [21].

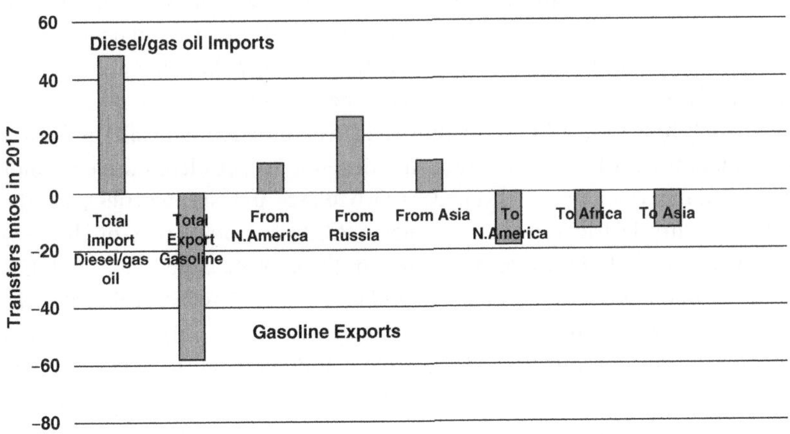

Figure 1.6 EU-28 Imports of Diesel/Gas Oil and Exports of Gasoline, Million Tonnes in 2017.

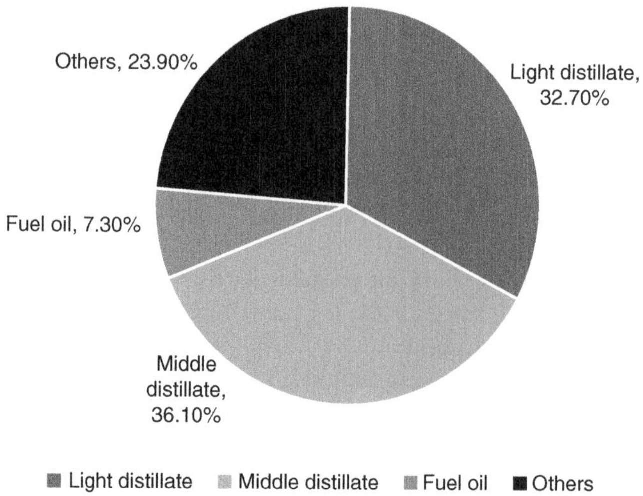

Figure 1.7 Distribution of Fuel Consumption by Refinery Description, Worldwide, 2018. Light distillate: aviation and motor gasoline with light distillate feedstock. Middle distillates: jet, heating kerosene, gas oils and diesel fuel, including some marine bunkers.[7] Fuel oil: marine bunkers (heavy fuel) and crude oil used as fuel. Others: refinery gas, LPG, solvents, bitumen, lubricants and other refinery products and losses.

Looking to the future, in a 2011 survey for the UK department of energy [19], the projection to 2030 was for a fall in the gasoline share of demand (to 10% of total petroleum products) and a corresponding rise in diesel fuel demand (to 40%), an increase in jet fuel consumption and continued fall in the demand for heavy fuels. The last two did occur but the continued rise in diesel demand was being limited by worries about its contributions to city pollution, inroads of biodiesel and decreasing energy intensity[8] as the new Asia Pacific economies mature.

In 2017, however, a break in the trend was underway, resulting from a combination of rising air pollution in cities, due to petroleum-driven motor transport, in particular diesel vehicles (particulates and NOx worries), and the rapid rise in electric vehicle technology. ACEA reported that of the new passenger cars sold in Western Europe, in 2016, 49.5% were diesel and 45.8 were petrol engine powered (2.1% hybrid electric, 1.5% rechargeable electric, 1.2% LPG) [22]. In June 2017, diesel passenger cars took only 39% of sales in Germany, and it was predicted that this could soon fall to 30% [23]; new passenger cars in Europe in 2018 showed a complete reversal to 36% diesel and 57% gasoline (electric rechargeable and hybrid 6%) [24]. As Yogi Berra said, 'It is difficult to make predictions, especially about the future'.

1.2.3 Catalytic Conversions

As already discussed, there is not sufficient distillate fuel after straight-run, atmospheric distillation to meet customer demand. This mismatch between what is in crude oil, that may be separated by distillation, and the needs of the market for transportation is resolved by cracking the molecules in the higher boiling fractions – so converting the low-value high-boiling gas oil into valuable gasoline and diesel. Vacuum distillates are hydrotreated to remove the sulphur that would poison the catalysts, fed into the crackers, and redistilled to give the lower boiling cuts needed for automotive fuels.

There are two main types of cracker [6]. In one, the Fluid Catalytic Cracker (FCC), hydrotreated, liquid vacuum gas oils are passed through a bed of zeolite catalyst at about 500°C, under low pressure (1 to 5 bar). The FCC produces hydrocarbons with a high degree of unsaturation (alkenes and aromatics) as well as branched lower alkanes and hydrogen (which is used in the hydrotreaters). The product is then distilled to produce streams that are valuable, high octane blend components for gasoline, which is the primary purpose of the FCC. In addition, there are smaller amounts of aromatic-rich kerosene and middle distillate fractions.

The other is the hydrocracker, in which vacuum gas oils are cracked at 350°C to 450°C in the presence of hydrogen. There are degrees of hydrocracking: mild hydrocracking uses hydrogen at a pressure of 35 to 70 bar and a nickel-molybdenum catalyst to remove sulphur, nitrogen, and destabilising olefin unsaturation. Conventional hydrocracking uses hydrogen at 85 to 140 bar and a nickel-molybdenum on silica-alumina-zeolite catalyst to remove most of the sulphur and most of the aromatics; at the same time, hydrocracking[9] and hydroisomerising occur to produce distillation streams that have high cetane numbers, so are suitable for ultra-low sulphur diesel fuel [6].

For light distillates, there is a further catalytic process. Catalytic reforming is used to convert low octane naphtha into gasoline components that are rich in iso-alkanes and aromatics, which have high octane numbers. Catalytic reforming is carried out by passing the naphtha over a platinum-based catalyst, under moderate pressure (5 to 25 bar) at 500°C [6].

1.2.4 Alkylation

Alkylates are the products of the sulphuric acid-catalysed reactions of short chain alkenes, usually butylene or propylene, with isobutene[10]. Alkylates are highly branched and so have high octane numbers – for example, the isobutylene/isobutane product is iso-octane which has a 100 octane number [6].

1.2.5 Coking

Finally, there are the atmospheric and vacuum distillation residues. These may be converted to distillable liquids by 'coking' whereby the distillation residues are heated at a high temperature so that carbon is extruded, as coke, from the high molecular weight, carbon-rich molecules leaving liquid products of lower molecular weight that contain more hydrogen. The process also leaves much of the sulphur and metals in the coke. The chemistry of the process is thought to be that as the already high molecular weight molecules condense together into poly-aromatic structures, low molecular weight alkanes, alkenes and aromatics are split off. Eventually, the poly-aromatics have graphite-like structures and contain low proportions of hydrogen. The product liquids are distilled into a full range of products: gases such as methane and alkenes such as ethylene, propylene and butylene (valuable chemical intermediates); aromatic-rich (hence high octane) gasoline components; aromatic-rich kerosene and gas oils that need hydro-treatment to be useful as jet and diesel blend components [6].

In delayed coking, the residue is heated to about 500°C, at several bar pressure in a furnace. It is then kept in a '(heat) soak drum' at about 450°C to 480°C until about 20% to 30% of the residual fuel is converted to coke plus lower boiling liquids, which are then redistilled. In a flexicoker, the residual fuel is passed through a fluidised bed, at about 520°C and low pressure (under 1 bar), with a flow of steam; in this process, the extruded carbon is burned off (only 2% left behind) providing the heat needed for the endothermic coking process. Visbreaking is a similar but milder process in which 15% to 25% of the residue is converted to lower molecular weight components that lower the viscosity, which is often necessary in heavy fuel blending. In the visbreaker, the residue is passed over heat-exchange tubes at 425°C to 450°C, depending upon the severity of cracking required; for example, at 450°C, 75% may be cracked to lower boiling materials [6,7]. Alternatively, the heated residue may be kept in a soaker drum for some time (as in delayed coking but at a lower temperature and shorter time).

1.3 Finished Fuels

As the boiling point range of the fuel is raised, the molecular weights and densities of the fuel components also increase giving rise to the terms 'light' and 'heavy', which refer to lower and higher boiling fractions, respectively. Finished fuels from a refinery are usually made up by blending a range of distillate streams (plus residual fuel for heavy fuels) – a range that can be several streams in a complex refinery (Figure 1.1). Blending enables all of a refinery's production to be used. The process is guided by the need to meet a range of specifications for the products. Specifications are set by various industry and government bodies to ensure that fuels meet a minimum quality to provide the performance needed by the customer and minimise environmental pollution.

1.3.1 Gasoline

Several refinery streams may be combined to produce gasoline that meets the local specifications – the most important two are octane number and volatility. Octane number is the gasoline quality that indicates whether the fuel will work effectively in the engine; it is specified either as a Research Octane Number (RON) of at least 95 or a Motor Octane Number (MON) of at least 85[11] (EN228 specifications) [25,26]. A minimum octane rating is necessary to ensure that gasoline used in an engine with standard compression[12] does not cause the engine to knock. Such knocking is caused by auto-ignition of the gasoline/air mixture lower down the engine cylinder as the flame front moves through the cylinder. This causes an out-of-time spike in pressure accompanied by a sharp sound (knock or ping) that can lead to engine damage (section 6.2.1). Auto-ignition is a free radical process which is inhibited by compounds that form relatively stable radicals, such as highly branched alkanes as 2,2,4-trimethylpentane[13] and alkyl aromatics such as toluene (see Chapter 6).

In order that gasoline burns evenly and completely in the combustion chamber, it must vaporise readily – the volatility of a gasoline should also be sufficient to ensure easy starting. Volatility is specified, for example in Europe and the USA (Table 1.3 [9,26]), and it depends upon the distillation properties of a gasoline, so the specifications are met by blending streams of different distillation ranges. Laboratory distillations of refinery products (by a standardised distillation procedure, ASTM D86) provide the cumulated weight percentage collected at each distillation temperature (see section 4.2.5). For the European EN 228 gasoline specification, distillation is presented as the amount evaporated (same as collected distillate) at 70°C, 100°C and 150°C or 180°C, and at the final boiling point (Table 1.3a)[14]. In the USA, ASTM gasoline specification D4814 distillation requirements are presented as the distillation temperature at which 10%, 50% and 90% of the distillate is collected, together with the end point – the temperature at which the last of the distillate is collected, the final boiling point (Table 1.3b). In general, the carbon numbers of the hydrocarbons that make up gasoline are mostly C_4 to C_{11}, with small amounts of C_3 and C_{12} [9].

Different classes of volatility are set to allow for climatic variations owing to region and season. Vapour pressure is a balance between the need to avoid vapour lock (vapour bubbles in the fuel lines, inhibiting the pumping of the fuel in hot weather) and to provide easy starting in the cold (which needs the lower boiling material). Since the vapour pressure is measured at a fixed

Table 1.3a EN 228, European volatility class gasoline specifications (the first, fourth and sixth of the six classes have been selected).

Volatility/ Distillation	Unit	Class A	Class B	Class C	Class D	Class E
Vapour pressure	KPa at 38°C	45–60	45–70	50–80	60–90	65–95
Ambient temperature	°C	>15	5 to 15	−5 to + 5	−15 to −5	<−15
% Evaporated at 70 °C	Vol %	20–48	20–48	22–50	22–50	22–50
% Evaporated at 100 °C	Vol %	46–71	46–71	46–71	46–71	46–71
% Evaporated at 150 °C	Vol % maximum	75	75	75	75	75
Final boiling point	°C, maximum	210	210	210	210	210
MON/RON		85/95	85/95	85/95	85/95	85/95

Table 1.3b ASTM D4814, American volatility class gasoline specifications (the first, fourth and sixth of the six classes have been selected).

Volatility/Distillation	Unit	Class AA	Class C	Class E
Vapour pressure	Max. Kpa at 38°C	54	79	103
Temperature for 10% evaporated	°C, maximum	70	60	50
Temperature for 50% evaporated	°C, maximum	77–121	77–116	77–110
Temperature for 90% evaporated	°C, maximum	190	185	185
End point	°C, maximum	225	225	225
Anti-Knock Index, (MON + RON)/2		Regular 87	Mid-range 89	Premium 91–94

temperature, this figure rises along the series of fuel classes to reflect the lower temperatures at which these classes of fuels are used; class E fuels would be used when the ambient temperature is low (winter) and class A when it is high (summer) (Table 1.3). The variation in the distillation temperatures of the

Table 1.4 The refinery components (Streams) that may be blended into the gasoline.

Component	Vapour Pressure, Kpa at 38°C	RON
Butanes, iso/normal	483/354	93/93
Pentanes, iso/normal	132/100	93/72
Light straight run (LSR) gasoline	76	66
Heavy straight run (HSR) gasoline	7	62
Light hydrocracker gasoline	88	83
Heavy hydrocracker gasoline	7	68
Coker gasoline	24	67
FCC light gasoline	95	92
FCC heavy gasoline	10	83
Reformate 94 RON	19	94
Reformate 98 RON	15	98
Alkylate C3'	39	91
Alkylate C4'	31	97
Alkylate C5'	7	90

lowest boiling 10% (USA data) shows that the higher vapour pressure is a result of including more lower boiling material in the gasoline. The same higher boiling blend components are used in all grades so the amounts of distillate at higher test temperatures vary little across the class series.

To meet both the requirements of octane and volatility, a refiner blends together a range of components that depend upon their availability in his refinery. The range of components that may be used have varied octane numbers and volatilities (Table 1.4). Gasoline fractions are known either as light and heavy gasoline or as light and heavy naphtha. These are mostly volatile enough to be fed into the engine as a vapour mixed with air after passage through the carburettor. Modern gasoline engines now use fuel injection which provides much better, electronic control of the fuel/air mixture; however, when the gasoline enters the combustion chamber, it has been almost completely vaporised in the inlet port. The light and heavy gasolines make up the bulk of the final blend but, when straight run, these have relatively low RONs [6] (Table 1.4). However, in a complex refinery there are other gasoline blend components that have the similar boiling ranges but higher RONs; such components come from distillations of FCC and hydrocracker products, and from the reformer and alkylation (sections 1.2.3 and 1.2.4).

Volatile alkanes – butanes and isopentane – help raise the RON of the blend towards the specification minimum of 95 but their use is limited by their volatility. Using the calculation for vapour pressure contributions to a gasoline blend [6], just 1.0 volume % of iso-butane increases the vapour pressure by 23 kPa, so it must be used sparingly.

Clearly, with this range of components, given the limitations of availability in a refinery, achieving a 95 RON may be quite difficult. There are, however, several alternative blend components with significantly higher RONs that can provide a boost to the RON of the blend. For example, methyl-tert-butyl ether (MTBE) was a much-favoured blend component having a RON of 115; a textbook example [6] shows that 7% of MTBE raises the RON from 88 to 95. Other components that can be used are short chain alkylated benzenes, such as toluene and xylene (RONs of 120 and 118), isopropanol and ethanol (RONs 118 and 109) – ethanol is a favoured component as it is readily made in large quantities from natural, renewable resources and, for example, blending 30% ethanol with gasoline (RON of 91) gives E30 which has a RON of 101 [27].

The major change to gasoline since 1990 has been the reduction of sulphur in most parts of the world. Sulphur dioxide is harmful if inhaled, causing lung diseases such as asthma and bronchitis. Initially, sulphur in gasoline was reduced to low levels of below 100 ppm (from up to 0.5% in the worst case) but other emission controls were also introduced to minimise unburnt hydrocarbons and carbon monoxide, which required catalytic convertors in vehicle exhaust systems. Such catalyst are very sensitive to deactivation by sulphur, so ultra-low sulphur specifications of 10 to 15 ppm maximum have been introduced widely. Sulphur is removed by hydrogenation of the fuel over a catalyst.

1.3.2 Middle Distillates

Fuels with boiling points higher than those of gasoline are generally known as middle distillates: jet fuel, kerosene, diesel fuel and heating oil. Higher boiling fractions may be blend components for some diesel and heating oils, or they may be blended into heavy fuels; various proportions may not leave a complex, modern refinery at all but are, instead, converted to the more valuable lower boiling fractions in the crackers. Where there is an excess of heavy distillates, there is a market for these as lower cost fuels for heating, power generation, off-road vehicles (earth moving and farm equipment), and marine fuels.

Middle distillates, in general, can suffer from a problem which does not arise with gasoline – that is their tendency to produce wax in the cold. This wax results from the higher molecular weight n-alkanes present in middle distillate fractions crystallising out of the fuel. The higher molecular weight, least soluble n-alkanes are also the highest boiling, so wax formation can be the factor that limits the use of high-boiling fractions in a fuel blend (see Chapter 4).

1.3.2.1 Jet Fuel

Jet fuel is a kerosene stream with more clear-cut specifications than those of general kerosene. The main international jet fuel, Jet A-1, distils to give 10% recovered at 205°C and an end point of 300°C; hydrocarbons in jet fuel have carbon numbers C_8 to C_{16} [28]. The upper boiling limitation is to ensure complete vaporisation in the jet engine and to ensure no wax formation in the wing tanks; the lower boiling specification is to remove the fire risk and vaporisation losses that were problems with the early, broad-cut jet fuel, which contained gasoline fractions [28]. It is often taken from the straight run atmospheric distillation, but hydro-treated streams may also be blended in. It is not heavily de-sulphurised, having a sulphur maximum specification of 0.3%, but mercaptans are neutralised by conversion to the less corrosive disulphides in the Merox© process [6]. There are a number of items specified to ensure the suitability of a particular fuel for handling, storage and use [28,29]):

- Conductivity and flash point to avoid the risk of vapour/air explosions initiated by a static spark.
- Freeze point to avoid wax separation at −40°C to −55°C in wing-tanks and viscosity to ensure no problems in pumping the fuel from wing-tank to jet engine.
- Reliable combustion relies upon energy content, distillation to ensure vaporisation in the engine, and smoke point and aromatics content to ensure that no soot is produced.
- Acidity and sulphur content are limited to prevent corrosion of the fuel system and corrosion of turbine metal parts a by sulphur oxides.
- Stability against the formation of fuel-line-blocking degradation products during storage.
- Lubricity of the fuel to prevent wear in the fuel pump.

Since a jet fuel is, in fact, burned in air, it does not have to meet an octane- or cetane-like requirement, though it does need a minimum calorific value, as do gasoline and diesel fuels. A sooty flame would provide a number of problems such as erosion of turbine blades, build up and blockage of the air inlets inside the jet engine, and changes to heat absorption by the jet engine walls. The other controls on handling and storage, such as stability, corrosion, lubricity, and prevention of static sparks, are similar to those for gasoline and diesel fuels. Wax separation in jet fuel is avoided altogether by limiting higher boiling components to meet a freeze point specification, which is set below the lowest temperature that a jet fuel is expected to meet.

1.3.2.2 Diesel Fuel

Diesel engines all use fuel injection as most of the fuel components (usually C_{12} to about C_{30} hydrocarbons) have low volatility. Fuel is sprayed into the combustion chamber through very fine orifices under very high pressure (30,000 psi or 200 MPa [8]). This occurs at top dead centre of the piston-in-cylinder cycle. At top dead centre, maximum compression of the air, taken in on the first down stroke of the piston, produces a temperature that is high enough to cause the injected fuel to auto-ignite. Auto-ignition, a problem to be avoided in gasoline engines, is essential to the operation of diesel engines – if it does not occur at the right time on every stroke, 'diesel knock' occurs, which reduces engine efficiency and causes excessive wear. Auto-ignition fuel qualities are reported as cetane numbers[15], which respond to fuel component chemistries in a fashion contrary to the response of octane numbers (data here drawn from different references [8,9] [30,31]):

- n-Alkanes have low auto-ignition temperatures and a low tendency to form relatively stable radicals, which would slow down auto-ignition. This is desirable for diesel engines with the standard for a cetane number of 100 being cetane (hexadecane).
- Branched n-alkanes, such as iso-octane, form relatively stable free radicals via H-atoms donated from tertiary C–H positions. This is bad for auto-ignition, so branched n-alkanes have low cetane numbers: 2,2,4,4,6,8,8-heptamethyl nonane, known as iso-cetane, is used as a low cetane standard having a cetane number of 15.
- Alkyl-aromatics can form relatively stable radicals on the carbon attached to the aromatic ring. These also have low cetane numbers – for example xylene has a cetane number of 30 compared with a RON of 118; 1-methylnaphthalene is defined as having a zero cetane number. The aromatic-rich FCC gas oils and kerosenes have low cetane numbers but can be hydro-treated to make them more suitable for diesel blends. Severe hydrotreating is needed to provide ultra-low sulphur diesel (ULSD, maximum 10 ppm sulfur), so separate hydro-treating of a FCC component may not be necessary.

In the aerosol of injected diesel fuel, some of the fuel vaporises and some remains as small liquid droplets which have the potential to produce fine particulates of unburnt carbon in the exhaust – particularly if the fuel contains a high proportion of high boiling components that contain hydrocarbons above C_{30} and multi-ring aromatics, such as alkyl naphthalenes. To minimise particulate emissions by road vehicles, boiling ranges and polycyclic aromatics contents are controlled by the current diesel specifications, especially in the

European specification EN 590 [26,32]. The main EN 590 specifications, for on the road diesel, are listed below (Table 1.5). An additional control of the emission of harmful particulates is provided by diesel particulate filters that are now fitted to diesel vehicles (see section 6.2.4).

Other properties that are measured to agreed specifications are viscosity and corrosiveness; cloud point, pour point, and cold filter plugging point to protect against waxing problems in cold weather; oxidative stability to prevent the formation of insoluble particles on storage; lubricity as a measure of ability of the fuel to lubricate the fuel pump; and flash point to be > 55°C for safe handling [8].

Components with distillation ranges within the diesel fuel specifications that may be combined into a diesel fuel may be selected from the following list, depending upon the refinery concerned – different degrees of conversion and different product demands affect the availability of the components:

- Kerosene.
- Light gas oil (LGO, No. 1 fuel in the USA).
- Middle gas oil (MGO; No. 2 fuel in the USA).
- Heavy gas oil (HGO, while too high a boiling range for EN 590 diesel, they were allowed in off-road and marine diesel fuels until recently).
- Fluid catalytically cracked gas oils (FCC-GO) have low cetane numbers so their use is limited unless they are hydrotreated.
- Hydrocracked gas oils (HCGO).

The boiling ranges of LGO, MGO and HGO can vary from refinery to refinery or the refinery may have only two gas oils which cover this boiling range.

Table 1.5 Selected specifications for on the road auto diesel oil, EN 590.

Property	Units	Minimum Value	Maximum Value
Cetane Index (CI)[16]		46.0	–
Cetane Number (CN)		51.0	–
Density at 15°C	kg/m^3	820	845
Sulphur content	mg/kg	–	10.0
Water and sediment	mg/kg	–	200
Polycyclic aromatics	% Wt.	–	4.5
Distillation, vol.% recovered, at:			
250°C	% V/V	–	<65
350°C	% V/V	85	–
95% recovered at:	°C	–	360

1.3.2.3 Heating Oils

Heating oils, also known as furnace oils, cover a wide range of boiling points and aromatic contents and often provide a sink for lower value refinery streams that cannot be used in other fuels. Such streams include aromatic middle distillates (also known as cycle oils), which are by-products of the FCC (after gasoline production), high boiling atmospheric, and vacuum distillates and residual fuels. Different burners for heating systems have varied tolerances for aromatics which, when burned, have a high oxygen requirement and a tendency to produce a sooty flame, or 'smoke'. Specifications limit the amount of aromatics through the use of the aniline point and smoke point tests [6]. Kerosenes with low aromatic content are more valuable as jet or diesel blend components; however, low aromatic/low sulphur grades of kerosene, which burn with little smoke, are also necessary for small, free-standing domestic paraffin heaters – these were widely used from about 1900 to the late 1980s when grades such as Esso Blue and Shell Pink were widely advertised.

Currently, in the UK, domestic heating oil (EN 2869 Class C2 [33]) is now kerosene that contains aromatic fractions – the difference between this and diesel fuel is clear from its smell. Oil heaters are now somewhat like mini-jet engines and have assured ventilation and exhaust systems. The older domestic, free-standing paraffin heaters relied upon a drip-feed system and no special ventilation arrangement. In larger heating facilities, and in general in Europe and North America, heating oil is a gas oil, somewhat similar to diesel fuel; however, it may contain more aromatic fractions, so it has a lower cetane number (45 minimum) and often has a higher cloud point. It is also used as off-road diesel in tractors, construction equipment, and boats. Heating oil usually has no duty and a low VAT, so it also contains a chemical marker and a dye for identification. In North America, heating oils (furnace oils) vary widely with region and may be a kerosene-like fuel ('No. 1 fuel') or a blend of No. 1 fuel and No. 2 fuel (see ASTM D396 [34] for fuel oil specifications).

While some off-road and heating oils had high sulphur contents (0.2% and 0.5% for Classes C2 and D) before 2010 [35], since then, off-road ('Red') diesel has a maximum of 10 ppm sulphur – gas oil specification EN 2869 Class A2 – as for modern on-road diesel. The class D gas oil specification now allows sulphur contents up to 1000 ppm (0.1%, EN 2869 Class D) for use as marine red diesel and static uses. In many parts of the world, various higher boiling gas oils, some of which contain residual fuels, are burned to raise steam in large-scale heating facilities or power stations.[17]

1.3.2.4 Marine Diesel Fuels and Power Generation

Marine diesel engines used a wide range of blends and specifications – a list of the specifications shows 4 distillate categories and 11 residual fuel categories

[36]. These contain high-boiling, vacuum-distilled fractions that are not found in on-road diesel fuel. Distillate heavy fuels are known as marine gas-oils, or just gas-oils. Residual fuel categories also contain the various residual fuels that are found in a refinery. Fuel storage vessels on ships are termed bunkers and the loading of ships fuels is 'bunkering' [37].

Atmospheric residue is now less commonly available because, in most refineries, it is vacuum distilled. Residual fuels, from the vacuum still or visbreaker, contain low solubility compounds known as asphaltenes. To guard against the possibility that the asphaltenes may separate out and to control their high viscosities, residual fuels are usually blended with aromatic light and heavy cycle oils. The sulphur contents of these fuels have been much higher than those of on-road diesels but have recently been sharply limited: in 2015 the maximum of 1.0% was reduced to 0.1% in European coastal waters and, in 2020, the maximum was reduced from 3.5% to 0.5% in the open sea [38]. At the time of writing, the whole range of marine fuel specifications have been or are being rewritten to meet the new limitations on sulphur contents and it appears that it may take some time for them to be followed everywhere [39]. Residual fuels may also end up in cokers – a more likely destination now that the worldwide sulphur specification for the heaviest marine fuels is limited to a maximum of 0.5%.

Heavy fuel oil (HFO) may be blended from any or all of these components [36]:

- Light and heavy gas oils (LGO and HGO);
- Light and heavy cycle oils (LCO and HCO);
- Waxy distillate (WAXD);
- Atmospheric residue (ARES);
- Vacuum residue (VRES);
- Visbreaker residue (VisRES).

Any of the distillates could be straight-run or by-product from the cracker or coker streams – they are more likely to be the low-value, by-product streams that cannot be used in the high-quality and high-value distillate products, such as on-road diesel. Cost is a particularly important factor for shipping which burns huge quantities of fuel in a highly cost-competitive business.

The large engines that are used to propel the marine world have been adapted to the production of electricity. For example, Wartsila, one of the major producers of marine diesel engines, including some of the largest of such engines, also produces power plants that use these engines [40]. There is a full range of diesel generators up to major power stations that may burn the heavy fuels also found in bunkers. Alternatively, these heavy fuels may be found in power plants in which the fuels are burned to produce steam to drive steam turbines for the generation of electricity.

1.3.4 Coal, Gas or Biomass to Liquids

The fuels discussed so far have come from crude oil. However, liquid fuels can also be obtained from other sources using the Fischer-Tropsch (FT) process. This was invented in Germany in the 1920s to converts a mixture of carbon monoxide (CO) and hydrogen (H_2) into a range of paraffins, at moderate pressures and temperatures over a range of catalysts [41,42]. The reactant mixture, known generally as producer gas, may be produced by blowing steam through a bed of burning carbon, coal or coke. Coal-rich countries without access to oil have made the earliest use of this Coal To Liquids (CTL) process: Germany, in the Second World War, and South Africa in response to oil sanctions in the 1950s and onwards. Sasol (founded in 1950 in Sasolburg, South Africa) still makes full use of this process. Now, methane (natural gas) is used as the preferred source of producer gas in the gas-to-liquids (GTL) process. Sasol International licenses its technology and provides FT conversion units to other countries with large coal reserves, such as China, or with large methane reserves, such as Nigeria [43]. Any carbon-containing fuel can be used as a source for the producer gas, and processes have been developed to use biomass such as straw and wood chips for Biomass To Liquids (BTL).

FT products are n-alkanes, so the lower boiling liquids have low octane numbers and need to be cracked and reformed to produce iso-paraffins and aromatic compounds that are suitable for use as gasoline. n-Alkanes have high cetane numbers, but those with middle distillate volatility and carbon numbers are waxes. These waxes are selectively hydrocracked and isomerised to produce liquid alkanes that have some alkyl branching while retaining a high cetane, making them suitable for diesel or jet fuel [44]. In Qatar, Shell Global have been responsible for the building of the world's largest GTL plant which produces low-sulphur naphtha suitable as chemical feedstock, kerosene for jet fuel and heating oil, gas oil for diesel fuel and paraffin waxes [45].

1.3.5 Biofuels

There is now a further fuel category: biofuel. European biodiesel contains Fatty Acid Methyl Esters (FAMEs), which are derived from natural sources such as soya, sunflower, rapeseed (canola), and palm oils [46]; also, used from cooking oil, recovered from major users such as fast food outlets. FAME has good cetane properties and low sulphur content (specified as EN 14214 [47]). Biodiesel containing FAME is described by its B# – for example, B10 contains 10% FAME and 90% petroleum diesel. Another, less common version is Hydrotreated (hydrogenated) Vegetable Oil (HVO), which has the advantage of being pure hydrocarbon, with a good cetane number. These hydrocarbons are mainly n-alkanes which contribute to the wax content of the fuel (see Chapter 4) so, as for GTL, they may need catalytic reforming to be converted into more soluble iso-alkanes.

For gasoline, alcohols and ethers are used as blend components under the label 'Oxygenates' [46]. Ethanol is the main bio-alternative, though consideration has also been given to butanol; both ethanol and butanol are made by fermentation processes. MTBE, methyl-t-butyl ether, was popular for a time, but its use was discontinued on concerns over drinking water contamination. Butanol is less volatile and more fuel-like (C_4H_9OH), with a higher proportion of hydrocarbon than ethanol (C_2H_5OH), but ethanol is currently accepted as the biological gasoline source of choice. The mixture of ethanol and petroleum-derived gasoline is sometimes referred to as gasohol and rated with an E# – for example, E10 refers to a gasohol containing 10% ethanol and 90% petroleum gasoline.

Biofuels are often blended into petroleum-sourced fuel, and the blend must meet the same specification requirements as 100% petroleum fuel. Some refiners and writers refer to biofuels as oxygenates and as additives. However, those in the additive industry consider them to be blend components, as distinct from performance-enhancing additives used at, usually, less than 1% of the fuel [48,49]. Blend components add volume and calorific value to a fuel while additives are added at such low levels that they do not add significantly to the fuel [50]. This distinction is worth remembering when looking at the statistics of quantities of fuel additives in use because some citations have included oxygenate blend components in their figures while others have not.

1.4 Fuel Additives – Value and Need

1.4.1 Value

Fuel additives are, in the most part, not sold as finished products to the end user, such as motorists, ship or truck operators and home heating consumers, but to the producers and marketers of bulk fuels [48]. Additive packages, at levels that are adequate to provide at least the minimum needs for fuel to meet specifications or consumer handling, are usually added at the refinery. Premium grades of fuel will contain higher levels of some components, such as the fuel detergent, and may be added either at the refinery or at the fuel distribution terminal [4] [18]. Relatively small amounts of fuel additives are sold directly to consumers in small packages with promises of improving their vehicles performance or protecting it from harm. Certainly, in the developed world, commercial fuels have to meet tight specifications so contain the required additives and more, for competitive reasons: commercially available fuels usually have no need for further additives.

Of the worldwide petroleum consumption total of 4662 million tonnes in 2018, 76%, or 3543 million tonnes per annum (mtoa), were liquid fuels [18]. Fuel additives are used at a range of total concentrations depending upon need and fuel grade. For gasoline and middle distillates these can range from 1 to 1000 ppm (parts per million, 0.0001 to 0.1%, by weight) of undiluted, active ingredient [48] – with a reasonable average being 500 ppm, a total that is usually a combination of additives. Examples of additive concentrations, or 'treat rates', in automotive fuels are found in a paper from the Technical Committee of Petroleum Additive Manufacturers (the ATC) in Europe [48] and, in general, in many patents. For example, for heavy fuels, tests on an asphaltene dispersant (stabiliser) were carried out at 700 ppm [51] or 2000 ppm [52] while a combustion improver was said to be used at preferred 10 to 30 ppm [53]. The reader is directed to relevant sections in this book for each additive type.

Assuming that all the world's fuel is treated with 500 ppm of additive (on average), this would indicate a potential total worldwide fuel additive use of 1.8 mtoa (in 2018). 'ATC estimates ... that the EU-27 market for fuel additives is over 200,000 tonnes per annum' in 2013 [54], which would be to treat about 15% of the world's liquid petroleum products, giving an estimated potential of about 1.3 million tpa. A published estimate of fuel additive volumes expected in 2016 is 26.5 million tonnes, but 94% of this is made up of ether oxygenates for gasoline [55] – more generally considered to be blend components and without which this estimate is for 1.6 million tpa for performance additives. The European ATC estimated a worldwide turnover of €7 billion in 2013 [54] while more recent figures are of $7.5 billion in 2017 and an outlook of $8.7 billion in 2023 [56].

1.4.2 Need

Fuel additives are needed to provide fuels with properties that they do not have after the refining processes, or to provide those properties that would cost much more to provide by a refining solution. These fuel properties are often seen as being essential such as the ability of the fuel to lubricate the fuel injection pump (lubricity), inhibiting autoxidation to prevent the formation of filter-blocking particulates during storage and the ability of diesel fuel to flow in cold weather (cold flow). It may be possible to solve some of these problems during refining but at much higher cost. For example, cold flow in diesel fuel can be provided by adding more kerosene; however, about 5% additional kerosene is needed for one degree centigrade of cloud point reduction while ten degrees of cold flow improvement may be obtained by the addition of 100 to 500 ppm of additive[19].

A problem with refinery solutions is that the improvement in one property may lead to the degradation of another property. Take lubricity, this is not a problem if sulphur is not removed at the refinery, but desulphurization of gasoline and diesel is now written into legally binding specifications as an environmental measure, almost worldwide. The process to remove sulphur also removes naturally occurring lubricity-providing compounds – lubricity additives replace this loss. With cold flow, more n-alkanes in diesel are good for cetane but they produce more wax and, hence, poor low temperature handling. In this case, a cetane improver can help cetane and a cold flow improver solves the low temperature handling [57]. The availability of these additives to meet the necessary performance specifications enables a refiner to optimise distillations and fuel component blends and so match production with demand more efficiently.

Competitive marketing makes good use of valuable features that a fuel additive may provide. For example, an anti-foaming agent enables quick filling of a fuel tank at the pump – especially useful for a 200-litre tank on a 40-tonne truck. The need for safe, trouble-free distribution favours the use of anti-foam, anti-static, biocide, and demulsifier additives. Detergents in automotive fuels are now seen as being essential for keeping critical parts of the fuel systems clean, particularly injector nozzles, which have extremely fine orifices through which fuel is sprayed. The use of fuel detergents reduces maintenance problems, fuel consumption, emissions and ensures a smoothly running engine.

Some advantages may seem small to the private motorist, such as the evidence-backed 3% fuel-saving claim for Shell Advanced Diesel (1988 [57]), but the value of this 3% to a truck fleet operator was €1,350 per year per 40 T truck (Goodyear's estimated value of a 3% fuel saving in 2008 [47,58]). Such savings are even more valuable to a shipping company – a 7000 teu[20] container ship typically burns 217 tonnes of fuel per day! [59]. On the other hand, the private motorist or householder with oil-fired heating would be terribly upset if their car or heating did not work because fuel deposits blocked the fuel lines. Most people would consider it unacceptable to permit any risk of failure in an aeroplane jet engine due to deposits or corrosion.

Leakages from corroded large storage tanks have been either a minor inconvenience or catastrophic, with terrible fire risk and outstanding environmental effects. A recent example is that of the scale of such problems is provided by the Norilsk oil spill starting on the 29 May 2020, when a fuel storage tank collapsed, spilling around 21,000 cubic metres of diesel fuel into the local river which is a tributary to other rivers downstream [60]. The company involved referred to softening permafrost as a cause but this tank had been identified years earlier as needing treatment because of extensive corrosion. At time of writing, it is reported that the Russian government is fining the company, Norilsk Nickel, $2bn as the cost of cleaning up the affected rivers [61]. So, a corrosion problem can not only cause damage to the environment but can also cost an errant company dearly.

1.5 The Application of Fuel Additives

Additives for some applications, such as detergents, are used in most fuels though specific details vary from fuel type to fuel type. Cold flow improvers may be used in all fuels except gasoline, but jet, diesel, marine-distillate, and marine-residual fuels all require different, specific cold flow improvers. Others are quite fuel specific: octane improvers are used only in gasoline and cetane improvers only in diesel. However, the physical chemistries relating to the modes actions of a range of detergents, lubricity improvers, cold flow improvers, anti-foam additives, and the rest, are usually the same across the full range of fuels. Also, the chemical entities of the various versions of a additive type are usually related. For these reasons, the approach taken in this book is to classify by additive type, since that best relates to additive chemistries and their modes of action.

An aspect of fuel additives that is sometimes overlooked is that they are used in a hydrocarbon medium. The active parts of additives are usually polar so will tend to aggregate in hydrocarbons. Consequently, the individual molecules need to carry substantial non-polar groups to confer solubility – they are, in fact, surfactants [62]. Surfactant chemistry is mostly concerned with aqueous solutions, in which the hydrocarbon segments of surfactant molecules aggregate leaving the polar groups in contact with the water [63,64]. A fuel additive may be added at only a few hundred parts per million to a fuel but the polar groups, having no solvation, must aggregate together – their dipole-dipole and hydrogen-bonded interactions outweighing the entropic advantage of separation. With the polar groups and water from the fuel on the inside and the hydrocarbon groups on the outside (the long tails), such aggregates are known as reverse or inverse micelles [65] (Figure 1.8). This arrangement is likely to affect the activities and interactions of different fuel additives, both in the dilute state in fuel and in the concentrated fuel additive package.

Additives are supplied to an oil company as a concentrated solution that is then, usually, automatically injected into a fuel stream – either during refinery fuel blending or further down the supply-line between fuel terminal and tanker. An oil company needing to use several fuel additives will not wish to use several different storage and injection systems but just one, for a concentrate into which all the required additives have been blended. Such concentrates can

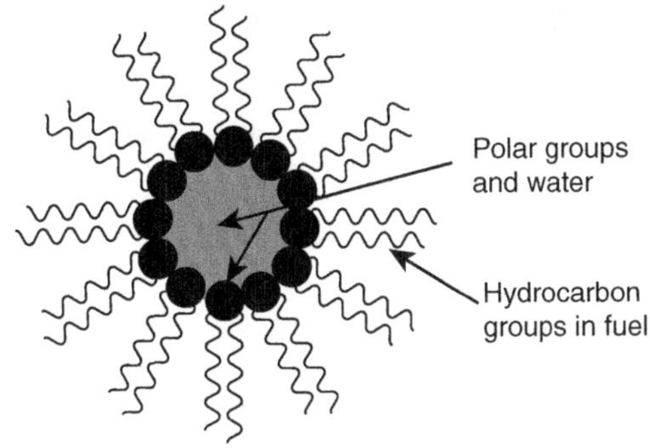

Figure 1.8 Diagrammatic Inverse Micelle.

suffer from physical instability, leading to undesirable phase separations with a different mix of additives in each separate phase – one objective of the development work of additives and their packages is to ensure the stability of the concentrates.

After decades of research, testing, development, and application of fuel additives, significant changes have occurred since the early 1990s. There has been a continual engineering push to improve engines for fuel efficiency and reduced emissions, aided by tightening of fuel specifications to help these improvements, backed by legislation. This progress has led to changes in fuels and more severe requirements in an engine. In addition, there has been the introduction of biofuels. These fuels, alcohols or esters (also known as oxygenates), provide significant levels of polarity and alter the solvency of the fuel, a factor that may affect the activity of the additives and which enables the fuel to absorb water. Other new fuels, such as GTL or HVO, are more petroleum-like and can provide high cetane numbers with little effect on the solvency or hydrophilicity of the fuel.

There are many fuel additive types available and their use may be optional or necessary, dependent upon fuel type. Also, some additive names vary depending upon the fuel in which an additive is used – for example, a friction modifier for gasoline is now termed lubricity improver when applied to diesel fuel. A summary of fuel additives and their applications to the different fuel types has been compiled as a guide (Table 1.6). It is worthwhile qualifying this list with a discussion by fuel type.

List compiled using references [7–9,50,57,66–72]. Heating oil is usually mentioned alongside diesel fuel but possible additives for heating oil are sometimes listed [73]. The Shell website [74] contains a list of additives that may be used in jet fuel (and 'Avgas'); the fuel additives sections of the Chevron [8,9], Lubrizol [68], Infineum [69] and Innospec [70] websites give descriptions of the problems that fuel additives solve in a range of fuels.

Table 1.6 Fuel additives and fuel type.

Additive Type	Gasoline	Jet Fuel	Diesel Fuel	Heating Oil	Heavy Fuel
Anti-icers	y	y	y	y	
Anti-static additives	y	Y	Y	y	
Biocides/Anti-foulant	y	y	y	y	y
Cold flow improver		(y)	y	Y	Y
Combustion improver	Octane improver		Cetane improver	Ferrocene	Y
Corrosion inhibitor	Y	y	Y	Y	Y
Particulate filter additives			Y		
De-/anti-foamers			Y	Y	
Demulsifiers/Dehazers	y		Y	y	
Detergent	Y		y		Y
Dispersant		y	y	Y	
Drag reducers	Y		y	y	
Dyes and markers	Y		y	y	
Emissions reduction	y		Y		
Friction modifiers	Y				
Lubricity additives		y	y	y	
Metal deactivators	y	y	y	Y	
Oxygenates/biofuel	Y		Y	y	
Re-odorant			y	y	
Silver corrosion inhibitors	Y				
Slag and ash inhibitors					Y
Stabiliser, antioxidants and metal deactivators	Y	y	Y	Y	y
Sulphur scavenger					Y

Additive use in jet fuel – also known as aviation turbine fuel – is particularly tightly controlled. Only specific additive types allowed and the individual, commercial compounds must be approved as additives for use in jet fuel (see Appendix D, Additives, in Aviation Fuel [29]). Jet fuels are specified internationally as Jet A-1 and, in N. America, Jet A; in addition, there are some national grades and several military grades. Here, use in Jet A-1 is taken as the major criterion: anti-oxidants and static dissipaters are required; anti-icing, corrosion and lubricity additives are used by agreement between jet fuel supplier and customer; metal deactivators and biocides are optional. Thermal stability additives are required in some special fuel grades that are used in USAF aeroplanes.

Both automotive fuels, gasoline and diesel, use most of the types of additives listed – especially detergents – with a few major exceptions. In gasoline, at one time octane improvers were widely used but, with the removal of lead anti-knock for environmental reasons, there is only a minor role for octane number improvers. Refiners usually depend upon blending together the right refinery components to attain the necessary octane number. Diesel fuel sometimes needs the addition of a cetane improver to meet the specification cetane number. Cold flow improvers are widely used in diesel fuel to modify the wax that separates in the cold, which is not a problem with gasoline.

Heating oil is often like diesel fuel but with a higher cloud point, so it also needs cold flow improvers. It often contains more aromatic compounds than does automotive diesel fuel and, since it is usually stored for many months before use, biocide and stability additives are needed. In some countries, such as the UK, domestic heating oil is usually kerosene – one containing much of the aromatic FCC by-product kerosene (also known as light cycle oil).

Heavy fuel, containing residual fuel, will have a relatively with high sulphur content (up to 0.5% from 2020) so has special problems of corrosiveness and may also need a hydrogen sulphide scavenger. While detergents may have some beneficial effect on the injectors, they are often added to stabilise the asphaltenes in the fuel, which may otherwise separate out on storage. When heavy fuels are burned in furnaces for steam generation in power stations, large heating installations or refineries, there is a need to control the build-up of ash in the form of a slag on the walls of the furnace, heat exchangers and exhaust systems, where it causes severe corrosion.

Certain problems apply to all fuels, such as the needs for corrosion inhibitors, biocides, and stabilisers, which are particularly necessary during long-term storage. Demulsifiers or dehazers are needed in any fuel that forms emulsions by agitation with any water that may be present in fuel tanks or lines. Distillate fuels usually need anti-static additives to prevent ignition from static sparks during the pumping of fuel through delivery hoses. Dyes and markers provide no technical advantage but are required to ensure tax is collected on motor fuel or that the correct grade of aviation gasoline is used.

1.6 Fuel Quality, Taxation, Dyes and Markers

1.6.1 The Need for Quality and Brand Recognition

Perhaps the earliest patent on the use of a dye to identify a fuel was in 1921, when there were concerns about the qualities of commercial motor fuels [75]. This early description of the problem to be addressed pointed to the fact that 'as is well known there are several grades of petrol' which may be of different qualities but were identical in appearance. Concern was that only the finest grades of petrol should be used in aeroplane engines, for there had been aeroplane crashes caused by engine problems attributed to the use of inferior grades of petrol. Not surprisingly, general concern was quickly raised by owners of the new internal combustion (IC) engines that powered their automobiles. To avoid the problem of being misled as to the quality of the petrol being supplied, it was proposed that an aniline dye should be added to the fuel to identify it as being of high quality. Some of the dyes suggested are still in use today, then referred to as Soudan 1, 2 or 3 or just 'Soudan[21] Red', to be applied at a low concentration of 2 ppm. Dyes are still used to visibly identify different grades of aviation gasoline that meet the varied requirements of different aircraft engines (Table 6.1). In 1931, invisible markers were proposed to enable a supplier to identify his grades of petrol or motor spirit; such markers would be detectable by some physical or chemical test [76].

Very soon in the history of internal combustion engine powered transport, the oil companies sold their motor fuels based on their trustworthy brands. As the private motor vehicle numbers expanded in the first two decades of the twentieth century, many gasoline-supplying companies appeared in the USA; the break-up of Rockefeller's Standard Oil company lead to the formation of six Standard Oil companies, Standard Oil of New Jersey became Esso, Standard Oil of New York became Mobil and so on, with others such as Gulf, Texaco and Atlantic-Richfield [5]. In the UK, the Anglo-American Oil company was another Standard Oil company, taking the name Esso in the 1930s. The Shell and BP brands also have their roots in this era. Brands took over as signs of quality so there was little need for dyes and markers to set them apart; there were very few patents on this subject until after the 1939–1945 war.

1.6.2 The Introduction and Growth of Fuel Taxation

In the early days of motor transport, motor fuel taxation was low and part of the total taxation of motor vehicles levied to pay for road building. In 1928, fuel tax amounted to 24% of the cost of petrol in the UK (at a price of 1s 6½d, now 8p, per imperial gallon) [77]. From 1937, UK motor taxes have been part of general taxation and, after the Second World War, began an inexorable rise to, now, 60% or more of the price paid at the pump. This is a universal picture, though European countries apply the highest rates with France, Britain and Germany at about \$3 per US-gallon, Asia-Pacific countries about \$2 per US-gallon and the USA about \$0.5 per US-gallon [78]. However, alongside of this, big exemptions were made for public transport, heating oils and off-road applications such as agriculture, mining, marine and construction. As a result, the big driver for the effective dying and marking of fuels across the western world became the protection of fuel taxation from the illegal use of fuels for which low rates of taxes are levied, such as the widely known 'Red diesel'. While dyes give an immediately visible identification of a low-taxed fuel, after its illicit removal such a fuel may pass as a fully taxed fuel. To counteract such a problem, further identification may be provided using markers that are not visible in dye-treated fuels but may be detected by some further physical or chemical analysis.

In the USA, Sudan Red and chemically similar red dyes have long been mandated to be added to heating oils and diesel fuel for off-road applications [50] to distinguish them from taxed motor diesel. More recently, the red dye has been applied to fuels with a high sulphur content to differentiate them from ultra-low sulphur automotive diesel fuels [79]. Red diesel was used in the UK and many European countries [80] for many years to indicate that it was for use in agricultural and off-road applications, for which there was a significant concession of reduced fuel taxation [81]. Around the world, shades of red are most widely used but yellow, blue and green are also used [81]. For example, in India there is concern about the quality of motor fuels, as indicated by the developments of their diesel and gasoline specifications (listed on the website of Indian Oil [82]) as these move to match the European specifications [83]. The Hindustan Petroleum Corporation has a fuel quality promise for its kerosene backed up by the use of a blue dye, which serves the dual purpose of identifying their quality kerosene and also enabling diesel or gasoline customers to check if their fuel has been adulterated by the blue-dyed kerosene [84]. The different grades of aviation gasoline are universally identified as red, purple, blue or green.

Taxes on motor fuels have risen over the years, particularly in European countries, since 1990 [85,86]. The difference in cost to a commercial transporter between fully taxed motor diesel (in the UK, 58p per litre of diesel fuel in 2012 [86]) and lowly taxed off-road red diesel (12p per litre) had made it worthwhile for some organisations to illegally remove the tagging dye from the cheaper fuel and share the cost saving with their customers[22]. The red dyes could be removed from the fuel by acid extraction, which leaves some acidity in the fuel and produces a potentially toxic waste that would be dumped illegally.

In 2002, following a proposal in 1995 to develop a suitable fuel marker to provide greater security [80], a European community mandate was made for the addition of solvent yellow 124[23] as the Euromarker to fuels for agriculture and marine use and for heating oil, at a concentration which does not add visible colour to the fuel [81]. Such fuels are often like highly taxed automotive diesel fuel for reasons of refinery efficiency. However, Solvent Yellow 124 can also be extracted from the fuel with aqueous acid as a bright red compound – the process by which it is detected (Figure 1.15 and Figure 1.16). To quote 'This has resulted in substantial losses of tax revenue' and many member states had pointed this out to the European Commission, leading to a detailed study and calling for new candidates in the search for a more secure fuel marker to work along with existing national dyes, in 2015 [87]. Clearly, the recognition of this major problem in Europe had not been lost on dye and marker manufacturers, whose patent activity in dyes, markers and their applications increased markedly in the 1990s, after a lull in the 1980s. In earlier years, dyes had been widely mentioned in patents for improved fuels but only as one of the many additives that come under the umbrella statement 'the fuel may also contain ...'.

An example of the lack of an effective answer to the need to mark low-tax fuels was the illegal tax evasion in Ireland involving the removal of dye from off-road diesel fuel. After an extended period of identification of the properties needed for secure markers and careful evaluation of candidates [87], new markers were introduced. The new technology has been demonstrated as successful in the UK, both in a 2016 news article [88] and in a 2018 Dow report [89] that provides some details of the efficacy of the introduction of the new fuel marker ACCUTRACE™ S10 in 2015 by UK and Irish authorities. The Dow report [89], which is a summary of official reports, shows a reduction of fuel-laundering plants operating in Northern Ireland from 38 in 2013/14 to 6 in 2016/17 and of hazardous fuel-laundering waste, that was collected, from 1237 tons to 148 tons on the introduction of this new fuel marker.

There had also been a burst of patent activity in the 1960s and 1970s in realisation of the need for reliable marking of fuels for tax reasons and the rapid growth in the volumes of motor fuels used. In this period, most of the patents were concerned with the solubility of the dye in hydrocarbon solvents, not only in fuel itself to ensure that the dye formed no deposits, but also in solvents such as xylene in order that the liquid solutions sold to refiners were as concentrated as possible while being stable. As with all fuel additives, the more concentrated the solution, the better the savings in cost of solvent, in transportation and storage of lower volumes, plus greater ease of injecting lower volumes into the fuel line exiting the refinery.

1.6.3 The Use and Chemistries of Fuel Dyes

Dyes have light absorption bands, due to electronic transitions, in the visible spectrum. Without an absorption band in the visible, compounds reflect all light so appear white. With absorption bands in the short visible wavelength of 400 to 430 nm, a dye would absorb in the violet end of the spectrum leaving an overall yellow appearance by the remaining reflected or transmitted light [90]. Similarly, the appearances of compounds having other absorption bands are well known (Table 1.7).

Dyes have particularly strong absorptions, giving colours at extremely low concentrations. Differences between light absorptions by similar solutions of different dyes can be in the order of powers of ten, so their efficiencies as colouring agents can differ widely which is then reflected in the cost required to produce a colouration. In practical terms, the dyes that are used in fuels will give a noticeable colour at extremely low concentrations – as low as 0.5 ppm in some cases. However, though they would be added at up to 10 ppm in order to ensure that the colour would be obvious to the eye and to take into account the possibility that the low-tax fuel may be used to adulterate a taxed fuel at, perhaps, 20% of the blend.

Initially, synthetic dyes were discovered and widely developed for colouring textiles, then inks and paints, well before they were used for colouring fuels – the first synthetic dye, mauveine, was invented by Perkin in 1856 and commercialised the following year. Diazo dyes were discovered and began their proliferation in the mid-1870s [91]. One of the earliest patents for the use of established dyes, in particular Soudan Red, for colouring fuel was in 1921 [75]; this patent was a claim for the use of any petrol soluble and water insoluble dye, particular aniline/azo dyes, for the purpose of identifying a fuel of adequate quality.

The issue of solubility in a concentrate, rather than in a fuel at 10 ppm, was realised as necessary for supplying a fuel dye in as little solvent as possible for

Table 1.7 Dye absorption bands and colour.

Absorption Band (wavelength, nm)	Appearance, Colour
430–480	Orange
480–550	Red
550–600	Violet
600–700	Blue
400–450 and 580–700	Green

easy storage and injection into fuel, at a much lower concentration. When interest in patentable aspects of dying fuels re-emerged in the 1960s, it was often of known dyes that were derivatised to ensure that they had high solubilities in organic solvents such as xylene. The simple azo dye structure of aniline diazotised and coupled with N,N-diethylaniline gave N,N-diethyl-4-phenylazo-aniline, now known as Solvent Yellow 56 (Figure 1.9), which adds a yellow colour to petrol at only 2 ppm. The N,N-diethyl-4-phenylazo-aniline version had increased solubility (35% to 50% in xylene) [92,93] over the pre-existing N,N-dimethyl-4-phenylazo-aniline dye, enabling a more concentrated commercial solution[24]. A further improvement, to enable a concentrate of up to 60% in xylene, was claimed for the azo dye from the coupling of the benzene diazonium ion with N,N,N′,N′-tetraethyl-1,3-diaminobenzene [94].

For use in oils and fuels, red dyes became widely used – these were the Sudan Reds, or Solvent Reds, that are azo and diazo compounds. These dyes have a diazo link to a naphthalene molecule and usually have a hydroxyl, alkyl ether or alkylamino substituent on the benzene or naphthalene group, such as 1-(2-methoxyphenolazo)-2-naphthol, known as Sudan Red G, or solvent Red 1 (Figure 1.10) [95].

Bis(diazo) compounds have extended delocalisation and are variations of early Sudan Reds. They have improved fuel colouring properties, such as Solvent Red 23, 1-(4-(phenyldiazenyl)-phenyl) azanaphthalen-2-ol (Figure 1.11). More recent variations on this structure were made to provide better fuel solubility and better solubility in an aromatic solvent, to allow higher concentrations in commercial solutions. A methylated variation, Solvent Red 26 (Figure 1.11) was in use over many years, in the USA (at 11 ppm), for tagging low tax heating oil

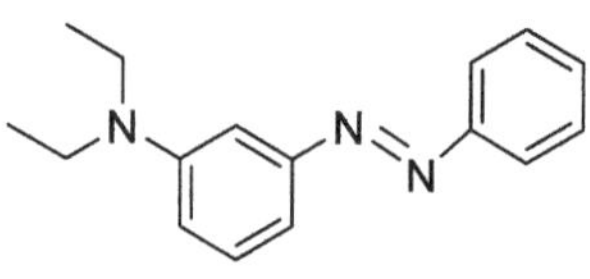

Figure 1.9 Solvent Yellow 56.

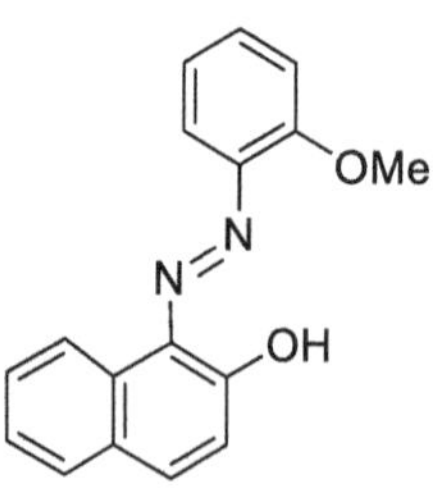

Figure 1.10 Sudan Red G.

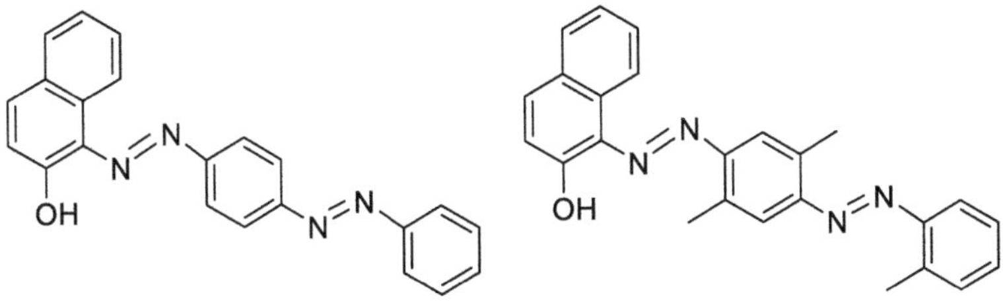

Figure 1.11 Sudan Reds, Solvent Red 23 and 26.

to prevent its being added to fully taxed diesel fuel [79]. A further variation, with a heptyl group on the naphthalene ring, solvent Red 164 [96], was introduced to tag high sulphur fuel when ULSD was introduced in the USA.

Such structures, widespread among the various versions of solvent red, are prepared by diazotising 4-amino-azobenzene and reacting it with 2-naphthol [97]. The various alkyl substitutions depend upon the alkyl anilines used to prepare the 4-amino-azobenzene and the alkyl naphthol used.

At the same time as there was a proliferation of azo dyes, a relative of an ancient red dye, madder, was brought into play as a fuel colourant. Madder contains alizarin, 1,2-dihydroxy-anthraquinone, but 1,4-dihydroxyanthraquinone is more easily synthesised from phthalic anhydride and hydroquinone, so this was the product, known as quinizarin or Solvent Orange 86 (Figure 1.12), used as a dye for fuel. However, the low solubility of quinizarin made its application difficult so the version alkylated with 2-ethylhexene was preferred for its good solubility in aromatic solvents (Figure 1.12) [98]. The 2-(2-ethylhexyl)-quinizarin may be used at very low concentrations in a distillate fuel, even as little as 0.2 ppm, when its colour is not apparent but it is readily extracted by alcoholic[25] aqueous caustic soda to give a deep purple aqueous layer [98]. In addition, quinizarins fluoresce under UV light so, with two methods of detection, they are among the earliest invisible markers.

Dyes that are green or blue are also based on the anthraquinone structure – the colour shift is obtained by amine substituents instead of the hydroxy substituents. For example, Sudan Blue II (Solvent Blue 35) is 1,4-bis(butylamino)anthracene-9,10-dione (Figure 1.13). Blue is in use in the republic of Ireland and Denmark [80] and is one of the dyes used to mark a grades of aviation gasoline [29].

A green dye is used in Norway for agricultural fuel [81] and in Italian gas oil [80]. The green dye used in fuel can be a combination of blue and yellow or Solvent Green 33, which is 1,4-bis(4-methylanilino) anthracene-9,10-dione (Figure 1.14).

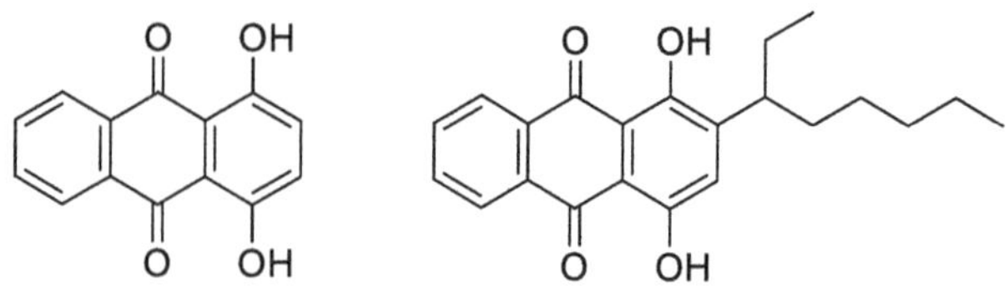

Figure 1.12 Quinizarin and 2-(2-Ethylhexyl)-Quinizarin.

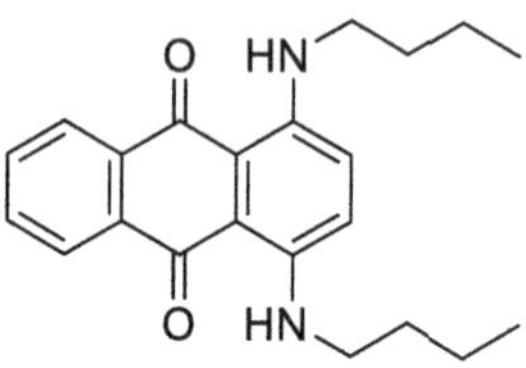

Figure 1.13 Solvent Blue 35.

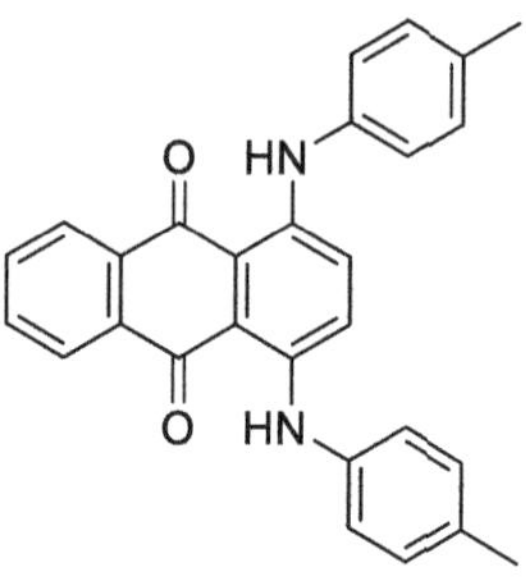

Figure 1.14 Solvent Green 33.

1.6.4 Invisible Fuel Markers

Solvent Yellow 124 (Figure 1.15) was mandated by the EU from 2002 for use as a common marker for fuels having concessionary low taxes [99]. Solvent Yellow 124 is N-ethyl-N-[2-(1-isobutoxyethoxy) ethyl]-4-(phenylazo)aniline, which is chemically stable, very oil soluble, easy to detect at very low levels and supposedly resistant to removal [80]. At the recommended concentration (6 ppm), Solvent yellow 124 is a colourless marker that is detected by extraction into dilute HCl, which hydrolyses the acetal group and protonates the product (Figure 1.16) to give an intense red colouration [99,100]. It is also detectable, quantitatively, in the presence of other commonly used fuel dyes, by its UV/visible absorption at 450 nm [101].

The use of Solvent Yellow 124 was not the first colourless marker (at its recommended concentration) to be used either alone or in combination with a

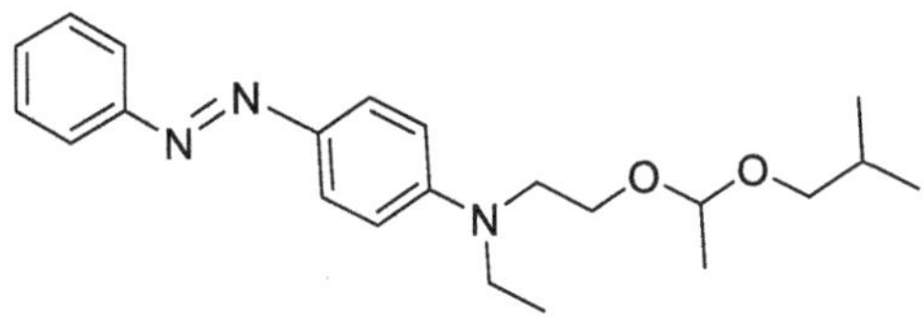

Figure 1.15 Solvent Yellow 124.

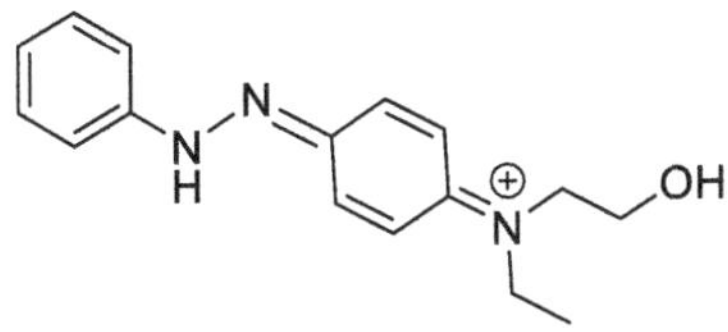

Figure 1.16 Hydrolysed Protonated Solvent Yellow 124.

visible dye. A variety of invisible markers, and means by which they should be detected, had been proposed in 1931 [76]: anthracene and its derivatives, which may be detected at low levels by their UV-fluorescence, and the indicator phenolphthalein, which turns bright pink when extracted into aqueous alkali. Furfural has been widely used in the past as a gasoline marker – it could be detected at extremely low concentrations by extraction with aniline in acetic acid to give an intense blue colour [98,102,103]. However, furfural fell out of use because of drawbacks: false positives due to low levels of contamination between fuels at the refinery, it decomposes on storage of the treated fuel and may be hidden by the coloration that fuels take up on ageing [102].

Another group of invisible markers that has been proposed [104] are those that rely upon a colour that develops on reaction with a diazonium salt. An invisible marker such as N-ethyl-N-(2-hydroxyethyl)-4-methylaniline is used to mark the fuel and is detected by contacting it with a diazonium salt of 1-aminoanthraquinone[26] in one of various forms (aqueous solution adsorbed onto titanium dioxide powder or onto a test paper) to give a dark blue-red colour (a wide range of options are described in the patent [104]).

The most recent development of invisible markers is from the Dow Chemical company. Dow® has been actively patenting many poly-aryls and their alkyl ethers, amine derivatives and others[27]. The basic concept is that the marker is a stable compound that behaves much like a fuel molecule, is not extractable but, being a compound that does not occur naturally, is readily detected by gas chromatography with a mass spectroscopy analyser [105–109]. The patents propose compounds such as 2,4,6-substituted-1,3,5-triazines, triphenylamines, bis-(aryl)-diaryl-ethers and trityl-phenol derivatives (such as those in Figure 1.17) – all of which may bear further substituents on any or all of the benzene rings. Their molecular weights can be varied to match the volatilities of

2,4,6-Triphenoxy-1,3,5-Triazine

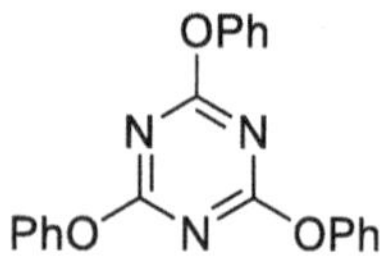

4,4'-Bis-Methylphenoxymethyl-1,1'-Biphenyl

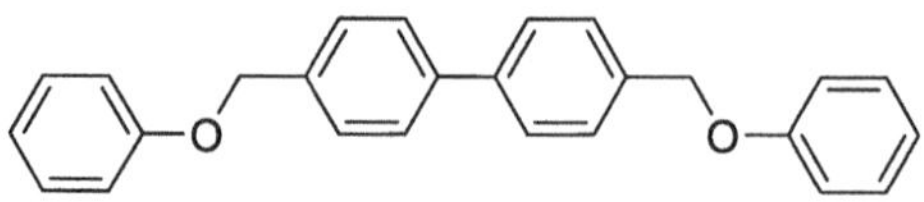

2-sec-Butyl-4-Tritylphenol

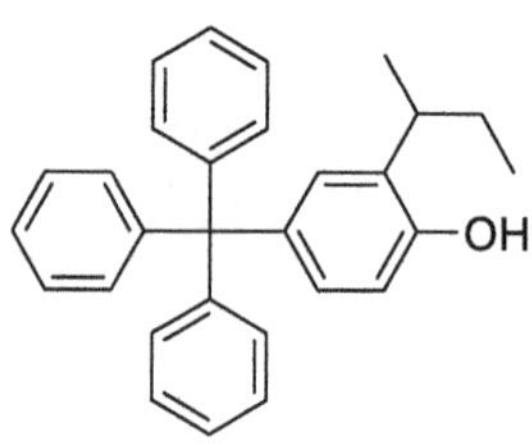

Figure 1.17 Some Markers in the Dow Technology. 2,4,6-Triphenoxy-1,3,5-triazine [105], 4,4′-bis(3-methylphenoxymethyl)-1,1′-biphenyl [106] and 2-(sec-butyl)-4-tritylphenol [108].

gasoline, kerosene or a middle distillate, as well as to provide a myriad of markers of different molecular weights.

The Dow® inventors have proposed that chosen combinations of these compounds could provide a wide range of fuel digital markers. Dow® has also patented a GC-MS method for detecting these compounds in petroleum fuels, singly or in mixtures, from their characteristic m/e mass spectroscopy signals to provide the digital markers [110]. For example, 2,4,6-triphenoxy-1,3,5 triazine in diesel fuel (7 ppm) was detected by its characteristics m/e signals by GC-MS at 264, 238, 145 and 121, which provide its digital signature [105]. The marker system that was reported to have been introduced in 2015 in the UK and Ireland, ACCUTRACE™ S10 [88,89], is probably an example of this chemistry and procedure. The digital marker system is also proposed to provide the identification of fuels in spills that, usually, end up contaminating water and are then untraceable.

1.7 Future Need for Fuel Additives

At the time of writing, the media are full of discussions about electric cars and trucks and their taking over from the traditional Internal Combustion Engine (ICE)- driven vehicles. One could be forgiven for the thought that this may mean the end of the market for petroleum liquid fuels and hence the demand for fuel additives. However, revolutions of this sort do not occur overnight. BP statisticians have made projections of the demand for the different fuels up to 2040 [111], based on current statistical data and trends. In summary, BP statisticians conclude that the demand for oil as a primary fuel will peak around 2030 and will be little changed between 2020 and 2040.

ExxonMobil have made similar projections [112]. Of total energy, electrical energy consumption is growing rapidly and is projected to continue to do so, while the demand for petroleum transport fuels is expected to peak around 2025 and remain steady until 2040. Oil demand for light-duty vehicles is expected to enter a slow decline from 2025, but this is balanced by the growth in the demand from commercial transport (trucks, aviation, marine and rail). The decline in light-duty use of oil is a combination of more fuel efficiency and a gradual displacement by electric vehicles, alongside an almost 50% increase in vehicle numbers between 2017 and 2040. Commercial transport energy demand is expected to increase by about 50% by 2040, driving the rise in oil demand from this sector, while the expectation is that 'Electrification plays a role in certain applications (e.g. short-haul trucks and buses) but electricity in commercial transportation grows slowly due to upfront costs, range limitations, payload requirements, and the pace of infrastructure development' [112]. The growth in the demand for energy and motor vehicles follows from the world's expanding population and its enrichment. At the same time, the energy efficiency of ICE vehicles will continue to grow, and their emissions will fall – as in the past, fuel additives will aid such developments.

It is worth remembering that over the most recent ten year period (2007–2017), that oil consumption has fallen by a small proportion in the west: little change in N. America, 1,105 Mtoe to 1,113 Mtoe, and a fall of 817 to 742 Mtoe in Europe[28]; while that in the Asia-Pacific region has increased dramatically, from 1,250 to 1,695 Mtoe [113]. In the ten years to 2018, the worldwide split between light and middle distillate fuels (mainly gasoline and diesel, respectively) has moved very little, from consumptions of about 33%/36% light/middle distillates in 2008 to about 31%/36% in 2018.

With respect to the electrification of the transport sector, the enormity of this is challenging: in 2017, for the EU-28, transport, mostly reliant upon oil products, took 31% of the total end use of energy (a total of 1,060 Mtoe) while electricity provided only 23% [14]; clearly, in order to convert the whole of transport's use of energy to electricity, the EU-28 would need to increase the production of electricity by 135%. While the richer countries currently have an excess of electricity-generation capacity over demand, it may be less when considering the reliance on sources that are productive for only part of the time (e.g. solar cells); this is allowed for by the 'de-rated supply' which has a relatively small margin over winter peak demand [114]. The effects of a major proportion of vehicles plugging-in on a winter evening – an event that may create double the current electricity demand – may be unwanted. However, while the proportion of electric vehicles is small, it is growing rapidly.

Many point to the rapid growth in the renewable energy sector, particularly wind-power and solar cells. However, these renewable power sources are, for years ahead, needed to replace the electricity lost by the closures of coal and oil-fired power stations. Combine these uncertainties with the current low level of investment in new oil production as major oil companies, such as Shell and BP, commit themselves to supporting the change to renewable, clean power, plus the growth in demand from the developing world, and the future decade or two may see a shortage of oil. This appeared to be Exxon's view [115]. It could be that, as it has before in its history [5], oil could still spring a few surprises.

Notes

1 The oil industry uses two measures of oil quantities, barrels per day and tonnes per annum. Switching between these is confusing, so quantities in this book are mostly in tonnes and tonnes per annum.

2 Meaning the vacuum distillation temperature being corrected to one atmosphere pressure.

3 °C, corrected to 760 mm Hg pressure, as a Normalised Boiling Point, NBP. For Kuwait export crude.

4 In Figure 1.2, the Kuwait example, the gasoline output would be a blend of the light naphtha and some of the heavier naphtha.

5 'Mtoe' is million tonnes of oil equivalent – one toe is a common unit to compare quantities of other fuels, which are converted to an amount of oil that would contain the same calorific value, one toe defined as containing 10 gigacalories [1].

6 Light distillate are gasoline and naphtha cuts; middle distillates are jet, kerosene, diesel and gas oil.

7 Heavy distillate fuels are included as middle distillates while residual fuels are included in fuel oil.

8 How much the economy depends upon energy as processes and engines become more efficient and economies rely more upon services and less upon manufacturing. Measured as how much energy is needed to produce a unit of Gross Domestic Product.

9 While in the FCC, large molecules are cracked into smaller, unsaturated molecules with the loss of hydrogen, in the hydrocracker, the free radical ends of the smaller molecules produced are trapped by hydrogen to give saturated hydrocarbons. Ranges of catalysts that include other metals are also used.

10 The butylenes, propylene and isobutene are collected as by-products of the FCC and the coker.

11 RON and MON are measured in a gasoline test engine which has variable compression, in comparison with mixtures of iso-octane (octane number 100) and n-heptane (octane number 0). The same engine is run under different conditions for RON and MON and some specifications take an average of the two.

12 Gasoline engine compression ratios are usually between 10 and 14 times.

13 Also known as iso-octane which is defined as the standard for a RON of 100.

14 Other properties that are specified as measures of the quality of gasoline fuels are gum content, oxidative stability, water tolerance, sulphur content (10 or 15 ppm maximum USA and EU) and corrosiveness [9].

15 Cetane numbers are measured on a standardised diesel engine with variable compression. The compression ratio is varied until the delay between fuel injection and ignition is 2 milliseconds; the cetane number is then defined by the mixture of cetane and iso-cetane that gives the same delay.

16 Cetane Index is a cetane measure that is calculated from the density and distillation of a fuel.

17 More detail on this in the chapter 5 on Corrosion Inhibitors, as such fuels are more corrosive than ULSD.

18 Large quantities of bulk fuels are transported by pipeline from refineries to terminals, close to centres of population, at which tankers are loaded for local delivery.

19 Finding an additional 50% of kerosene would disrupt a refiner's ability to match the demand barrel.

20 TEU – Twenty-foot Equivalent Unit, a standard shipping container 20 feet long, 8 feet wide.

21 Early references spell this as Soudan [75] and recent documents (such as in Wikipedia [79]) use the spelling Sudan, though the most recent nomenclature format for non-aqueous soluble dyes is 'Solvent #', so Soudan (or Sudan) Red or Sudan 3 or Solvent 23 are the different names for the same dye.

22 In Germany, 2012, red fuels carried no duty [116] but the duty on motor fuels was 65.5-euro cents per litre on gasoline and 47 cents on ULSD [78].

23 Solvent Yellow 124 turns red if extracted from fuel into dilute hydrochloric acid.

24 The N,N-diethylaniline products were covered in both patents published in 1969 but the first application for BASF's patent [92] was in January 1967 (in Germany) while for ICI's patent [93] it was in September 1965.

25 Methanol or ethanol preferred.

26 The diazonium salt of 1-aminoanthraquinone is exceptionally stable at ambient temperatures, compared with diazotised aniline which decomposes above 10°C.

27 Initial patent applications were made by Angus Chemical company, then a subsidiary of Dow®, and Rohm and Haas, which was acquired by Dow in 2009, and were granted under Dow's name.

28 These are the figures for the whole of Europe, including Switzerland, Turkey, the Ukraine and 'other Europe' while the figures in section 1.2.2b were for the European Union only.

References

1 BP Statistical (2018). Review of world energy. BP.com. https://www.bp.com/content/dam/bp/en/corporate/pdf/energy-economics/energy-outlook/bp-energy-outlook-2018.pdf (Cited: 13 November 2018).

2 British Coal Corporation (1994). The role of fuel additives to control environmental emissions and ash fouling. european commission technical coal research. catalogue number CD-NA-14893-EN-C.

3 Libutti, B.L. and Pall, R.L. (1978). Criteria for selection of coal additives. Drew Chemicals, now part of Ashland Specialty Chemical Co. https://web.anl.gov/PCS/acsfuel/preprint%20archive/Files/23_1_ANAHEIM_03-78__0206.pdf (Cited: 4 November 2014).

4 Richards, P. (2014). *Automotive Fuels Reference Book*, Third edition. Society of Automotive Engineers, Warrendale, P.A. ISBN 9780768006384.

5 Yergin, D. (2009). *The Prize, the Epic Quest for Oil, Money & Power*. The Free Press: New York.

6 Fahim, M.A., Al-Sahhaf, T.A., and Elkilani, A.S. (2010). Fundamentals of petroleum refining. Elsevier B.V.

7 Speight, J.G. (ed.) (2007). *Chemistry and Technology of Petroleum*, 4th edition. Hoboken: CRC Press, Taylor and Francis.

8 Chevron. (2007). Diesel fuels technical review. chevron.com. https://www.chevron.com/-/media/chevron/operations/documents/diesel-fuel-tech-review.pdf (Cited: 26 November 2012).

9 Motor Gasolines Technical Review. (2009). chevronwithtechron.com. http://www.chevronwithtechron.com/products/documents/69083_MotorGas_Tech_Review.pdf (Cited: 26 November 2012).

10 Simple, A. Guide to oil refining. ExxonMobil.com. http://www.exxonmobil.com/UK-English/files/simpleguideoilrefining.pdf (Cited: 15 October 2014).

11 Fawley Refinery (2019). ExxonMobil.co.uk. 9 May https://www.exxonmobil.co.uk/Company/Overview/UK-operations/Fawley-refinery (Cited: 4 April 2020).

12 Parkash, S. (2003). Refining processes handbook. Elsevier, ISBN075067721X.

13 Stratiev, D., Dinkov, R., Petkov, K. (2010). et al. Evaluation of crude oil quality. *Petroleum & Coal* 52 (1): 35–43. ISSN13377027.

14 Energy balance sheets, (2017) DATA. 2019 Edition. ec.europa.eu/eurostat/documents. August https://ec.europa.eu/eurostat/documents/3217494/10077623/KS-EN-19-001-EN-N.pdf/59b44e6f-ff33-488b-a85f-9c4f60703afc (Cited: 29 March 2020).

15 Cooper, J. (ed.). (2019). Refining products for our everyday life, Statistical Report 2019. fuelseurope.eu. https://www.fuelseurope.eu/wp-content/uploads/FuelsEurope-Statistical-Report-2019-2.pdf (Cited: 29 March 2020).

16 Commission, European. (2017). Oil and petroleum products – A statistical overview. ec.europa.eu/eurostat.

17 Short-Term Energy Outlook Data Browser. (2019). U.S.energy information administration. https://www.eia.gov/outlooks/steo/data/browser (Cited: 31 March 2020).

18 BP Statistical Review of World Energy (2019). 68th edition, Oil. bp.com/statisticalreview. June 2019. https://www.bp.com/content/dam/bp/business-sites/en/global/corporate/pdfs/energy-economics/statistical-review/bp-stats-review-2019-oil.pdf (Cited: 30 March 2020).

19 Cuthbert, N., Leavens, J., Kennaby, D. et al. (2011). *Developments in the International Downstream Oil Markets and Their Drivers: Implications for the UK Refining Sector*. UK Government. June. https://www.gov.uk/government/publications/developments-in-the-international-downstream-oil-markets-and-their-drivers-implications-for-the-uk-refining-sector (Cited 17 October 2014).

20 Trends in New Car Registrations. (2013). acea website. 1 January http://www.acea.be/uploads/statistic_documents/20140225_Diesel_1312_1.pdf (Cited: 19 October 2014).

21 Domm, P. (2013). US becoming ‘Refiner to the world’ as diesel demand grows. CNBC website. 7 August http://www.cnbc.com/id/100943620 (Cited: 19 October 2014).

22 Trends in fuel type of new cars between 2015 and 2016. (2017). ACEA, european automobile manufacturers association. 15 February https://www.acea.be/statistics/article/trends-in-fuel-type-of-new-cars-between-2015-and-2016-by-country (Cited: 30 March 2020).

23 German Carmakers Exhausted. (29 July 2017). *The Economist*, 424(9051), 57. No ISBN with this magazine.

24 The Automobile Industry Pocket Guide 2019-2020 (2020). www.ACEA.BE. https://www.acea.be/uploads/publications/ACEA_Pocket_Guide_2019-2020.pdf (Cited: 2 April 2020).

25 Unleaded Petrol BS EN 228:2008.Mabanaft. (2011). http://www.mabanaft.co.uk/Mabanaft/en/services/grosshandel/Mabanaft_Limited/sicherheitsdatenblaetter.php (Cited: 21 October 2014).

26 Olivares, E. and Foukes, E. (December 2012). EU: Fuels: Diesel and Gasoline. Transport Policy.net. http://transportpolicy.net/index.php?title=EU:_Fuels:_Diesel_and_Gasoline&oldid=8446 (Cited: 21 February 2013).

27 Theiss, T., Alleman, T., Brooker, A. et al. (2016). *Summary of High-Octane, Mid-Level Ethanol Blends Study*. Oak Ridge National Laboratory. Report: ORNL/TM-2016/42 for US DOE.

28 Chevron Global Aviation – Aviation Fuels Technical Review. (2006). cgabusinessdesk.com. https://www.cgabusinessdesk.com/document/aviation_tech_review.pdf (Cited: 21 October 2014).

29 ExxonMobil Aviation: World Jet Fuel Specifications. (2005). ExxonMobil web site. http://www.exxonmobil.com/AviationGlobal/Files/WorldJetFuelSpecifications2005.pdf (Cited: 21 October 2014).

30 Balaban, A.T. (1983). Topological indices based on topological distances in molecular graphs. Pergamon Press Ltd, UK, *IUPAC, Pure & Applied Chemistry* 55 (2): 199–206.

31 Petroleum and Coal, The Chemistry of Petroleum Products. Purdue University Online Chemistry Courses. http://chemed.chem.purdue.edu/genchem/topicreview/bp/1organic/coal.html.

32 Sulphur Free Diesel BS EN 590:2009. (2009). Mabanaft.co.uk. http://www.mabanaft.co.uk/Mabanaft/en/services/grosshandel/Mabanaft_Limited/sicherheitsdatenblaetter.php (Cited: 21 October 2014).

33 Kerosene BS EN 2869:2010Class C2. Mabanaft Ltd. http://www.mabanaft.co.uk/Mabanaft/en/services/grosshandel/Mabanaft_Limited/sicherheitsdatenblaetter.php (Cited: 24 October 2014).

34 ASTM D396.2002 US Fuel Oil. (June 2002). Public.Resource.Org.Inc[US]. https://law.resource.org/pub/us/cfr/ibr/003/astm.d396.2002.pdf (Cited: 27 October 2014).

35 Engineering Toolbox. (2003). Classification of Gas Oil. engineeringtoolbox.com. https://www.engineeringtoolbox.com/classification-gas-oil-d_165.html (Cited: 6 April 2020).

36 Vermeire, M.B.. (June 2012). Everything you need to know about marine fuels. http://www.chevronmarineproducts.com/docs/Chevron_EverythingYouNeedToKnowAboutFuels_v3_1a_DESKTOP.pdf (Cited: 23 October 2014).

37 In Use Marine Diesel Fuel, EPA420-R-99-027. US EPA web site. (August 1999). http://www.epa.gov/otaq/regs/nonroad/marine/ci/fr/dfuelrpt.pdf (Cited: 21 October 2014).

38 Benwell, S. and Corradi, M. (3 December 2019). Heading for IMO 2020. Inevitable change for marine industry stakeholders ahead. Insight. https://www.infineuminsight.com/en-gb/articles/fuels (Cited: 6 April 2020).

39 Stone, A. (7 April 2020). New ISO standards expected to improve the quality of marine fuels. F+L Magazine.

40 Wartsila. (2016). Power Plants Solutions 2016. Wartsila.com. (Cited: 12 February 2017).

41 Fischer, F. and Tropsch, H. Process for the Production of Paraffin-hydrocarbons with more than One carbon atom. GB Patent 255,818. Filed 26 March 1926.

42 Process for the Production of Paraffin-Hydrocarbons with more than One Carbon. US Patent 1,746,464. Filed 1 April 1926.

43 Exciting Deverlopment at Escravos. (April 2015). https://www.sasol.com/about-sasol/group-functions/technology/news (Cited: 7 April 2020).

44 Schrauwen, F.J.M. (2003). Shell middle distillate synthesis (SMDS) process, Ch.15.3. In: *Handbook of Petroleum Refining Processes*, 3rd e. (ed. R. Meyers), 15.26–15.39. McGraw-Hill Professional. ISBN-10:0071391096

45 Shell Global, Gas to Liquids. Discover the processes at Pearl GTL. Shell.com. 13 October 2014. http://www.shell.com/global/future-energy/natural-gas/gtl.html.

46 ACEA (2019). *Worldwide Fuel Charter*, 6th e. Gasoline and Diesel. Fuel.acea.be. 28 October. https://www.acea.be/publications/article/worldwide-fuel-charter-2019-gasoline-and-diesel-fuel (Cited: 7 April 2020).

47 Wikipedia Contributors (2008). *EN14214 – European Standard for FAME*. Wikipedia. http://en.wikipedia.org/w/index.php?oldid=502082859 (Cited: 2 November 2014).

48 ATC Europe. (2004). Fuel additives and the environment. atc-europe.org. 1 January http://www.atc-europe.org/public/doc52.pdf (Cited: 1 October 2013).

49 Henly, T.J., Kulinowsky, A.M., and Roos, J.W. (1999). The role of fuel additives in formulating clean transportation fuels. In: *Designing Transportation Fuels for a Cleaner Environment* (ed. J.G. Reynolds and M.R. Khan), 61–77. London: Taylor and Francis Group.

50 Reynolds, R.E., Kasperson, A., Herwick, G. et al. (2007). Changes in diesel fuel. scribd. http://www.scribd.com/doc/220519094/Changes-in-Diesel-Fuel (Cited: 26 November 2012).

51 Hertel, G., Feustel, M., Bystrom, H. et al. (1997). Heavy oils having improved properties and an additive therefor. US Patent 6,488,724. Filed USA, 27 October Clariant GmbH.

52 Bernasconi, C., Faure, A., and Thibonnet, B. (1986). Homogenous and stable composition of asphaltenic liquid hydrocarbons and additive useful as industrial fuel. US Patent 4,622,047. Filed USA, 6 January.

53 Thunker, W., Lohmann, G., Marschewski, A. et al. (1997). Use of Ferrocene. US Patent 5,746,784. Filed USA, 14 April.

54 Fuel Additives: Use and Benefits, ATC Doc. (2013). 113. atc-europe.org. http://www.atc-europe.org/public/Doc113%202013-11-20.pdf (Cited: 2 February 2014).

55 Global Demand for Fuel Additives to Surge by 2016. (2013). Freedonia. 21 January Oil & Gas Journal, ProQuest, 23.

56 Fuel Additives Market. May (2019). https://www.marketsandmarkets.com/Market-Reports/fuel-additives-market-723.html (Cited: 8 April 2020).

57 Garrett, T.K. (1994). Chapter 2. Fuel quality and additives. [book auth.]. In: *Automotive Fuels and Fuel Systems*, 2 (T.K. Garrett), 12–24. London: Diesel. Pentech Press.

58 Factors Affecting Truck Fuel Economy. (2008). goodyear.eu.com. https://www.goodyear.eu/pl_pl/images/truck-fuel-economy_tcm1330-81819.pdf (Cited: 8 April 2020).

59 How much fuel does a container ship burn? (2009). Answers.com. October (Cited: 21 June 2013). Answers.com > Wiki Answers > Categories > Cars & Vehicles > Miscellaneous Vehicles > How much fuel does a container ship burn?

60 Wikipedia Contributors. (2020). Norilsk oil spill. the free encyclopedia. 27 June (Cited:8 July 2020). Page version ID: 964715460.

61 Seddon, M. (2020). *Cleaning up Russia His Norilsk Nickel with Record $2bn Fine for Arctic Rivers Diesel Spill.* Financial Times (UK). 8 July.

62 Jao, T.C., Passut, C.A., and Zoller, U. Applications of Surfactants in Lubricants and Fuels. Marcel Dekker A.G., 2009, Surfactant Science Series, 141, 331–344.

63 Tadros, T.F. (2007). Ch.1 General principles of colloid stability and the role of surface forces. colloid stability. In: *The Role of Surface Forces*, 1, 1–22 (Ed. T.F. Tadros), 17–27. Weinheim: Wiley-VCH. ISBN: 9783527314621.

64 Cosgrove, T. (2010). *Colloid Science, Principles, Methods and Applications*, 2nd e. Blackwell Publishing Ltd. ISBN: 9781444320190.

65 Strestha, L.K. and Aramaki, K. (2011). Structure of nonionic surfactant micelles in organic solvents. [book auth.], In: *Self-Organised Surfactant Structures* (ed. T.F. Tadros), 17–57. Wiley-VCH.

66 Owen, K., Coley, T.R., and Weaver, S.R. (1995). *Automotive Fuels Reference Book*, 2nd e. Warrendale, P.A.: Society of Automotive Engineers. ISBN1560915897.

67 Song, C., Hsu, C.S., and Mochida, I. (2000). *Chemistry of Diesel Fuels*. Taylor & Francis. ISBN1560328452.

68 Lubrizol Corporation. Fact sheets and technical presentations. lubrizol.com. http://www.lubrizol.com/FuelAdditives (Cited: 25 September 2013).

69 Fuel Additives. Infineum website. http://www.infineum.com/Pages/fuel.aspx (Cited: 2 November 2014).

70 Innospec Fuel Specialties. Fuel additives product summary. http://www.innospecinc.com (Cited: 4 January 2013).

71 Thiel, C.Y., Hayden, T.E., and Kaufman, B.J. (2003). Fuel Additives Strategies for Enhancing the Performance of Engines and Engine Oils. ChevronTexaco, OroniteTechnology. http://www1.eere.energy.gov/vehiclesandfuels/pdfs/deer_2003/session1/2003_deer_kaufman.pdf (Cited: 2 November 2014).

72 Clariant, G. Refinery. http://www.oms.clariant.com/en-us/refinery/Pages/terminal-fuel-additives.aspx (Cited: 27 January 2013).

73 Germanaud, L., Eydoux, F., Doucet, C. et al. (2005). Use of additives to improve the odour of hydrocarbon compositions and hydrocarbon compositions comprising such additives. EP 1,591,514 Europe, 26 April.

74 Aviation Jet Fuel Additives Shell website. http://www.shell.com/home/content/aviation/products/fuels/aeroshell_performance_additive101 (Cited: 2 November 2014).

75 Knight, L. (1921). Improvements relating to petrol and other liquid fuels. GB 170075.

76 Ferguson, H. (1931). Improvements to and relating to motor fuels or like combustible liquids. GB 361310.

77 Wikipedia Contributors (2018). Hydrocarbon Oil Duty. Wikipedia. 11 February https://en.wikipedia.org/w/index.php?title=Special:CiteThisPage&page=Hydrocarbon_Oil_Duty&id=825016998 (Cited: 14 May 2018).

78 Wikipedia Contributors. (2020). Fuel Tax. Wikipedia. 31 March https://en.wikipedia.org/w/index.php?title=Fuel_tax&oldid=948415809 (Cited: 8 April 2020).

79 Wikipedia Contributors. (2018). Sudan III. Wikipedia. 15 March https://en.wikipedia.org/w/index.php?title=Sudan_III&oldid=830595803 (Cited: 13 May 2018).

80 Czarnocka, J., Zborowski, U., and Kesik, A. (2011). Dyeing and marking ystem in European Union and Poland. *CHEMIK* 65 (3): 200–203.

81 Wikipedia Contributors. (2018). Fuel Dyes. Wikipedia. 16 April retrieved 10 May 2018, https://en.wikipedia.org/w/index.php?title=Fuel_dyes&oldid=836747660 (Cited: 10 May 2018).

82 Indian Oil Diesel meets the requirements of IS 1460: 2005 (5th Revision) (2016). Indian oil. 17 December https://www.iocl.com/Products/HighSpeedDiesel.aspx (Cited: 20 December 2016).

83 Chia, B.P. (2016). *India Jumps to Bharat VI*. Insight, Infineum International Ltd. 12 October.

84 Hindustan Petroleum Corporation. (2014). Fuel promise. http://hindustanpetroleum.com/aboutretailgoodfuelpromise. (Cited: 20 May 2018).

85 Fuelling Oil Demand. (2011). What happened to fuel taxation in Europe? www.transportenvironment.org. April (Cited: 13 May 2018).

86 Johnson, P., Leicester, A., and Stoye, G. (2012). *Fuel for Thought*. The what, why and how of motoring taxation. Institute for Fiscal Studies. May https://www.racfoundation.org/media-centre/fuel-for-thought (Cited: 13 May 2018).

87 Elordui-Zapatarietxe-Del-Aguila, S. and Emteborg, H. (2017). *Evaluation of the Performance of the Short-Listed Candidate Markers regarding the Technical Requirements*. JRC Science Hub, JRC 107206. Publications Office of the European Union. ISBN 978-92-79-70178-8.

88 Fuel Laundering Problem Effectively Eliminated, says Revenue (2016). 30 September https://www.irishexaminer.com/breakingnews/ireland/fuel-laundering-problem-effectively-eliminated-says-revenue-757253.html (Cited: 13 May 2018).

89 Fuel Marker Demonstrates Success in United Kingdom & Ireland. April (2018). https://www.dow.com/en-us/fuel-markers/news (Cited: 13 May 2018).

90 Elvers, B. (2014). Dyes, general survey. In: *Ullmann's Encyclopedia of Industrial Chemistry – Fine Chemicals*, 11, 7th e (ed. B. Elvers). Wiley-VCH. ISBN 9783527334773

91 Waring, D.R. and Hallas, G. (2005). *The Chemistry and Application of Dyes*. Plenum Press. ISBN-139781468477177.

92 Hagelloch, G., Kunert, F., and Pape, G. (1968). Organic solvent solutions of monoazo dyes. GB Patent 1,142,239. Filed 17 January.

93 Berrie, A.H., Bebbington, S.T., Senior, R.K., and Hoop, G.M. (1966). Coloration process and aminoazo compounds for use therein. GB Patent 1,153,144. Filed 21 July.

94 Smith, M.J. and Buckland, D.J. (1970). Improvements in or relating to hydrocarbon-soluble monoazo dyestuffs. GB Patent 1,309,903. Filed 22 April.

95 Zeidler, G., Dehnert, J., Hansen, G. et al. (1974). Oil-soluble azo dye with n-substituted 2-naphthylamine as coupling component. US Patent 4,315,756. Filed 5 December.

96 The use of CI Solvent Red 164 as a penetrant dye in the detection of cracks in metal components. 14 February (2008). http://www.hse.gov.uk/aboutus/meetings/iacs/acts/watch/140208/solventred164.pdf (Cited: 13 May 2018).

97 Orelup, B.(1967). Liquid azo dye composition and process therefor. US Patent 3,690,809. Filed 24 October.

98 Turner, R.J., Booth, D.L., and Berry, R.D. (1973). 2-(2′Ethylhexyl)-Quinizarin. US Patent 3,883,568. Filed 14 May.

99 Contributors, W. (2018). Solvent yellow 124. Wikipedia. 1 March https://en.wikipedia.org/w/index.php?title=Solvent_Yellow_124&oldid=828184049 (Cited: 3 June 2018). Page version ID: 828184049.

100 Lewis, D.M. (2005). Hydrocarbon markers. EP 1,580,254. Filed 21 March.

101 Linsinger, T., Koomen, G., Emteborg, H. et al (2004). *Validation of the Draft Community Reference Method for the Determination of Solvent Yellow 124 in Gas Oil (Euromarker)*. European Commission Directorate, Joint Research Centre, Dutch Customs Laboratory Amsterdam. https://ec.europa.eu/taxation_customs/sites/taxation/files/resources/documents/taxation/excise_duties/energy_products/legislation/euromarker_report.pdf online link has been changed to 911 https://publications.jrc.eceuropa.eu/repository/handle/JRC27

102 Orelup, R.B. Colored petroleum markers. (1986). US Patent 4,735,631. Filed 16 October.

103 Foley, W.M., Sanford, G.E., and McKennis, H. (1952). The mechanism of the reaction of aniline with furfural in the presence of acid. *Journal of the American Chemical Society* 74 (21): 5489–5491.

104 Dyllick-Brenzinger, R., Raulfs, F.W., and Schlosser, U. (1993). Anilines as markers for mineral oils. US Patent 5,627,077. Filed 9 November.

105 Banavall, R.M. and Stephens, W. (2007). Chemical markers. US Patent 7,858,373. Filed 22 January.

106 Green, G.D. and Sedo, R.J. (2011). Biphenyl benzyl ether marker compounds for liquid hydrocarbons and other fuels and oils. US Patent 9,005,314. Filed 4 October.

107 Green, G.D. and Swedo, R.J. (2012). Ortho-phenylphenol compounds as markers for liquid hydrocarbons and other fuels and oils. US Patent 8,961,624. Filed 7 May.

108 Butterick, R., Green, D.G., and Swedo, R.J.(2013). Tritylated alkyl aryl ethers. US Patent 9,422,493. Filed 1 July.

109 Green, G.D., Smith, W.E., and Swedo, R.J. (2015). Tetrarylmethane ethers. US Patent 2017/0,073,290. Filed 21 April.

110 Gras, R.L., Luong, J.C., and Smith, W.E. (2014). Analytical methodfor detecting fuel markers. US Patent 9,689,859. Filed 2 May.

111 BP Energy Outlook (2018) edition. bp.com/energyoutlook. 2018. https://www.bp.com/content/dam/bp/en/corporate/pdf/energy-economics/energy-outlook/bp-energy-outlook-2018.pdf (Cited: 13 November 2018).

112 2019 Outlook for energy: A perspective to 2040. exxonmobil.com/energyoutlook. 2019. https://www.exxonmobil.co.uk/-/media/Global/Files/outlook-for-energy/2019-Outlook-for-Energy_v4.pdf (Cited: 28 March 2020).

113 BP Statistical Review of World Energy (2019). 68th edition. bp.com/statisticalreview. June 2019. https://www.bp.com/content/dam/bp/business-sites/en/global/corporate/pdfs/energy-economics/statistical-review/bp-stats-review-2019-oil.pdf (Cited: 30 March 2020).

114 Energy & Climate Intelligence Unit. (2020). Will britain keep the lights on? https://eciu.net/analysis/briefings/uk-energy-policies-and-prices/will-britain-keep-the-lights-on (Cited: 9 April 2020).
115 Brower, D. (2020). FT big read. Energy. The Financial Times. 29 October.
116 Durham, R. (2012). Germany fines red diesel yachstmen. Yachting Monthly. 1 March http://www.yachtingmonthly.com/news/germany-fines-red-diesel-yachtsmen-3896 (Cited: 13 May 2018).

2

Fuel Stabilisers

Antioxidants and Metal Deactivators

2.1 Introduction

Fuel stability generally means the stability of distillate fuels against degradation initiated by autoxidation. All hydrocarbons suffer varying degrees of autoxidation when in contact with the atmosphere. Autoxidation is a slow process at ambient temperatures, but fuels may be stored for extended periods, as happens with heating oils and fuels stored on military bases. First formed products of degradation are hydroperoxides (ROOH) which decompose on heating – a condition met with by all fuels, at some point in their use – or at ambient temperatures if catalysed by dissolved metals, particularly iron. Both the formation of hydroperoxides and their decomposition produce radical chain reactions that perpetuate the autoxidation process.

The term as applied to heavy fuels refers to its stability with respect to the separation of asphaltenes from the rest of the fuel. This is dealt with using detergents (Chapter 3).

The decomposition of hydroperoxides leads to the formation of aldehydes, ketones, carboxylic acids and alcohols. These reactive oxygenates form polymeric materials that may separate from solution first as gums and then as particles that block sensitive parts of fuel systems, such as fuel filters and fuel injectors. Such blockages, partial or complete, lead to a degradation of performance such as irregular engine operation, lower fuel efficiency and higher emissions.

Problems of deposits and gums were first observed in gasoline engines when carburetted engines drew fuel through a filter under low, negative pressure. These filters were overly sensitive to being blocked by deposits, which also interfered with the operation of the carburettors. In the 1930s, patents for the prevention of gum formation in gasoline showed an understanding of the problem as one of oxidation.

Fuel Additives: Chemistry and Technology, First Edition. Robert D. Tack.

Jet fuel then attracted much interest, as any interference with the smooth running of a jet fuel system could have disastrous consequences. Jet fuel is not only burned in the jet engine but is also circulated as a coolant around critical parts of the jet engine, heating up the fuel thus accelerating its autoxidation.

Similar problems of fuel stability apply to heating oils, diesel fuel and biofuels such as biodiesel, containing fatty acid methyl esters, and gasohol (blends of petroleum gasoline and ethanol), both of which are more susceptible to autoxidation than are hydrotreated hydrocarbon fuels.

Stability additives, or stabilisers, are antioxidants – often combined with Metal Deactivating Additives (MDAs) – whose purpose is to inhibit the oxidation of fuels [1,2]. Despite the use of antioxidants and MDAs, fuels absorb oxygen and some oxidation products are inevitable, especially over long storage periods and during heating. Unwanted products of autoxidation and thermal degradation may still form deposits in critical parts of an engine or burner; fuel detergents are then needed to keep systems clean (Chapter 3).

2.2 Detailed Problems

The supply of oxygen from air dissolved in fuel is limited, but normal operation enables more oxygen to be absorbed as fuel is churned in the fuel tank (shown in experiments on B10 biodiesel [3]). While the vapour pressure of gasoline keeps most air out of the storage tank, air may be drawn in during refuelling. Jet fuel oxidation derives from peroxides formed by the oxygen dissolved in the fuel during the normal transportation and pumping of fuel; the Jet Fuel Thermal Oxidation Tester (JFTOT) takes into account that there is little further exposure to oxygen and the test is run with limited oxygen availability [4]. In jet fuel stability studies, it was clear that the supply of oxygen is, as expected for all fuels, an important factor, so it is likely that differences in results by different workers result from the different testing conditions. For example, some static diesel fuel aging tests are carried out on fuel in sealed containers at a fixed temperature. Such tests give different results from those in a diesel vehicle, in which the fuel in the tank is churned in the presence of air and some of the fuel is heated up near the injection pump and the injector before being returned to the tank[1].

2.2.1 Oxidative Stability of Jet Fuels

The oxidative stability of jet fuels is addressed in two parts [5]. The first is that shared with the other distillate fuels – the absorption of oxygen and degradation during storage. This is, however, not generally a problem with civil aviation jet fuels, Jet A and A-1, because they are usually used within a few weeks of production. For the military, however, a substantial quantity of jet fuel may be stored for quite long periods – a year or more – in order that it is ready for immediate use on the outbreak of hostilities. Small airports, with a low turnover of fuel, may also have aviation fuel in store for a substantial period.

Storage instability is caused by the reaction of oxygen with the less stable hydrocarbons in fuel by a free radical process, forming hydroperoxides. Hydroperoxides are moderately stable below 90°C so build up in a fuel over long periods of storage. In the 1960s, hydroperoxides were found to degrade seals in the jet engine fuel systems, when present at more than 16 ppm, causing their embrittlement [6]. Hydroperoxides also lead to poor thermal stability because they decompose when the fuel gets hot. Where there is a concern, antioxidants should be added to the fuel to inhibit the formation of hydroperoxides on storage – where specified, the concentration of peroxides is limited to a maximum of 8 ppm in jet fuels. The use of phenolic antioxidants is required in some US military jet fuels such as grades JP-4 and JP-5 but is optional in IATA specifications for Jet-A and A1, though these fuels must meet the oxidative stability specifications [6]. Because the formation and decomposition of peroxides are catalysed by metals such as copper and iron, which can be dissolved into fuel by the action of any acids present, metal deactivators (MDAs) are often used in combination with antioxidants in jets fuels; MDAs are optional in most grades and required in some military grades [6].

Consideration of thermal stability is particularly important for jet fuels because jet fuel is used as the heat exchange fluid to remove heat from the engine oil, hydraulic fluid and the air conditioner. In high MACH aircraft, fuels may experience temperatures as high as 250°C, certainly at temperatures well above those at which hydroperoxides decompose [7,8]. As the fuel gets hot, the decomposition of hydroperoxides causes degradation of fuel to form gels, gums and particulates; other free radical polymerization processes, involving various alkenes, may also contribute to deposits.

Such gums and particulates can deposit:

- on fuel filters, impeding the flow of fuel
- in injector nozzles, disrupting the spray pattern, causing an uneven distribution of fuel burning in the jet engine
- in engine control systems, interfering with fuel flow and control of the engine
- on heat exchangers, reducing their efficiency and impeding fuel flow

While antioxidants help fuel storage stability, they may not be effective in improving thermal stability under the extreme conditions experienced in military aircraft. Problems of deposits caused by thermal instability may not become apparent for hundreds or thousands of hours of operation, so the JFTOT (ASTM D 3241/ IP 323) [5] was developed as an accelerated test to correlate with problems seen in reality. The JFTOT looks at filter blocking and deposits on an aluminium tube at 260°C (see section 2.3). Specifications are set based on the JFTOT for most jet fuels around the world.

2.2.2 Oxidative Stability of Gasoline

Antioxidants are added to gasoline to inhibit the reaction of less stable components with oxygen to form peroxides and gums [9]. Unstable components – alkenes and alkyl aromatics – are products of cracking in the FCC, added to gasoline blends to boost their octane numbers. Problems of deposits and gums were recognised in the 1930s as resulting from autoxidation, leading to the introduction of antioxidants [10–12]. They are needed in almost all gasolines but especially in those with high alkene content. Hydroperoxides, the initial autoxidation products, can degrade antiknock quality, cause fuel pump wear and attack plastic or rubber fuel system parts. Soluble gums can lead to engine and injector deposits and insoluble gums can plug fuel filters. As with other fuels, oxidation on storage is relatively slow, but modern fuel-injected vehicles subject the fuel to higher temperatures and fuel recirculation produces more oxygen-exposure. While antioxidants reduce reactions with oxygen, gasoline still has a tendency to form deposits when it is heated to high temperatures, as it is in the fuel injectors or inlet ports of hot engines; these deposits are controlled by the use of detergents (Chapter 3).

2.2.3 Oxidative Stability of Diesel Fuel

Diesel fuel attracts the same military attention for long-term storage as does jet fuel. The US Navy has diesel fuel, now ULSD, stored in readiness for possible combat operations for one to three years, in many different climates [13]. It is accepted that peroxides in diesel, at concentrations greater than the specification maximum of 8 ppm, attack O-rings and seals. Hydroperoxides initiate fuel autoxidation and thermal oxidative instability when subjected to short-term high-temperature stress – as in the region of diesel injectors and pumps, from which excess fuel is recycled to the tank. Peroxides lead to the formation of gums and sediments. Investigations for the US Navy reported that ULSD had greater resistance to the formation of peroxides at temperatures prevailing in diesel vehicles (at less than 90°C, compared with the 150°C used in some tests) than had high sulphur, straight-run diesel fuel. However, some results of more stressed testing indicated the opposite [1]. Consequently, it was reasonably concluded that both higher sulphur diesel and ULSD tend to form peroxides so need antioxidants for long-term storage [13].

Another effect of fuel degradation in diesel engines is lacquering of the injector needle and those parts of the injector system controlling the needle – 'lacquering' is a term used to differentiate it from deposits building up in the nozzle of an injector [14]. When unstable compounds in a diesel fuel, particularly a biodiesel, become heated in the fuel injector (presumably also for gasoline in GDI engines) they can produce, initially, a thin varnish or lacquer on the injector needle and associated systems. Eventually, the thickness of this layer builds up and interferes with the reliable operation of the injector causing uneven running, higher fuel consumption, loss of power and higher emissions. Lacquering had been recognised in 1956 [15] as needing prevention using antioxidants and MDAs, combined with a fuel detergent [14].

2.3 Tests of Oxidative Stability

There are several tests of distillate fuels' risks of oxidative degradation, or that check how much oxidative degradation may have taken place on storage. Some are fuel specific, some look at the results of degradation and some measure resistance to oxidation. Simply storing fuel samples for long periods to test their stability is impractical; the tests that are routinely used accelerate the oxidative process by using higher temperatures and may also use oxygen under pressure. Here, short descriptions are given of the ones that often appear in specifications and the technical and patent literature.

2.3.1 Jet Fuel Stability Tests

Jet fuel can become hot when carrying out its role as coolant, especially in military aircraft, so it must meet a thermal stability specification. The time under which harmful deposits may build up in an active jet engine may be many months, so thermal stability of jet fuel undergoes a test procedure that speeds up the process. The test used is the jet fuel thermal oxidation tester (JFTOT [5]) in which fuel (from a reservoir of 600 ml) is pumped over an aluminium tube[2] which is heated to 260°C; it then passes through a 17 micron filter, to collect any deposits that are formed, and back to the reservoir, for 2.5 hours. During the test, the pressure drop across the filter is measured and should not rise above a maximum of 3.3 kPa[3]. At the end of the test, the tube is visually rated for any deposit or varnish that has coloured the aluminium surface; for standard Jet A, it is specified that this rating should be less than 3, quoted as 'no peacock or abnormal colour deposits allowed' [6]. Time to the pressure drop of 3.3 kPa, or the pressure drop at the end of the 150-minute test, is noted. This test has also been used at varying time periods and temperatures in studies of jet fuel stability [7].

Gum in jet fuel is determined in tests that are based on those used for gasoline (below). Existent gum, that which has already formed on storage, is measured by evaporating 50 ml of jet fuel in a stream of superheated steam, for 30 minutes, at 235°C. The total residue from this test should be no more than 70 g/m^3 in Jet A or Jet A-1 [6]. Potential gum, as measured for gasoline, may be obtained but is not generally specified for jet fuel; however, an engine manufacturer does specify a maximum of 140 g/m^3 for arctic grade kerosene, oxidised as in ASTM D873 (under gasoline).

Hydroperoxides[4] themselves, formed in the initial autoxidation process on storage, may be determined by titration with sodium thiosulphate [16,17]. This is not generally specified except for grade JP-5 jet fuel in Japan, where there is a maximum allowed of 8 ppm by weight of peroxides. France has an optional test for peroxide in for Jet A-1 and a maximum of 2 mEq/dm^3, if determined [6].

2.3.2 Gasoline Stability Tests

Gum present in gasoline is measured by evaporation of 50 ml of fuel at 155°C, aided by a controlled flow of air for 30 minutes, or until the weight of residue is steady (test ASTM D381 and equivalents EN 6246, IP 131). The residue is weighed then washed with heptane and reweighed. Results for unwashed gum (also known as existent gum) and washed gum may be quoted as milligrams per 100 ml of fuel[5], as stated in the ASTM method, or g/m^3 in European specifications. For top grades of gasoline, in Europe, the maximum amount of existent gum allowed is 50 g/m^3 (EN 228).

A measure of the oxidative stability of a fuel is its induction period. An oxidation induction period is the time from initial exposure to oxidative conditions, to the start of continuous take-up of significant amounts of oxygen. During the induction period, any natural or added antioxidants are effectively inhibiting oxidation; when all the anti-oxidants are oxidised, the fuel is no longer protected and it is rapidly oxidised. The test method used is that a sample of gasoline is placed in a bomb with oxygen, under an initial fill pressure of 690 kPa, at 100°C (test EN ISO 7536/ASTM D525, IP 40 [9]). The pressure in the bomb is monitored continuously until the pressure falls by 14 kPa over a 15-minute period, followed by another fall of 14 kPa/15 minute, as oxygen is consumed. The end of the first of these 14 kPa pressure falls is called the 'breakpoint', and the time to the breakpoint is taken to be the induction period. Induction periods for European unleaded gasoline should be more than 360 minutes (EN 228). This process, which goes back to the 1930s [12], has been automated in the PetroOXY equipment, which uses a 5-ml sample of gasoline and oxygen at an initial pressure of 700 kPa, at 100°C [18].

Aviation gasoline specifications require an oxidative stability test known as 'potential gum' [6] (test ASTM D873). For this, an oxidative induction period test is combined with the gum measurement procedure: 100 ml of gasoline is heated at 100°C for 4 hours in a bomb, charged with 690kpa (100 psi) of oxygen. The oxidised gasoline is filtered to collect the insoluble gum, and the soluble gum is obtained by evaporating the filtrate (ASTM D381 method). Combining the weights of both soluble and insoluble gums gives the potential gum, which should be less than 60 g/m^3 for aviation gasoline [6]. The same test applied to motor gasoline should yield less than 50 g/m^3 of potential gum [19].

2.3.3 Diesel Fuel Stability Tests

There are many tests that measure the susceptibility of diesel fuel to reaction with oxygen. Simple storage stability testing involves storing fuel samples over long periods and then examining the samples for oxidative degradation. For example, in ASTM D4625, several samples of diesel (400 ml each) are stored in glass bottles at 43°C; one sample is examined at the start and other samples are taken after periods of 4, 8, 12, 18 and 24 weeks. The samples are filtered, washed with a tri-solvent (toluene/methanol/acetone), dried and weighed [20].

Such tests are not practical for fuel quality control measurements on a continuous basis, so oxidative stability tests are accelerated using elevated temperatures and may also take place in oxygen. One diesel fuel oxidative stability test is carried out at atmospheric pressure to produce a potential gum: oxygen is bubbled through a diesel fuel sample (350 ml) for 16 hours, at 95°C. After that, it is filtered and the residue is washed with the tri-solvent, dried and weighed (test EN ISO 12205). The maximum amount for insolubles so obtained is specified as 25_g/m^3 (EN 590). Another accelerated diesel fuel storage test is carried out at 150°C, in air, for 180 minutes; after finally filtering the fuel, the filter paper is rated by the reflectance of the filter pad, which should have a minimum of 80% reflectance (deposits are dark in colour so the more deposit the lower the reflectance) (ASTM D6468) [20].

Diesel fuel may also be stability tested under pressurised oxygen, as for gasoline potential gum or induction time. ASTM D5304 stores a sample (100 ml) of diesel fuel at 90°C, in a bomb charged with oxygen at 690 kPa for 16 hours; insoluble materials are filtered off, washed, dried, and weighed as in D4625 [20]. As for gasoline, the induction time may be measured in the automated PetroOXY equipment (5 ml of fuel and oxygen at an initial pressure of 700 kPa) but for diesel or biodiesel, at 120°C [18].

Biodiesel, or fatty acid methyl ester (FAME), produces volatile acids when undergoing autoxidation. These acids (mainly acetic and formic) are used to detect the degree of oxidation of biodiesel when exposed to air. In the Rancimat test, air is passed through a small sample (3 g for 100% biodiesel, EN 14112, and 7.5 g for biodiesel /petroleum diesel blends, EN 15751) held at 110°C. The air stream carries volatile acids over into a vessel containing water, the conductivity of which is monitored. When the conductivity rises sharply, that is taken to indicate the time at which oxidative degradation has started and the time from start to this point is the induction period. Induction periods for biodiesel blends should be not less than 20 hours (EN 590) and not less than 6 hours for 100% biodiesel (EN 14214 [21]).

2.4 Stability Additives: Antioxidants and Metal Deactivators

2.4.1 Antioxidants

There are two main types of antioxidants used in fuels, hindered phenols and phenylenediamines. Variations have been developed to make them soluble in hydrocarbon fuels by ensuring that they carry sufficient alkyl groups. Hindered phenols, such as 2,6-di-tert-butyl-4-methylphenol, also known as butylated-hydroxytoluene (BHT), are widely used, inexpensive antioxidants. The use of antioxidants in commercial jet fuels (Jet A and A-1) is optional and of low importance because of their short storage lives. Military grades, however, often require them to be added, especially when the fuel contains hydrotreated components, because they may be stored for long periods. Antioxidants approved for use in jet fuel are the hindered phenols 2,6-di-tert-butylphenol, 2,6-di-tert-butyl-4-methylphenol and 2,4-dimethyl-6-tert – butylphenol [6].

Phenylenediamine antioxidants were thought to produce deposits so do not appear in the list of approved antioxidants for military jet fuel. Aviation gasoline, however, permits the use of the N,N'-di-isopropyl-para-phenylenediamine or N,N'-di-sec-butyl-para-phenylenediamine, as well as the hindered-phenol antioxidants [22]. These same antioxidants are also used in gasolines and diesel fuels [2,23].

As already noted, the problems due to autoxidation of gasoline were observed early in the twentieth century. The earliest patents for antioxidants to overcome the oxidative instability of motor gasoline appeared in the 1930s. Initially, simple amines were proposed, such as diethanolamine and dibutylamine, 1929 [11]. Later, the para-aminophenol system was discovered by Kodak, notably the dibutyl ester of para-hydroxyphenyliminodiacetic acid as published in 1936 [12] – this discovery was demonstrated using the induction-time oxidative stability test described above. Kodak went on to discover the effectiveness of N,N '-di-sec-butyl-para-phenylenediamine (patent application in 1946 [24]) as a stabiliser for use in motor fuels[6] and aviation gasoline – a stabiliser that is still widely used today [23] under commercial names such as Tenamene 2 and Topanol M [25].

N,N'-di-sec-butyl-para-phenylenediamine is prepared by a route that is an interesting combination of standard synthetic procedures [26] (Figure 2.1). Chlorination of benzene (extracted from petroleum distillates, particularly after the FCC) at room temperature, over a catalyst of ferric chloride, yields mainly chlorobenzene. Nitration and separation from the ortho isomer yield 4-nitrochlorobenzene, which is aminated by ammonia at elevated temperature and pressure – the nucleophilic displacement of chloride is made possible by the electron-withdrawing effect of the nitro-group. Reduction by iron and hydrochloric acid, or hydrogen and a catalyst, gives para-phenylenediamine. Finally, the para-phenylenediamine is alkylated by reductive amination of methyl-ethyl ketone (2-butanone) with hydrogen and a copper oxide catalyst [24].

Also, in the 1930s, the antioxidant, gasoline-stabilising nature of phenols was first discovered. Initially, the use of simple phenols, such as catechol, naphthol and phenol ethers, were recommended as anti-oxidant stabilisers for gasoline to inhibit the formation of gums [27]; similarly, phenol alkylated by a refinery gas that contained ethylene, propylene and butylene were also recommended [28]. More specifically, hindered phenols in use today seem to have been first discovered in

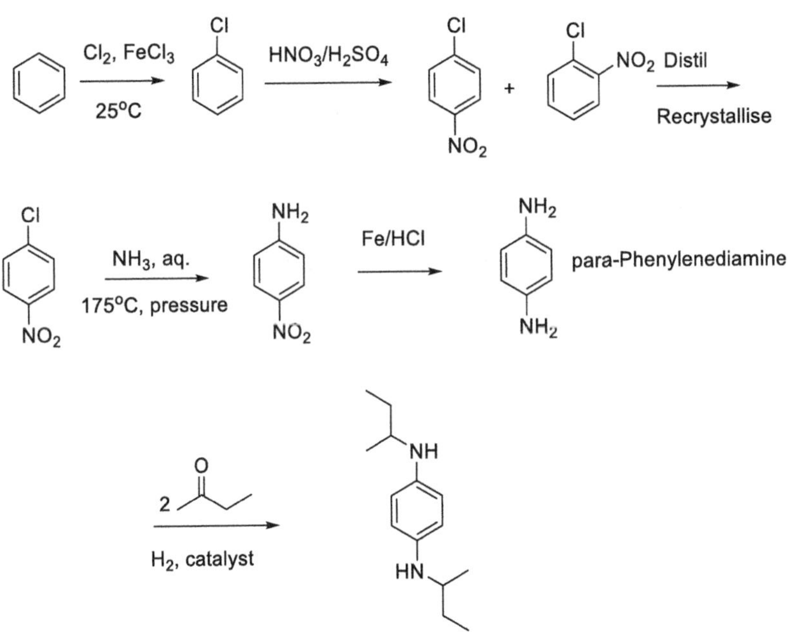

Figure 2.1 A Commercial Synthesis of N,N'-di-sec-butyl-para-Phenylenediamine.

the 1940s: a 1944 invention [29] claimed a range of poly-alkylated-phenols, that included 2,6-dimethyl-4-tert-butylphenol and 2,4-dimethyl-6-tert-butylphenol (close relatives to the widely used BHT [23]), as anti-oxidants to stabilise gasolines that contain amines[7]. The synthesis of BHT provides the reason for the name, butylated hydroxytoluene: para-cresol (4-methylphenol) is butylated with isobutylene under a slightly raised pressure, catalysed by sulphuric acid or aluminium phenate, followed by a water wash and recrystallization from aqueous ethanol, the yield depending upon reaction conditions (Figure 2.2) [30].

Both the para-phenylene diamine and hindered phenols were considered to be in common use as stabilisers for gasoline [31] and for diesel [15] in the 1950s. The only further antioxidant to enter the scene is di-tert-butylhydroquinone (DTBHQ) that has been claimed to show advantages when combined with BHT as a stabiliser for FAME [32]. Patents literature suggests that the main interest in DTBHQ is in biofuels [33] – particularly, in several Chinese patents. In one route, hydroquinone may be made using the same process as that used for the manufacture of phenol, but with 1,4-di-isopropylbenzene as the starting material. The dimethylbenzyl tertiary C-H positions readily lose hydrogen atoms to form free radicals that react with oxygen to form hydroperoxides. Acid or metal catalysed rearrangements to oxygen – the Baeyer-Villiger reaction – yield the phenolic hydroxyls and acetone as a by-product [30]. Acid-catalysed alkylation with isobutylene gives the 2,5-di- tert-butyl-hydroquinone [34] (Figure 2.3).

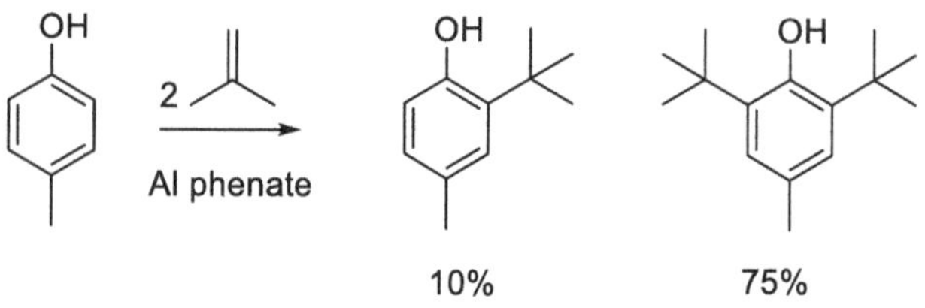

Figure 2.2 Production of BHT.

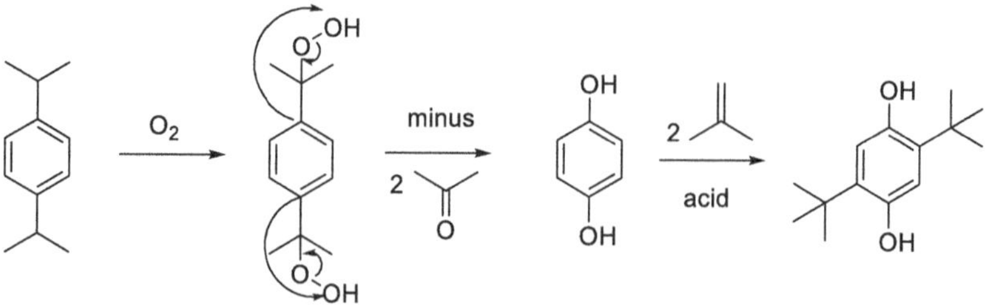

Figure 2.3 Preparation of 2,5-Di-tert-butylhydroquinone.

2.4.2 Metal Deactivators (Mdas)

Iron and copper can catalyse the initiation of free radical oxidation. They are dissolved into fuels by the action of any acids that the fuel may contain. Carboxylic acids are end products of oxidation, and fatty acids are added to low sulphur fuels to provide lubricity and anti-corrosion activity; acids may also be present (at low proportions) in acid-derived additives such as detergents. Biodiesel, fatty acid methyl esters (FAME), contains small quantities of carboxylic acids that result from hydrolysis side reactions during its preparation from fats and oils; specifications for such biodiesels allow for acid numbers of up to 0.05 or 0.08 mg KOH/g fuel[8] (EN 14214 or ASTM D6751). Ethanol, as used in gasohol (petroleum-derived gasoline plus ethanol), may contain small proportions of acetic acid [35].

To inhibit their initiation of oxidation, these solubilised metals are deactivated by the addition of the MDA N,N'-Disalicylidene-1,2-propanediamine (Irgamet 30) to aviation fuels, diesel, gasoline and heating fuels [2,6,23]. This MDA coordinates strongly with copper, zinc, and iron impurities, inhibiting their abilities to catalyse oxidation. A range of related compounds, that produced from salicylaldehyde reacted with 1,2-ethylene-diamine, were first described as effective co-additives to antioxidants, for preventing the oxidative degradation of gasoline, in a 1937 patent [36]. They showed that these compounds do not act as antioxidants but as compounds that deactivate copper ions, preventing their ability to catalyse the oxidation of gasoline. The inventors of this patent also described the product of salicylaldehyde reacted with 1,2-propanediamine as being the preferred MDA for vanadium in gasoline, containing certain cracked stocks, in a 1940 patent [37].

The preferred MDA contains two Schiff base groups, which are formed readily by the reaction of salicylaldehyde (two molar equivalents) with 1,2-propanediamine (one molar equivalent) and the removal of water (Figure 2.4). This chelating structure is well suited to strongly coordinate with iron and copper ions, forming stable complexes at extremely low concentrations, such as less than 6 ppm in aviation fuels [6] and 1 to 10 ppm in diesel fuels [15].

2.4.3 Thermal Stability Additives

Jet fuels, especially military jet fuels, may be exposed to the high temperatures that decompose hydroperoxides that may have formed during storage or during refining processes. The JFTOT test (described in section 2.3.1) gives some indication of the severity of the conditions that some military jet fuels may experience. To cope with such severe conditions, JP-8 military jet fuel may be treated with a thermal stability additive to raise the maximum temperature at

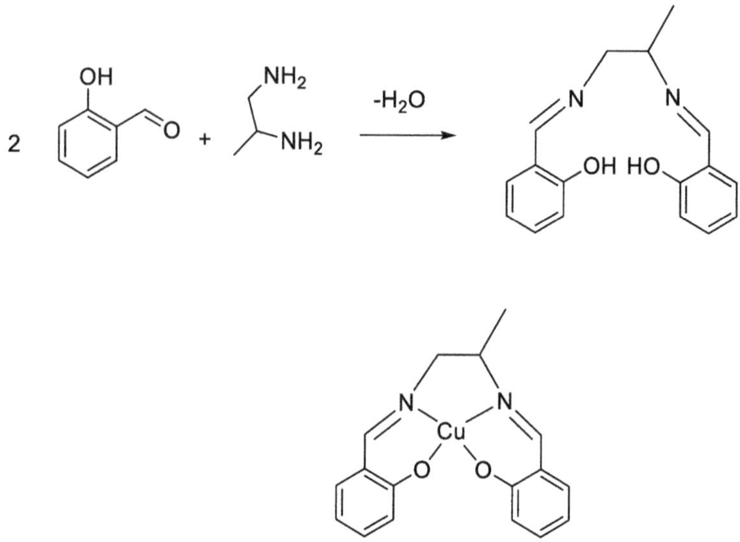

Figure 2.4 The Preparation of N,N'-Disalicylidene-1,2-propanediamine and Coordination with Copper Ions.

which it is stable from 325°F to 425°F (218°C); fuel treated in this way is called 'JP-8 + 100' [7]. The thermal stability additive used is a combination of a hindered phenol antioxidant, MDA, and a particular detergent/dispersant, known as Spec-Aid 8Q462 (listed in additives approved for use in jet fuel [6]). Spec-Aid 8Q462 is discussed in Chapter 3.

2.5 Mechanisms

Autoxidation of fuels leads to the formation of deposits on surfaces and particulates that interfere with the operations of valves, injectors, pumps and filters. These, in turn, cause deterioration in the performance of an engine, making its power output irregular and increasing noise and emissions. Reactions of oxygen with hydrocarbons in various applications all follow the same oxidative mechanisms; these mechanisms have been studied at great length in polymers [38,39], lubricating oils [40] and foods [41] as well as in fuels [13] and biofuels [32].

2.5.1 Hydrogen Atom Abstraction from Hydrocarbon Molecules

The process of deposit formation from fuel starts with the addition of oxygen to hydrocarbon free radicals, forming hydroperoxides. In common with many free radical processes, the reactions comprise an interplay of initiation, propagation and termination. All these processes involve hydrogen atom abstractions – including termination by antioxidants. Which hydrogen atom is abstracted, and whether or not one is abstracted, depends upon the relative strengths of the bonds to hydrogen that are broken and formed. So, first, the strengths of relevant carbon–hydrogen and oxygen–hydrogen bonds, as represented by their bond dissociation energies, should be considered. The bond dissociation energy represents the energy required to break a bond, or the energy released when that bond is formed.

While there are various values for exact bond dissociation energies, obtained by different workers using different methods, experimentally obtained figures for carbon–hydrogen and oxygen–hydrogen bonds, relevant to fuels (petroleum and biofuels), indicate which hydrogen atom is abstracted and which bond will form (Table 2.1).

Table 2.1 C–H Bond Dissociation Energies.

Structure	Description	C–H Bond Dissociation Enthalpy, kJ/mole (average of refs.)
HO—H	Water, O–H	498 [42]
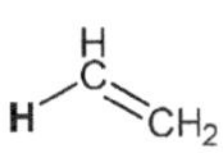	Vinylic C–H	463 [43]
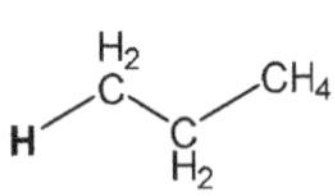	Primary alkyl C–H	422 [43,44]
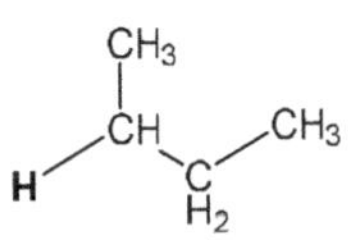	Secondary alkyl C–H	411 [43,44]
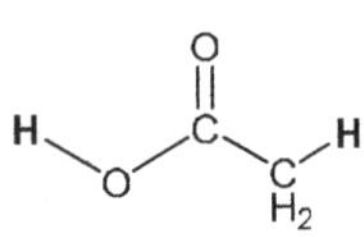	Carboxylic, alpha C–H Carboxylic O–H	405 [45,46] 469 [43]
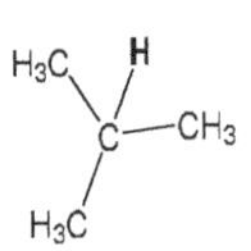	Tertiary allkyl C–H	404 [43,44]
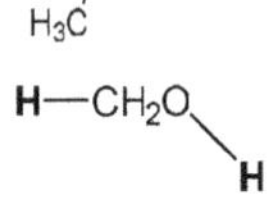	Methanol O–H Methanol, C–H	438 [43] 402 [43,45]

Continued

Structure	Description	C–H Bond Dissociation Enthalpy, kJ/mole (average of refs.)
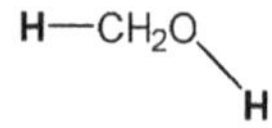	Ethanol, 1-C–H	387 [45,47]
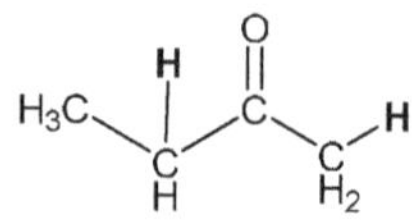	Ketone, alpha secondary CH–H Ketone, alpha primary CH_2–H	386 [45] 411 [45]
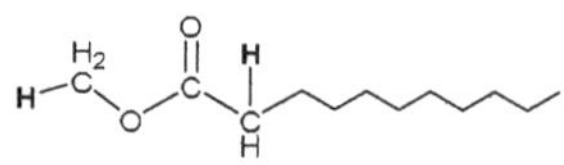	Methyl decanoate, methyl C–H 2-position methylene C–H	423 [48] 386 [48]
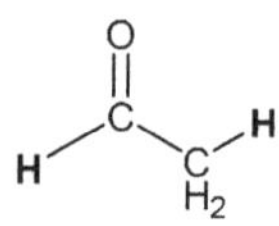	Aldehyde, alpha C–H Aldehyde, O = C–H, C–H	385 [43,45] 374 [43,46]
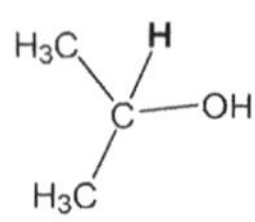	Isopropanol, 1-C–H	372 [47]
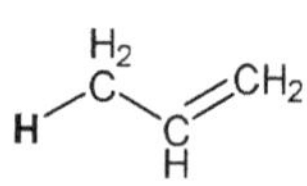	Allylic primary C–H	367 [43,44]
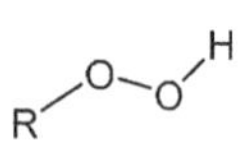	Alkyl hydroperoxide O–H Alkyl hydroperoxide O–O	368 [43] 197 [43]
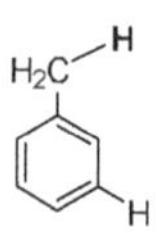	Benzyl C–H Phenyl C–H	351 [49] 469 [50]
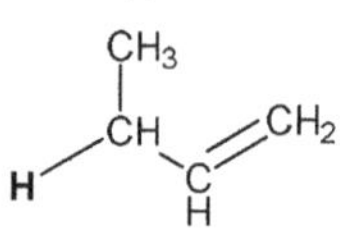	Allylic secondary C–H	345 [44]
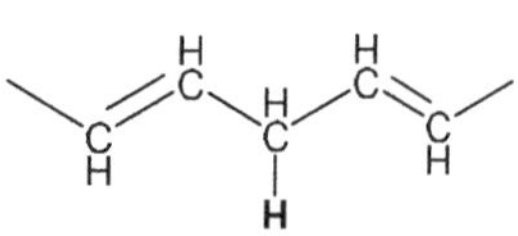	Bis-allylic C–H	319 [44,45]
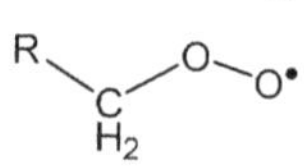	Peroxy radical O–O Peroxy radical C–O	272 [43] 159 [43]

2.5.2 Initiation

Initiation is the first formation of a hydrocarbon free radical, R ·, which takes place most readily if the radical is stabilised, such as in benzylic or secondary allylic positions. Initial radical formation may be catalysed by light, heat or metals with variable oxidation states – the last two of these are most relevant to fuels. Ferrous cations react with oxygen to form a ferric cations and peroxide radical/anions, which add protons to form hydroperoxide radicals (Figure 2.5). The hydroperoxide radical abstracts a hydrogen atom (a gain of 377 kJ/mole[9]) from a hydrocarbon benzylic C–H (cost of 351 kJ/mole) or allylic C–H (cost of 345 kJ/mole) to form a hydrocarbon free radical and hydrogen peroxide. Hydrogen peroxide is readily decomposed by metal ions to hydroxide ions and the very reactive hydroxy radicals – the formation of the second oxygen–hydrogen bond in water provides an energy gain of 498 kJ/mole, which ensures that it will abstract a hydrogen atom from any hydrocarbon molecule that it meets, as all carbon–hydrogen bond dissociation energies are lower than this (Table 2.1). This overall process is well understood because it is of concern in the oxidative spoiling of foods [41]. When a fuel is heated, any peroxides that it may contain – having been formed by contact with air at any stage from the refinery through to end use – will undergo homolysis, thus initiating further free radical reactions (see propagation).

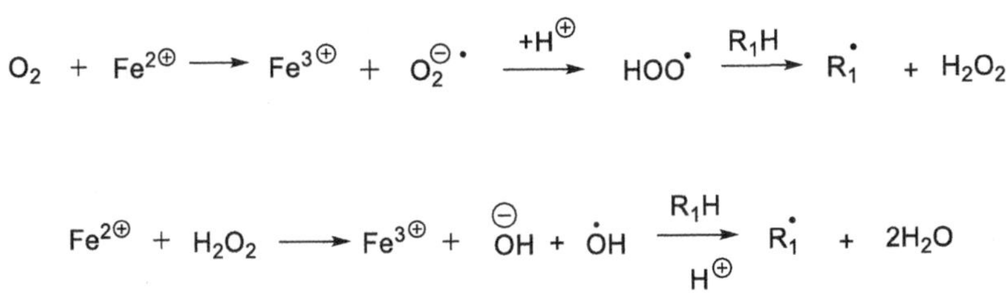

Figure 2.5 Free Radical Autoxidation: Initiation in the Presence of Soluble Iron.

2.5.3 Propagation

The first formed alkyl (or aralkyl) radical interacts with oxygen to form a peroxy radical (ROO ·) (Figure 2.6). This is a rapid reaction with no activation energy because the ground state of dioxygen is paramagnetic, having two unpaired electrons as a 'diradical' [51], and radical/radical combinations need no activation energy [52]. The peroxy radical then abstracts a hydrogen atom from another hydrocarbon molecule forming a hydroperoxide and another alkyl free radical.

The ease of hydrogen atom abstraction depends upon the dissociation energy of the carbon–hydrogen bond – the lower the bond energy the more easily the hydrogen atom is removed. Compare the dissociation energies of a secondary alkyl carbon–hydrogen bond at 411 kJ/mole, with that of a benzylic carbon–hydrogen bond, which is 351 kJ/mole and that of secondary allylic carbon–hydrogen bond, the most labile of these three, at 345 kJ/mole (Table 2.1). The gain in energy from forming the oxygen–hydrogen bond of an alkyl hydroperoxide from its radical, 368 kJ/mole, is more than that lost by breaking benzylic or allylic carbon–hydrogen bonds, but less than that lost by breaking a secondary alkyl carbon–hydrogen bond. Because a carbon–hydrogen bond needs to be broken, this step does have an activation energy, so it is slower than the combination with oxygen.

At ambient temperatures, once initiated, there is a continuous chain reaction of alkyl, allyl or aralkyl radicals reacting with oxygen to produce peroxy radicals, which react with other molecules to yield a hydroperoxide and another hydrocarbyl radical (top line of Figure 2.6). Cracked stocks contribute to the problem of deposit formation because thermal cracking produces alkenes, with allylic carbon–hydrogen bonds, and alkyl aromatic compounds, with benzylic carbon–hydrogen bonds – both of which lose a hydrogen atom relatively easily, enabling this free radical autoxidation process. Hydrotreating, as used in the removal of sulphur from gasoline and diesel fuels, saturates most of the alkenes and multi-ring aromatics that may be present in straight run and cracked fuels. This increases their stabilities under long-term storage at ambient temperatures [13]. At increased temperatures and with increased

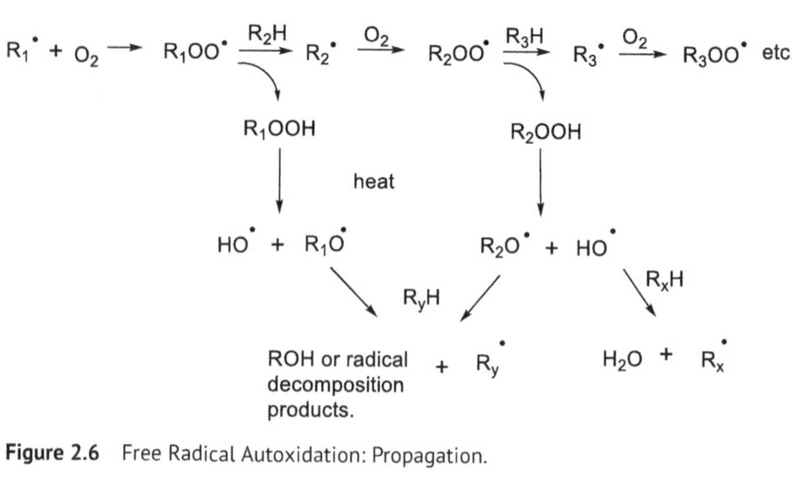

Figure 2.6 Free Radical Autoxidation: Propagation.

oxygen pressure (the short-term tests), hydrotreated fuels do form peroxides, but this is preventable by the use of antioxidants (BHT or PDA).

Bis-allylic hydrogens are particularly easily lost since such C-H bonds have the lowest dissociation energies in fuels at 319 kJ/mole (Table 2.1). These are found in methyl linoleate (methyl cis,cis-9,12-octadecadienoate), which is present at 30% to 70% in a number of vegetable oils, such as those from sunflowers, canola, soya, corn and peanuts [53], that are used to produce fatty acid methyl esters (FAMEs), used as biodiesel. As a result, the methyl linoleate is the major contributor to the greater oxidative instability of FAME and derived biodiesels, as is also the case in foods [43]. The position alpha to the ester carbonyl group has a carbon–hydrogen bond dissociation energy of 386 kJ/mole (as for methyl decanoate [48]), which is greater than that of the hydroperoxy carbon–hydrogen bond (368 kJ/mole) and the carbon–hydrogen bond of the methoxy group of the FAME ester (423 kJ/mole [48]) is much higher. Thus, a peroxy radical will most easily abstract a hydrogen atom from the bis-allylic position, less easily from the secondary allylic positions (C_8 and C_{14} in the linoleate, C_8 and C_{11} in the oleate) and with difficulty from other carbon–hydrogen positions in FAME molecules.

Hydroperoxides are stable at moderate temperatures (below about 90°C) so can build up in a fuel at ambient temperatures, causing premature ageing of seals. However, as motor fuels experience heating in the area of the fuel injectors and pumps, and military jet fuels in their capacity as cooling agents, they decompose by homolytic cleavage of the oxygen–oxygen bond [8] (lower part of Figure 2.6), which has a dissociation energy of only 197 kJ/mole. The alkoxy and hydroxy radicals formed can abstract hydrogen atoms from any fuel carbon–hydrogen positions to form alcohols and water, as they have high oxygen–hydrogen bond energies of 438 and 498 kJ/mole, respectively. Two further alkyl radicals are also produced that may form more peroxides, in the presence of adequate oxygen supply, thus causing divergent chain reactions.

Alkoxy radicals may also decompose by eliminating water or another alkyl radical to give aldehydes or ketones and new radicals (Figure 2.7); aldehydes are easily oxidised by peroxides or oxygen to give carboxylic acids as further, more stable decomposition products.

2.5.4 Termination

In the absence of antioxidants, termination of radical chains can occur by the interaction of pairs of radicals. This may be the combination of two peroxy radicals, followed by breakdown to oxidation products, contributing to deposit formation, or a peroxy radical with an alkyl radical to give a dialkyl peroxide.

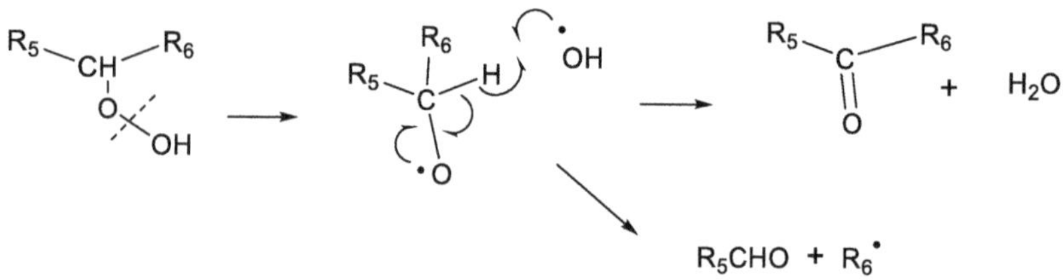

Figure 2.7 Two Peroxy Decomposition Pathways.

Two alkyl radicals may combine to form an alkane, or exchanging a hydrogen atom to form an alkene and an alkane. Given the extremely low concentrations of active radicals during autoxidation, these are slow reactions relative to being trapped by an antioxidant.

2.5.5 Formation of Difunctional Molecules during Autoxidation

It is entropically favourable for peroxy radicals to find a hydrogen atom in the same molecule, if one is present such as on the third carbon, leading to 1,3-difunctional molecules; or, in hot fuel as the peroxide decomposes to an alkoxy radical, leading to a 1,4-difunctional molecule, such as a hydroxy-acid (Figure 2.8). When multiple hydroperoxides are formed on the same molecule, either randomly or by a back-biting mechanism, multi-functional alcohols, aldehydes, ketones, and carboxylic acids are formed as they decompose.

Although the oxygenated compounds are present at low concentrations, they are brought together either as inverse micelles by association of their polar groups in the non-polar, non-solvating hydrocarbon fuel, or by adsorbing together onto any metal surfaces[10]. Heating such compounds, on hot metal surfaces or in hot fuel, causes polymerisations to polyesters, poly-aldols/ketals and Claisen-type condensation products. Such polymerisations are the steps for the formation of deposits, varnishes and particulates – the control of which is the reason for adding detergents (Chapter 3).

2.5.6 Mechanisms of Antioxidant Action

It is clear, then, that the formation and build-up of hydroperoxides must be prevented to stabilise a fuel against the formation of deposits, at any stage. Antioxidants are added to fuels to trap peroxy radicals, with which they react more rapidly than do most hydrocarbon molecules. Trapping peroxy radicals, as soon as possible, prevents the chains of reactions that lead to the continuous production of hydroperoxides (Figure 2.6). On donating hydrogen atoms to peroxy radicals, antioxidants form relatively stable radicals with

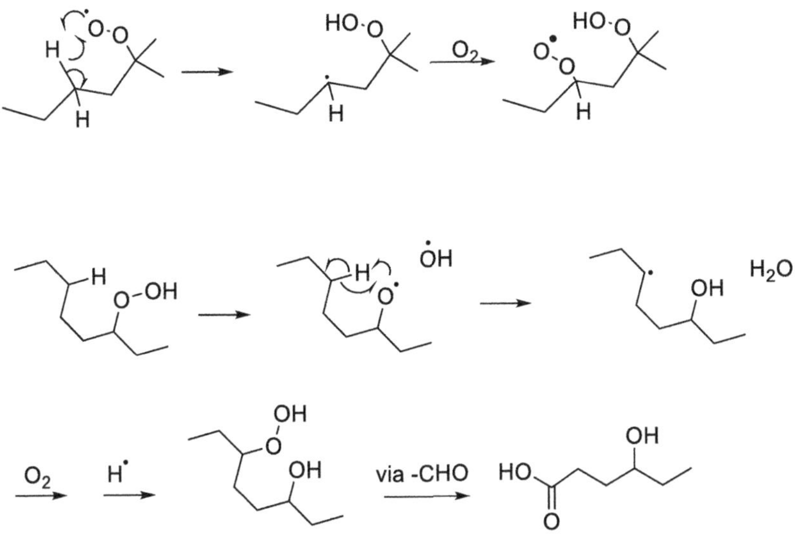

Figure 2.8 Formation of Difunctional Molecules during Autoxidation.

long lifetimes, which enables them to trap other radicals, as formed in the propagation stages, forming non-radical products. Through various pathways, each antioxidant molecule may absorb several free radicals. There are three types of antioxidants in general use: hydroquinones, hindered phenols and phenylenediamines.

Hydroquinones readily donate hydrogen atoms to free radicals, peroxy radicals, because the resulting hydroquinone radicals are stabilised by delocalisation across the molecule. The stabilised free radicals can combine with further free radicals, such as alkyl or aralkyl radicals, forming para-hydroxyphenol ethers (top row Figure 2.9). Alternatively, they may react with other hydroquinone radicals, disproportionating to form quinones and regenerating hydroquinones (middle Figure 2.9). Quinones, themselves, are also free radical traps: they can readily add free radicals (peroxy or alkyl) to form para-hydroxyphenol-ether radicals[11], which can combine with further free radicals to form non-radical di-ethers (bottom Figure 2.9). This is an example of the first stage of radical entrapment leading to a molecule that is also able to trap free radicals – a process that is necessary for antioxidant molecules to have high efficiencies as radical traps [54,55].

The range of reactions of antioxidant radicals, that follow their formation, are often complex – as shown, in part, for BHT (Figure 2.10) [56]. The first H atom may be abstracted from either the hydroxyl or the para-methyl group, leading to two routes, each of which can trap further radicals. Route-A leads to a quinone – neutralising two peroxy radicals per molecule – and route B to a stilbene quinone

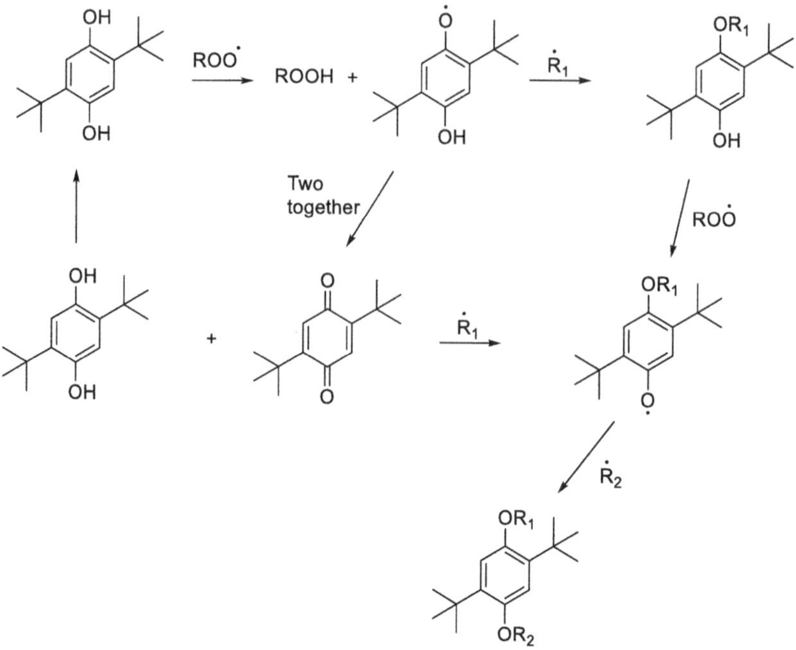

Figure 2.9 Process of Radical Trapping by 2,5-Di-t-Butyl-Hydroquinone.

– neutralising three radicals – as non-radical products [38]. Both product quinones can react with further free radicals (for example, as in Figure 2.9), increasing further the number of free radicals trapped per initial molecule of hindered phenol. As the product molecule becomes larger, the radical entrapment chemistry becomes more complex. The process of antioxidant product molecules further acting as radical traps, as demonstrated here by the hydroquinone and BHT, is common to most of the effective, organic antioxidants.

Para-phenylenediamines are also effective peroxy radical scavengers, said to be more effective than the hindered phenols, but can lead to colouring of the fuel [38]. The chemistry proposed for these antioxidants in action is also complex, involving hydrogen atom donation (Figure 2.11) or formation of N-oxides and, eventually, a degree of oligomerisation to extensively delocalised oligomers [57]. Under the more severe conditions of US military jet fuel thermal stability testing, it was this oxidative polymerisation of N,N-di-sec-butyl-para-phenylenediamine (PDA) that lead to deposits from the additive itself, resulting in such antioxidants being withdrawn from use in military jet fuels [13].

All three of these antioxidants, or close relatives, are in use in one fuel type or another. The action of an antioxidant may be summarised by its readily donating a hydrogen atom to a free radical, to produce a stable molecule. At the same time,

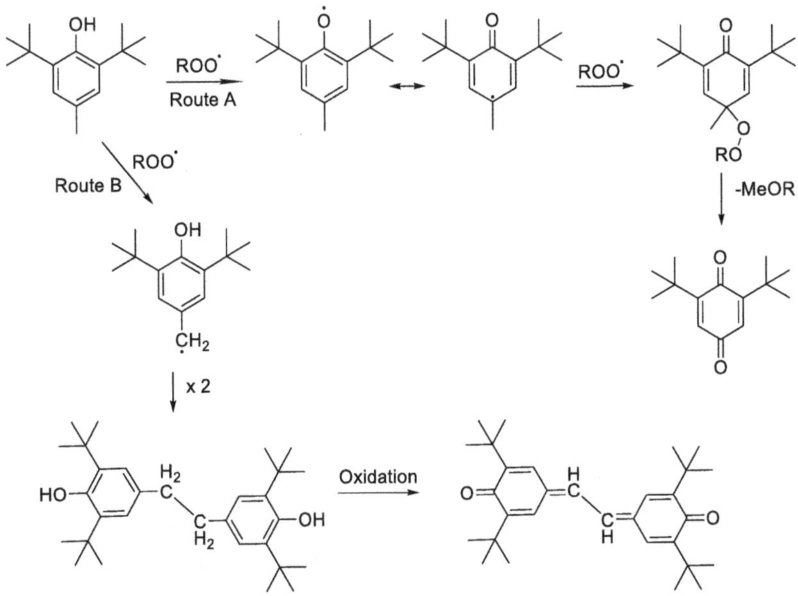

Figure 2.10 Some BHT Antioxidant Activity Pathways.

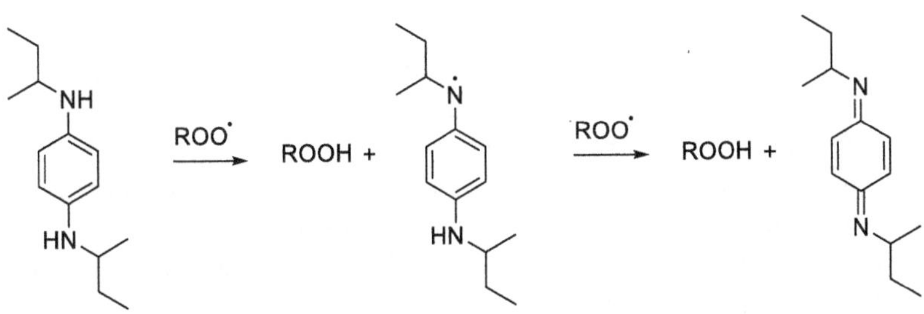

Figure 2.11 Initial Reactions of para-Phenylenediamine Antioxidants with Peroxy Radicals.

the antioxidant forms a relatively stable free radical which, after donating a further hydrogen atom to neutralise another free radical, is oxidised to a stable molecule which has extended conjugated unsaturation. This product is also able to neutralise further free radicals. The ability to do this enables such an antioxidant to be effective at low concentrations: recommended treat rates are up to 24 ppm in jet fuel [6] and in aviation gasoline [22]. Treat rates recommended for gasoline and diesel fuel are quite broad at 8 to 100 and 20 to 200 ppm, respectively [2], though their actual uses are at the lower ends of these ranges, such as 8 to 40 [10] or 20 ppm [29] in gasoline and, preferably, 10 to 50 ppm in diesel or biodiesel [14].

Notes

1 Diesel fuel pumps pump excess fuel, under pressure, to the region of the injectors. Excess fuel from around the pump and the injector equipment is collected in a by-pass arrangement and fed back to the tank.
2 Dimensions: 16 cm long, the test part between 5 cm shoulders is 6 cm long and 3 mm diameter.
3 Stated as 25 millimetre of mercury, mm Hg, in the test document.
4 Often referred to by the general term peroxides.
5 Such measures will be given as g/m3, i.e. parts per million by weight per unit of volume, throughout for consistency and comparability. However, test methods and specifications often quote a weight of milligram per 100 ml as that is the quantity used for the test, 1 milligram per 100 ml translates to 10 g/m^3 (10 ppm).
6 The patent [24] describes hydrocarbon motor fuel as finished gasoline (American English), being a blend of various refinery gasoline streams, but including the possibility of alcohols and ethers. In the UK, the term 'motor fuel' refers to both petrol and diesel fuel. European documents also use the term 'motor spirit' for gasoline [58].
7 Aromatic amines that were thought to act as anti-knock additives, either alone or in combination with TEL; such amines would give brown discolouration to gasoline as they were oxidised.
8 Biodiesel specification EN 14214 states acid number maximum of 0.5 mg KOH/g, which is $(0.5/40) \times 10^{-3}$ mole/g. Molecular weight oleic acid is 282, so this would be present at $282 \times (0.5/40) \times 10^{-3}$ g/g fuel, i.e. 0.35%.
9 For hydrogen peroxide, the O–H bond dissociation energy is 377 kJ/mole [59].
10 As do lubricity and anti-corrosion additives, see Chapter 4.
11 Radicals that may also have been formed by loss of a hydrogen atom from the para-hydroxyphenol ether.

References

1 Owen, K., Coley, T.R., and Weaver, S.R. (1995). *Automotive Fuels Reference Book*, 2^{nd}edn. Warrendale, P.A.: Society of Automotive Engineers. ISBN 1560915897.
2 Fuel Additives: Use and Benefits, ATC Doc. 113. atc-europe.org (2013). http://www.atc-europe.org/public/Doc1132013-11-20.pdf (Cited: 2 February 2014).
3 Engelen, B., Schmidt, M., Baldini, L. et al. (2013). Laboratory oxidation stability study on B10 Biodiesel blends. Report no. 10/13. CONCAWE and DGMK: CONCAWE Brussels, ISBN 9782875670243.
4 Heneghan, S.P. and Zabarnick, S. (1994). Oxidation of jet fuels and the formation of deposits: Butterworth-heinemann Ltd. *Fuel* 73: 35–43.

5 FTR-3. (2006). Chevron global aviation – Aviation fuels technical review. cgabusinessdesk.com. https://www.chevron.com/-/media/chevron/operations/documents/aviation-tech-review.pdf (Cited: 16 April 2020).

6 ExxonMobil. World Jet Fuel Specifications (2008). http://www.exxonmobilaviation.com/AviationGlobal/Files/WorldJetFuelSpec2008_1.pdf (Cited: 17 February 2014).

7 Heneghan, S.P. and Harrison, W.E. (1997). JP-8+100: The development of high thermal stability jet fuel. *6th International Conference on Stability and Handling of Liquid Fuels*. Vol. 1, 271–283. DOE, Vancouver.

8 West, Z.J., Adams, R.K., and Zabarnick, S. (2011). Homogeneous catalysis of liquid-phase hydroperoxide decomposition in hydrocarbons: ACS publications. *Energy & Fuels* 25: 897–904.

9 FTR-1. (2009). Chevron corp. motor gasolines technical review. chevronwithtechron.com http://www.chevronwithtechron.com/products/documents/69083_MotorGas_Tech_Review.pdf (Cited: 24 September 2013).

10 ATC Europe (2004). Fuel additives and the environment. atc-europe.org. 1 January http://www.atc-europe.org/public/doc52.pdf (Cited: 1 October 2013).

11 Calcott, W.S. and Lee, I.E. (1929). Gum inhibitor for gasoline. US Patent 1,940,445. Filed USA, 7 August.

12 Clarke, H.T. and Bean, F.R. (1936). Gasoline antioxidant. US Patent 2,122,408. Filed USA, 26 June.

13 Willauer, D.H., Hardy, D.R., Morris, R.E. et al. (2007). Potential for peroxide and gum formation in ultra-low-sulfur diesel fuels. In: *Chemistry Division, Navy Technology Centre for Safety and Survivability*. Washington: Naval Research Laboratory. NRL/MR/6180-07-9087.

14 Arondel, M., Dequenne, B., and Rodeschini, H. (2012). Additive compositions for improving the lacquering resistance of higher-grade fuels of the diesel or biodiesel type. US Patent 2015/0,033,617. Filed USA, 17 December.

15 Bailey, D.R. and Daniels, P.H. (1956). Improvements in and relating to diesel fuels. GB 790,978. Filed 20 July.

16 Peroxide Value, Cd 8-53 (1997). Official methods and recommended practices: American oil chemists' society.

17 Lin, C.-Y. and Chiu, -C.-C. (2009). Effects of oxidation during long-term storage on the fuel properties of palm oil-based biodiesel: ACS. *Energy & Fuels* 23: 3285–3289.

18 Petrotest -Oxidation Stability PetroOXY (2009). petrotest.com. 1 October http://www.laboaragon.com/docs/marcas/petrotest/OxidationStability-PetroOXY.pdf (Cited: 24 July 2017).

19 Gupta, A.A., Puri, S.K., Chand, S. et al. (2004). Antioxidant composition for motor gasoline. US Patent 2005/0091,914. Filed USA, 20 August.

20 Chevron (2007). Diesel fuels technical review. chevron.com. https://www.chevron.com/-/media/chevron/operations/documents/diesel-fuel-tech-review.pdf (Cited: 26 November 2012).

21 Karavalakis, G. and Stournas, S. (2010). Impact of antioxidant additives on the oxidation stability of diesel/biodiesel blends: ACS. *Energy & Fuels* 24: 3682–3686.

22 JIG. (2016). Product specifications bulletin – Aviation fuel quality requirements for jointly operated systems. jigonline.com. Issue 29, October. http://www.jigonline.com/wp-content/uploads/2016/10/Bulletin-96-AFQR-Oct-2016.pdf (Cited: 21 June 2017).

23 Richards, P. (2014). *Automotive Fuels Reference Book*, ThirdEdition. Warrendale, P.A.: Society of Automotive Engineers. ISBN 9780768006384.

24 von Bramer, H. and Ruggles, A.C. (1946). Improvements in motor-fuels. *GB Patent* 627 (880): Filed 23 February.

25 ChemicalBook. (2016). N,N'-Di-sec-butyl-p-phenylenediamine(101-96-2). chemicalbook.com. http://www.chemicalbook.com/ProductMSDSDetailCB8421520_EN.htm (Cited: 3 August 2017).

26 Booth, G. (2014). Nitro compounds, aromatic. In: *Ullman's Encyclopedia of Industrial Chemistry – Fine Chemicals*. Vol. 3. (ed. B. Elvers), 990. Weinheim: Wiley-VCH.

27 Universal Oil Products Company. (1935). Improvements in or relating to the stabilisation and refining of cracked hydrocarbon distillates. GB Patent 459,923. Filed 18 July.

28 Cantrell, T.L. (1936). Manufacture of phenol-olefin reaction products useful as gasoline gum inhibitors. US Patent 2,149,759. Filed 4 September.

29 Shell Development Company. (1944). Stabilisation of gasoline and of amines for addition thereto. GB Patent 587,314. Filed 29 June.

30 Fiege, H., Voges, H.W., Hamamoto, T. et al. (2014). Phenol derivatives. In: *Ullman's Encyclopedia of Industrial Chemistry – Fine Chemicals*. Vol. 3 (ed. B. Elvers), 1037–1098. Weinheim: Wiley-VCH.

31 Shell Refining and Marketing Company Ltd. (1951). Fuels for internal combustion engines and compositions for the production of such fuels. GB Patent 708,006. Filed 8 June.

32 Salley, S.O., Teng, H., deGuzman, R.C. et al. (2010). Effect of antioxidants on the storage stability of soybean-oil-based biodiesel. *Energy & Fuels* 24: 2028–2033.

33 Poirier, M. (2012). Dye-stable biofuel blend compositions. WO Patent 2013/062840. Filed 18 October.

34 Yu, T., Zhang, Y., and Jun, Y. (2004). Preparation of 2-tert-butylhydroquinone from 2,5-di-tert-butylhydroquinone. 2 May Chemistrymag.org. Vol. 6, 33.

35 DuMont, R.J., Kulinowski, A.M., and Cunningham, L.J.,. (2007). Methods and compositions for reducing deposits in engines combusting alcohol-containing fuels. US Patent 2008/02020561. Filed USA, 1 March.

36 Downing, F.B. and Pederson, C.J. (1937). Stabilization of organic substances. US Patent 2,181,121. Filed 29 December.

37 Downing, F.B. and Pederson, C. (1940). Stabilisation of nonviscous cracked petroleum distillates. US Patent 2,284,267. Filed 3 August.

38 Antioxidants, P. (2004). *Kirk-Othmer Encyclopedia of Chemical Technology*, 5th edition. Vol. 3. 145–159. NJ: John Wiley a& Sons Inc.

39 Dexter, M., Thomas, R.W., and King, R.E. (2014). Antioxidants. [book auth.]. In: *Encylopedia of Polymer Science and Technology*, 4th edn. (ed. H.F. Mark, A. Seidel, and M. Peterca),682–720. N.Y.: John Wiley & Sons.

40 Liston, T.V. (1992). Engine lubricant additives, what they are and how they function. May, *J.Soc.Tribologists and Lubrication Engineers* 48: 389–397.

41 Skibsted, L.H. (2010). Understanding oxidation processes in foods. In: *Oxidation in Foods and Beverages and Antioxidant Applications: Woodhead Publishing Ltd*, Vol 1 (ed. R.J. Elias, D.J. McClements, and E.A. Decker), 3–31: Great Abington, Cambridge: Understanding Mechanisms of Oxidation and Oxidation Activity.

42 Chandra, A.K. and Uchimaru, T.J. (2000). A DFT study on the C-H bond dissociation enthalpies of haloalkanes: Correlation between bond dissociation enthalpies and activation energies for hydrogen abstraction. *Journal of Physical Chemistry A* 104 (40): 9244–9249.

43 Blanksby, S.J. and Ellison, G.B. (2003). Bond dissociation energies of organic molecules: ACS. *Accounts of Chemical Research* 36: 255–263.

44 Senosiain, J.P., Han, J.H., Musgrave, C.B. et al. (2001). The use of quantum methods for a consistent approach to combustion modelling: Hydrocarbon bond dissociation energies. *Faraday Discussions* 119: 173–189.

45 Cherkasov, A., Jonsson, M., and New, A. (2000). Method for estimation of homolytic C-H bond dissociation enthalpies. *Journal of Chemical Information and Computer Science* 40: 1222–1226.

46 Korth, H.-G. and Sicking, W.J. (1997). Prediction of methyl C-H bond dissociation energies by density functional theory calculations. *Journal of the Chemical Society. Perkin Transactions* 2: 715–719.

47 Laarhoven, L.J., Mulder, P., and Wayner, D.D. (1999). Determination of bond dissociation enthalpies in solution by photoacoustic calorimetry. *Accounts of Chemical Research* 32: 342–349.

48 Oyeyemi, V.B., Dietrich, J.M., Krisiloff, D.B. et al. (2015). Size-extensivity-corrected multireference configuration interaction schemes to accurately predict bond dissociation energies of oxygenated hydrocarbons. *Journal of Physical Chemistry A* 119: 3429–3439.

49 Genco, J.I., Duke, F.R., Griffel, M. et al. (1956). *The C-H Bond Dissociation Energy in the Methyl Group of Toluene*. 746. ISC: Ames Laboratory, Iowa State College.

50 van Scheppingen, W., Dorrestijn, E., Arends, I. et al. (1997). Carbon-oxygen bond strength in diphenyl ether and phenyl vinyl ether: An experimental and computational study. *Journal of Physical Chemistry A* 101: 5404–5411.

51 Lewis, M. and Waller, G. (1982). *Advancing Chemistry*. Oxford University Press. ISBN 0199140839.

52 Norman, R.O.C. (1968). *Principles of Organic Synthesis*. Methuen & Co. Ltd. SBN 4164719502.

53 Contributors, W. (2017). Sunflower oil. 30 August https://en.wikipedia.org/w/index.php?title=Sunflower_oil&oldid=797936369. Page Version ID: 797936369 (Cited: 31 August 2017).

54 Denisov, E.T. (1980). Inhibitor regeneration in oxidation. In: *Developments in Polymer Stabilisation*Vol. 3. (ed. G. Scott), 1–20. Amsterdam: Elsevier Applied Science Publishers Ltd. Ch.1.

55 Scott, G. (1981). Mechanism of antioxidant action. In: *Developments in Polymer Stabilisation*Vol. 4. (ed. G. Scott), 1–22. London: Elsevier Applied Science Publishers Ltd. Ch.1.

56 Pospisil, J. (1979). Chain breaking antioxidants in polymer stabilisation. In: *Developments in Polymer Stabilisation*Vol. 1. (ed. G. Scott), 1–37. Springer Science B.V. Ch.1.

57 Pospisil, J. (1984). Aromatic amine anti-degradants. In: *Developments in Polymer Stabilisation* Vol. 7. (ed. G. Scott), 1–64. Elsevier Applied Science Publishers Ltd. Ch.1.

58 Eurostat. Energy Yearly Statistics (2005). *[Prod.] European Commission*. Europe: Office for Official Publications of the European Communities. SSN18307833.

59 H2O2 Dissociation – Heat, Free Energy and Equilibrium Constant (2020). USP technologies. 19 April http://www.h2o2.com/technical-library/default.aspx?pid=9&name=Technical-Library.

3

Fuel Detergents

3.1 Introduction

All fuels are, to a greater or lesser extent, exposed to air during storage, while being pumped from large tanks to tankers, then to smaller tanks. It will also be shaken as it is transported by road. On exposure to air, fuel-hydrocarbons absorb oxygen and form peroxides. When the per-oxygenated hydrocarbons in fuel get hot, particularly in contact with metals, they degrade. Degradation products are moderately small proportions of the fuel, but they are polar, so they aggregate and adsorb onto metal surfaces; as these aggregates grow, they produce particulates and further deposits on the metal surfaces. The hot areas of the fuel systems in which this occurs are also those critical to the operation of the engines or furnaces, such as the fuel injectors of gasoline or diesel engines, and the burners of jet engines or heating systems. These may become partially blocked, affecting fuel flow and spray patterns that are critical to efficient burning of fuel. Also, inline fuel filters get blocked, impeding the flow of fuel. Older automotive gasoline engines and many small aircraft, with rotary gasoline engines, use carburettors, rather than fuel injectors, to provide the air/fuel-vapour mix for the engine to burn; deposits building up in carburettors impair their operation.

In jet engines, other critical items are the narrow pipes of heat exchangers through which fuel is pumped. Thermal degradation can take place in these hot pipes, producing deposits that build up over time, limiting fuel flow and the heat exchanger efficiency.

Residual fuels, components of heavy fuels (HF), contain asphaltenes which can separate on storage and settle, forming viscous layers on the bottom of storage tanks. When such heavy fuels are burned in furnaces, they produce large amounts of effluent which block flues and coat heat exchanger surfaces. In all cases, combustion inefficiencies caused by the build-up of deposits not

Fuel Additives: Chemistry and Technology, First Edition. Robert D. Tack.

only reduce efficiency and power output, but also increase exhaust emissions: soot, carbon monoxide, hydrocarbons and particulates.

As their name suggests, the purpose of fuel detergents is to keep fuel systems clean. They prevent polar decomposition products from associating and from settling onto metal surfaces – even cleaning up existing deposits. Much attention has been paid to identifying the nature of deposits in internal combustion engine fuel systems, around the inlet valves of gasoline engines, and in gasoline and diesel engine fuel injectors. Deposits in the inlet port of gasoline engines have been found to contain simple hydrocarbons (fuel and oil), carbon black (soot), metal salts and oxides, and a polymeric fraction that, it was suggested, could have come from lube oil additives (section 3.5.1). In diesel injectors, deposits were identified as fuel oxidation products, sodium salts of fatty acids (from the lubricity additive) and sodium chloride (which some refineries use as a fuel drier); in addition, polymeric materials have been identified that infra-red spectra suggest could be derived from fuel or lubricating oil additives. Polymers derived from autoxidised fuel additives are not easily identifiable but may be chemically linked with the other deposit components – however, tests for fuel stability produce deposits on storage, in the presence of air or oxygen, in the absence of fuel additives (see Chapter 2). In addition to hydrocarbon sources, zinc oxide and zinc carboxylates have also been found in diesel injector deposits.

3.2 Detailed Problems

3.2.1 Gasoline Engines

Carburettors were used as fuel-metering devices up until about 1990, but they have since been replaced by fuel injectors. Carburettor operation was sensitive to the build-up of deposits in the carburettor itself. Fuel injectors give better control of fuel delivery – however, small amounts of deposits can critically interfere with the operation of fuel injectors [1,2]. Deposits can build up in the injector body and can also lead to 'coking' (burnt-on deposits) of the nozzles. Such deposits degrade the fuel spray pattern leading to lower fuel efficiency, reduced power and increased exhaust emissions [2–4].

There are two main arrangements for gasoline fuel injectors: Port Fuel Injectors (PFI) and Direct Injection (DI). With PFIs, fuel is injected just before, or onto, the intake valve, to be carried into the upper cylinder (combustion chamber) with the air stream. PFI deposits, from fuel, can build up on the Intake Valve (IVDs), the inlet port and in the injector, affecting the flow of air and fuel into the combustion chamber, where deposits may also form (CCDs) (Figure 3.1) [2].

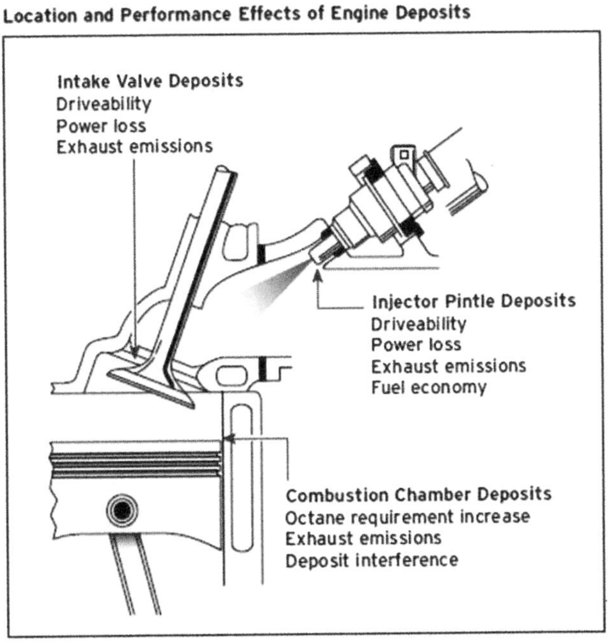

Figure 3.1 Port Fuel Injector. © 2019 Chevron. All rights reserved.

In DI, fuel is injected directly into the cylinder [2,3] – the arrangement is like that for diesel injection (Figure 3.2) with the addition of a space for the spark plug, above the injection region. When using DI, only air passes through the intake valve so deposits in the injector are the main problem. Fuel may also cause deposits in the upper part of the cylinder ('Upper Cylinder Deposits') which, over years, causes Octane Requirement Increase (ORI, need for a higher-octane fuel as the engine ages) [2]. Deposits in the carburettor fuel systems, in PFIs, on the intake valve, in the intake port and in the upper cylinder, appear to respond best to different detergent chemistries [4]. Before 2010, PFI was the most common system in automobiles, as gasoline DI was more expensive and was found only on more expensive cars (e.g. Porsche, Mercedes); however, since then, DI has become widespread in production-line vehicles.

New problems appear from time to time as results of automotive technology improvements introduced for environmental reasons. Exhaust Gas Recycle (EGR) is a widely used process by which a proportion of the exhaust gas is recycled into the air intake as inert gas to reduce NOx emissions [2]. Perhaps unsurprisingly, this process also introduces deposits in the fuel inlet systems and injectors of modern engines.

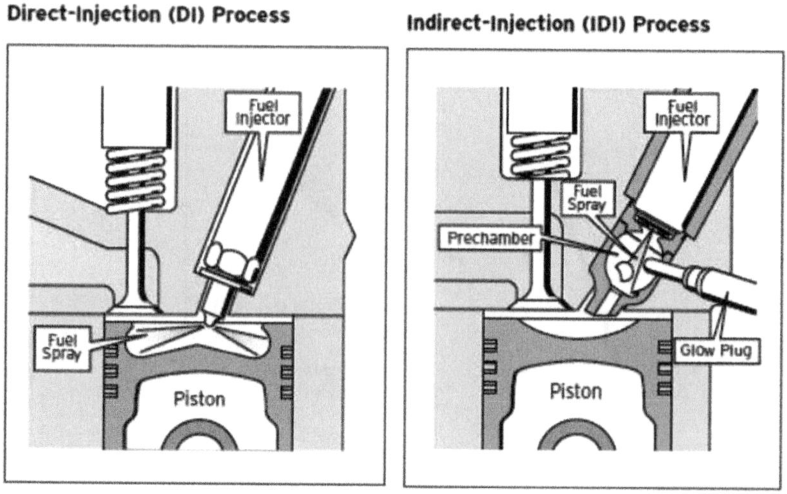

Figure 3.2 Direct and Indirect Diesel Injection Processes. © 2019 Chevron. All rights reserved.

3.2.2 Fuel Injector Deposits in Diesel Engines

Most diesel engines now use Direct Injection (DI) of fuel into the combustion chamber; the alternative indirect injection (IDI) of fuel, into a swirl pre-chamber above the combustion chamber, was developed for early passenger vehicle diesel engines (mid-1980s to mid-1990s, some of these may still be around) (Figure 3.2).

The advantage of IDI was quieter engines, but they had lower fuel economy than DI engines. DI engines have since been developed to overcome their disadvantages. In the push for greater fuel economy and lower emissions, diesel fuel injection technology has progressed significantly over the last two decades. In the most recent technology (at time of writing), injector nozzles are extremely small (a few microns) and injector bodies, housing the pintle[1], have very narrow clearances. With electronic controls, the injection system, operating under pressures of up to 2,000 bar, may inject fuel many thousands of time per minute, in quantities of 1 to 40 mm^3 (1 to 40×10^{-3} cc) over 1 to 2 millisecond [5,6]. Given such stringent requirements, even small amounts of deposits in the injector body and in the nozzles can interfere with the optimum operation. Deposits disrupt the spray pattern and atomisation of fuel (Figure 3.3), leading to incomplete combustion, poorer fuel economy, reduced engine performance and increased emissions.

Diesel engines are used in a wide range of applications, so have a wide range of sizes and injectors. They are used in almost all trucks', railway locomotives' and ships' engines, and in oil-fired electricity power plants. Larger engines in

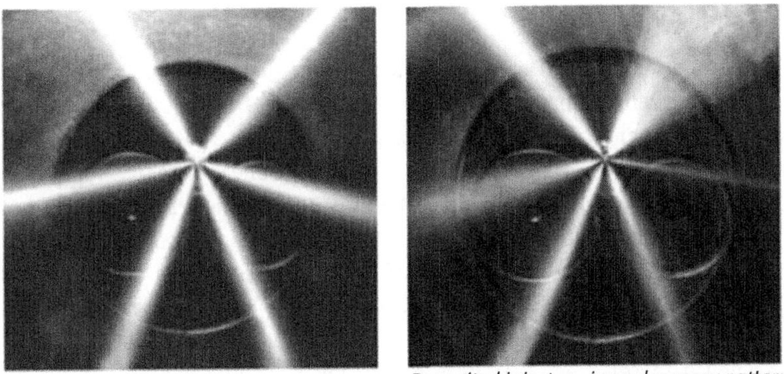

Figure 3.3 Diesel Injector Spray Pattern – Interference by Deposits.

commercial vehicles have larger injectors, which should allow for more deposits and particulates before blockages occur. However, larger engines also have a larger throughput of fuel and commercial vehicles are often running continuously (12 to 24 hours a day). While trucks and trains use the same high grade of fuel as passenger cars, ships and power plants use lower-grade fuels.

Marine vessels use a wide range of engines burning a wide range of fuels, varying from distillate fuels, containing various proportions of higher boiling fractions than used on land, through to blends containing various proportion of residual (usually vacuum residual) fuels. Residual fuels contain black asphaltenes (from which pitch is derived), various metal compounds and high sulphur levels, so are prone to readily produce particulates and much by way of emissions. Engines that burn residual fuels are two-stroke engines that use huge quantities of this much less expensive fuel. For example, on each stroke, the largest marine diesel, the Wartsila-Sulzer RTA96-C, injects 160 g of fuel per cylinder of 1,820 litre displacement on full load, at 102 rpm [7]. Compare this with a 2 litre, four-stroke, four cylinder car engine that may inject up to 0.04 cc of fuel, into a 0.5 litre displacement cylinder, at up to 3,000 rpm. So, while the large engine injectors are less sensitive, they deal with much larger quantities of particulate-rich fuels than does a passenger car.

Some grades of marine fuel are so called hybrid fuels – blends of distillate fuel and residual fuels. In such fuels, the asphaltenes from the residual fuel often separate out and the fuel is termed unstable. Such fuels are becoming more common in the now that the cap of marine fuel sulphur contents is lowered from 3.5% to 0.5% (section 1.3.2.4) – part of this reduction in sulphur is met by blends of residual fuel with higher proportions of low sulphur distillates, increasing the demand for asphaltene stabilisers [8,9].

3.2.3 Heating Oils

Heating oils which, like diesel fuels, are middle distillates may be of a high grade – some refiners provide heating oils from the same fuel component pool as used for diesel but using less kerosene. In the USA, most middle distillate fuel is pipelined as two fractions: No. 1 fuel (kerosene) and No. 2 fuel. These are blended to give the diesel fuel or heating oil that meet regional, seasonal temperature specifications (see section 4.2.2). In Europe, heating oil is more refinery-sourced and contains more of the aromatic streams (FCC by-products) that are unsuitable for diesel fuel, because they have low cetane numbers. In the UK, domestic heating oil is aromatic-rich kerosene.

As world diesel specifications move to low aromatic contents, following those in the EU and USA, heating oils take up more of the aromatic streams. Aromatic components have lower oxidative stability than have the aliphatic components preferred for diesel, so are more prone to oxidation and ensuing deposit and particulate formation on extended storage – heating oil may be stored for several months before being burned. Although heating equipment burners are less sensitive than fuel injectors, they too need to atomise fuel effectively to burn efficiently, with minimum soot, so are intolerant of deposits and particulates. Such deposits will eventually lead to failure of the burners and blockage of fuel filters (usually in the middle of winter!).

High heat output (≥ 1 MW) industrial boilers[2] may burn heavy fuels containing residual distillate bottoms [10]. In Europe, just over half of the residual fuel (11.7% of total refinery output, fuel oil Table 1.2) is used for non-transport uses, which include heating plants and electricity generation. In Russia, about 30% (50 million tonnes) of refinery output was said, in 2003, to be residual fuel, most of which was used either in heating plants or for electricity generation [11]; recent Russian statistics (January 2019) indicate that residual fuel oil still constitutes 27% of production[3] [12]. Burning residual fuel of worsening quality (as more atmospheric residue is vacuum- distilled), for its cheap calorific value, provides a number of problems: bottom deposits form and accumulate on fuel storage; combustion is less efficient, causing unburnt carbon and ash which form difficult-to-remove deposits on heat exchanger surfaces; polluting, toxic stack-gas emissions (burner effluent gases passing up a flu or stack) are exhausted into the atmosphere [11]. The problems of ash fouling, slag and corrosion, are reduced by the use of additives [13].

3.2.4 Jet Engines

Jet engines and jet fuel are paid a great deal of attention because failure can have catastrophic consequences. Jet fuel is pumped from the wing tanks not only to the burners, but also through the narrow pipes of heat exchangers and back to the tanks [14,15]. The metal heat exchanger pipes are used to cool engine oils, hydraulic fluids, and air conditioning systems. While jet fuel passes through these narrow pipes, it becomes quite hot. As a guide to the temperatures involved, the thermal stability test for jet fuel, the JFTOT (section 2.2.1), is operated at 260 °C, for the most widely used commercial Jet A-1 fuel [16]. Jet fuel in military jets may experience temperatures, at the nozzle, of 163 to 218 °C [15]. Jet fuel that has been in storage, which enables oxygen absorption, can form varnishes, gums and particulates that can:

- Disrupt spray patterns in the injector (burner) nozzles of the jet engine causing smoke, build-up of further deposits on other parts of the jet engine, and reduced efficiency.
- Interrupt the fuel flow by blocking fuel filters and fuel lines.
- Reduce heat exchanger efficiency.

However, storage of jet fuel occurs only in small airports and in military airfields – the USAF requires particular grades of jet fuel that are produced by some refineries, not continuously but in large batches, which may be stored, for months or years, until used. In addition, military jet fuel is more thermally stressed than is commercial fuel. To cope with the extended storage and high thermal stresses, military jet fuels are usually treated with specific, stabilising detergents, in particular SEC-AID 8Q405 [17,18].

For normal, commercial operations at major airports, deposits are not a problem because the commercial jet fuel has a short life cycle: it is produced in a refinery in a continuous process, run through a pipeline to the airport (such as Heathrow from Fawley [19]), loaded onto the various airliners and burned, possibly all in the same day. In addition, there is a water reaction drawback to using the detergent used in military jet fuel: at airports, water is removed from jet fuel by passing it through a continuous water separator, the effectiveness of which is compromised by the presence of the jet fuel detergent [20].

3.3 What Detergents Do

The problems of thermal decomposition products, just described, were clearly calling out for solutions as they became necessary to improve fuel efficiency, reduce emissions and reduce maintenance problems. Fuel additives were needed that, in low concentrations ('treat rates'), prevent these problems. Refinery operations, led by European specifications, have improved the qualities of automotive fuels significantly over the last 20 years, probably to their practical limit. Additives were, and still are, needed to keep fuel systems clean – especially narrow pipes (fuel lines), filters, injectors, inlet ports, burner jets and heat exchanger surfaces. Taking a page out of the domestic washday book, they were christened 'detergents'. Various fuel detergent additives have been developed to prevent the above-described problems, caused by deposits and particulates. Most interest has been in the effects of deposits on motor vehicles and airplanes, since problems in these are almost immediately made obvious by the engine no longer running smoothly.

Tests, in which deposits were formed in the absence of additives, have been devised for gasoline and diesel engines over many years, improving to meet developments in engine technology. One type of test is to measure performance in field trials, for example: running a passenger car for 750 km with fuel containing no fuel additive, then for a further 750 km and another 1500 km with fuel containing the detergent additive. At each stage, fuel flow (and fuel usage) through the injector would be measured, and smoothness, noise of the engine when running and deposits formed would be noted [21].

Another test method is to run an engine on a test bed in an engine test laboratory, measuring power output, fuel usage and emissions continuously, and finally checking the engine for deposits: e.g. for gasoline intake valve deposits (IVDs, using CEC method F-04-A-87) in a full engine [22] or in a single cylinder test engine [23]. Inlet valve deposits in a gasoline engine may be quite severe but are controllable using a suitable detergent (Figure 3.4) [24,25].

Diesel deposits are obvious on the outside of the injectors and on the removable injector plungers, when untreated fuel is used, compared with no deposits from fuel containing detergent additives (Figure 3.5) [24–30]. Injector nozzles have even been sliced open to observe and analyse critical deposits [29]. In other tests, fuel spray patterns have been photographed (Figure 3.3) to demonstrate the adverse effects of deposits and advantages of detergent additives [26].

Figure 3.4 Typical Gasoline Intake Valve Deposits, with and without Detergent. (LAC Lowest Additive Concentration.) © 2019 Chevron. All rights reserved[1].

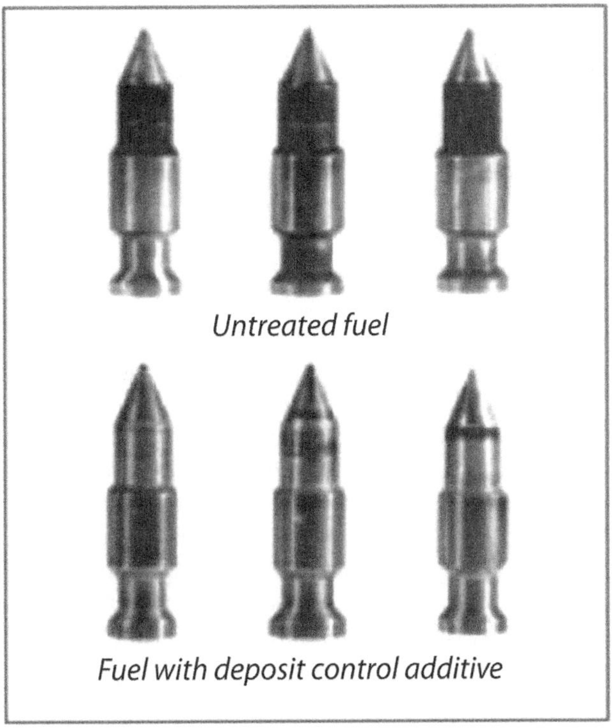

Figure 3.5 Typical Deposit Levels on Diesel Fuel Injector Plungers Sensitive to Deposit Formation. © 2019 Chevron. All rights reserved.

Advantages provided using detergents have been used to differentiate those fuels, with more effective detergency, from the average. For example, Shell V-Power 'Helps to restore engine condition by helping to remove performance robbing deposits from key fuel system components such as intake valves and/or fuel injectors' using DYNAFLEX Technology – enriched with powerful cleaning technology – that helps 'restore engine cleanliness by removing deposits that have already formed' [31]. Similar claims also come from other major oil companies (such as BP [32]).

Jet engine deposits have been subjected to special study, such as that to develop a test that will estimate a fuel's tendency to form deposits inside hot metal tubes (the JFTOT [33,34]) – part of a range of tests to be passed well before considering a test in an airplane. A more advanced test has also been developed to follow the oxidation of jet fuel and its inhibition by additives [15]. The problem of poor filterability of jet fuel after storage, due to resinous particle formation, has also received special attention [35].

Problems with heating oils, usually containing large proportions of FCC components, are filter blocking, corrosion and colour deepening caused by sediment (often referred to as 'sludge') that forms on storage. Such problems were identified as a concern in 'domestic furnace fuel' in the mid 1950s [36,37] and the problem still remains in need of treatment [38]. Early tests were to store the test fuel for three months then filter and measure filter time and amount of sediment. Now, there are ASTM storage tests in which sediment formation is accelerated by keeping fuel samples hot [39,40]. The use of fuel detergents was shown to significantly reduce filtration time and the amount of sediment that could be separated; the detergents also prevented colouring and tendency of the aged fuel to emulsify (hence accumulate) water [36].

A major problem experienced with heavy fuels, that contain residual fuel, is separation of asphaltenes during storage. Some settle, creating fuel stratification and handling problems; and all agglomerated asphaltenes cause serious filter blocking problems. This can be a greater problem in hybrid fuels, whose presence may grow with the incoming reduction in the sulphur cap to 0.5% for marine fuels [9,41]. An accepted laboratory test for this is to dissolve asphaltenes in toluene and add pentane and note whether the asphaltenes separate (ASTM D4370-32); fuel detergent additives can prevent such separation [42]. While field tests are difficult for heavy fuels used in naval vessels, there is a report of a heavy fuel detergent being tested by its effect on the filterability of heavy fuel, after by-passing normal pre-separation of deposits; to run the test, it was necessary to insert a manual pre-injector filter [43] – this filter needed changing and cleaning when it became blocked by asphaltenes. In this test, the filter needed cleaning every 4 minutes to keep the engine running on untreated fuel, while the filter to the engine running on treated fuel needed cleaning only every 40 minutes – once fuel, treated with a suitable detergent, had fully circulated the fuel lines [43].

3.4 The Chemistries of Fuel Detergents

3.4.1 General Background

Detergents have, as a basic structure, a polar, hydrophilic head-group attached to a non-polar hydrocarbon group. In aqueous systems of the kitchen, there are two kinds of dirt that are found on cooking and eating utensils, and on clothes: polar substances and fatty/oily materials. The polar substances are either partially burnt, carbonaceous deposits or those found in the ground (clay, sand, metal carbonates and sulphates). To remove burned-on deposits, caustic materials (caustic soda in oven cleaner, washing soda in dishwasher formulations) are used to break down the poly-ester/poly-acetal/carboxylic acid structures found in semi-burnt food, and detergents are used to disperse the particles that are released. For fatty/oily deposits, the non-polar part of the detergent, the hydrocarbon tail, interacts with the deposit and the polar end of the detergent interacts with water, allowing wetting of the detergent-covered oily deposit. The detergent then forms a stable emulsion of micelles, with the oil-plus-hydrocarbon tails on the inside and the polar head-groups on the outside, in contact with water [44,45].

Detergents used in hydrocarbon (fuel) solutions act analogously to those in water, except that they form 'inverse' or 'reverse' micelles, with the polar head groups in the core and the hydrocarbon tails on the outside, dissolved in the fuel [46]. In order to remove deposits of semi-burnt fuel or oil, or to prevent their precursors forming deposits, the fuel detergent needs a polar head group that interacts with the deposit and a sizeable hydrocarbon tail to solubilise the detergent plus deposit material into the fuel.

There are limitations to structures possible for fuel detergents that are not problems for detergents in aqueous systems. Fuel detergents need thermal stability and should not emulsify water into fuel. Water is present in fuel tanks, condensing into the fuel from the atmosphere or by contamination. A detergent that would stabilise water into fuel, by forming an emulsion, would attract more water. Eventually, the emulsion of water and fuel (and any dirt that it contains) would form a sludge that settles and accumulates in the fuel tank but is easily disturbed. This limits the use of highly polar head groups, such as the metal sulphonates and sulphates used in aqueous systems.

Since fuel system deposits are, mostly, caused by oxidation and interaction of the fuel with hot metal surfaces, there are similarities with deposits formed in crankcase lubricating oils. This suggested that lubricating oil additives might be of use. Deposit problems in the combustion zones of internal combustion engines are controlled using lubricating oil detergents. Lubricating oil detergents go back to the 1930s and are now, most usually, oil soluble calcium or magnesium salts of alkyl- (more than C_{20}) benzene sulphonic acids, of oligomerised alkylphenols or of alkyl salicylates [47]. Such salts bear a resemblance to detergents used in aqueous systems, such as sodium dodecyl benzene sulphonate, which has shorter alkyls for water solubility. However, if used in fuel, metal compounds would lead to a rapid build-up of additive deposits in fuel intake systems and particulates of metal oxides in the exhaust. With these considerations, lubricating oil detergents were not chosen for use in automotive fuels.

However, lubricating oil detergents – which may also contain colloidal dispersions of metal carbonates or oxides (for acid neutralisation) – are used in heavy fuels, which usually have high sulphur levels, to neutralise the sulphur oxides of combustion. They are also used to increase the melting temperatures of slag, deposits formed from the ash, making it less likely to melt and adhere strongly to surfaces of flues and heat exchangers, and cause their corrosion [48]. These detergents are, preferably, calcium or magnesium alkylbenzene sulphonates or carboxylates [48,49].

Lubricating oils, containing detergents, can keep the region around the combustion zone relatively free of deposits. However, they do not, alone, keep those deposits in the oil. Deposits separate out in the crankcase and other cooler parts of the engine, along with an emulsion of the water that condenses into the oil in a cold engine, during short runs and stop start conditions. Separating deposits and emulsions make frequent oil changes a necessity as they increase the viscosity of the oil and gradually block oil ports. As a result, at the oil change, it was necessary to run the engine with a 'flushing oil' before putting in the new oil. These problems were overcome in the 1960s by the introduction of 'dispersants', otherwise known as ashless detergents, that were designed to disperse crankcase sludge into the oil. While a number of potential dispersant molecules were proposed, those that dominate this application are variants around the reaction products of Poly-IsoButylene[4] Succinic Anhydrides (PIBSAs) with poly-ethylene amines[5], such as tetra-ethylene-pentamine (TEPA, $H_2N(CH_2CH_2NH)_4H$), or with polyols such as penta-erythritol; another group of dispersants are Mannich reaction products of poly-amines with alkyl phenols and formaldehyde [4,50,51].

Dispersants prevent sludge deposits forming solid masses inside the cooler parts of the engine and keep all deposit-forming material in suspension, to be removed with the oil at the next oil change. Being ashless, and effective in lubricating oils, dispersants were identified as potentially useful for keeping fuel systems clean, especially in those hot regions in the fuel injectors and inlet ports, or in and around burner jets. When used in fuels, lubricating oil dispersants are called detergents.

An earlier appearance of fuel detergents came as amines, which were deployed to keep carburettors clean. Judging by the occurrence of carburettor cleanliness patents, this must have been a considerable concern prior to the 1990s – those who can remember running mass produced cars with carburettors will attest to their sensitivity to deposits. The earliest claim seems to be one for simple, fat-derived amines, that came from the Armour Industrial Chemical company in the 1950s – though Armour claimed their amine for use in all middle distillates (diesel, jet, domestic and industrial heating oils), the testing was in domestic heating/furnace oils, containing 50% of cracked components [36–38]. Alkyl- and poly-alkyl-amines – usually bearing a primary amine group – have been widely claimed and used as gasoline detergents, initially for cleaning carburettors. In the general literature on gasoline of the late 1980s, the main concern was with carburettor cleanliness and the gasoline additives then commonly in use to control this were poly-isobutylene amines (PIB amines), PIB- succinimides (PIBSA-PAM), poly-ether amines and imidazolines [24,50,52].

Deposit control additives for gasoline engines follow a history of [53]:

- Mid 1950s: Alkenyl-succinimides, first from 1-alkenes, then from PIB
- About 1970: PIB amines
- About 1980: Poly-ether amines

At the time of these references (1989 to 1995), the use of indirect, port fuel injectors was becoming widespread in gasoline engines because they provided better control, giving lower emissions and better fuel economy than carburettors. However, fuel injectors are also susceptible to problems caused by deposits, so the need for gasoline detergents continued. Different detergent chemistries have different strengths, each being more able to prevent, or remove, the different deposits more or less effectively [4]:

- Poly-ether amines are better at controlling CCDs.
- PIB-amines and Mannich bases are better at controlling IVDs.
- PIBSA-PAM detergents are more effective at controlling PFI deposits and, possibly, direct injector deposits.

Commercial detergent packages often contain combinations of the different deposit control additives.

Diesel engines have always used fuel injectors but, around the same time as fuel injectors were replacing carburettors in gasoline engines, the more fuel-efficient diesel engine powered passenger cars became popular. The smaller diesel engines used in passenger cars are much more sensitive than truck engines to relatively minor problems with fuel flow or injector fouling. Car drivers, having a lot of choice in the autocar market, do not like noisy vehicles with uneven running, poor starting and smoky exhausts. However, both types of vehicle had experienced a growing need for fuel economy in response to the increases in prices of fuels and crudes over the last 40 years (crude from less than 20 $/barrel in the 1980s to 110 $/barrel 2012 [54]) and reduced emissions (enforced by legislation) — detergent additives are essential for continued injector cleanliness and efficient operation. Currently (April 2022) the price of crude has risen to 100 $/barrel, having fallen to below 20 $/barrel two years ago, but the demands of fuel specifications and trouble-free vehicle operation are unlikely to follow the same volatility. While fuel detergents are similarly listed for diesel fuel and gasoline [4,50], their preferred additive chemistries, once similar, have now diverged.

Most of the work on fuel and lubricating oil detergents has, by far, been in automotive applications, gasoline and diesel vehicles. Heating oil and marine applications have, mostly, been derived from this work. Jet fuel applications are much smaller in volume and are rather specialised.

3.4.2 Detail

3.4.2a Poly-IsoButylene, PIB

The earliest fuel detergents relied on alkyl groups derived from fats, which are relatively short (C_{16}-C_{18}). Stabilisation of dispersions in nonpolar media (oil) is mainly steric [45,55,56] – the oil soluble groups of the dispersant, dissolved in fuel, prevent the protected particles from agglomerating (coming together) by the entropic unfavourability of multiple chains (on different particles) occupying the same space. Thus, the larger the stabilising, fuel-soluble group, the better it is at stabilising the dispersion. Longer, straight chain alkyls would have poor solubilities, because of their tendencies to crystallise (see Chapter 4), so branched-chain, low molecular weight polymers are used, such as PIB [57]. Isobutylene is readily available in the oil industry from the steam cracker treatment of the petroleum gas streams (mainly C_2, C_3 and C_4 alkanes), as part of the C_4 stream, after removal of the more valuable ethylene and propylene. This C_4 stream contains about 2% ethylene and propylene, 10% 1-butylene, 10% to 15% 2-butylene and 15% to 30 % isobutylene, with the balance being a mixture of butane and iso-butane. Cationic polymerisation of the butylene mixture is selective for the isobutylene, especially at low temperatures. The process uses Lewis acid catalysts, such as aluminium trichloride[6], the soluble diethyl-aluminium chloride or boron trifluoride. These strong acids catalyse the formation of the tertiary cation from isobutylene (initiation), which adds other isobutylene molecules in a rapid, stepwise process (propagation, Figure 3.6).

PIB, made from the mixed butylenes, does contain a modest proportion of 1- and 2-butylene-derived units – an irregularity that probably further improves solubility. The molecular weight of the PIB produced depends upon how fast the chain grows (propagation) compared with how quickly it is terminated. Termination, at temperatures for the low molecular weight PIB, is mainly by chain transfer, with proton transfer to an isobutylene molecule (Figure 3.6).

Chain transfer is temperature dependent so, at about −100°C, high molecular weight PIB, or butyl rubber, is produced. At temperatures of around 0°C, the PIB produced has molecular weights of hundreds to a few thousand. These low molecular weight PIBs are preferred for fuel and lube oil detergent applications as they are very soluble in fuel and oil over a wide range of temperatures; they form good dispersants by providing effective steric barriers and they have low viscosities (high molecular weight polymers are too viscous for application as fuel detergents). Describing their use as PIBSA-based diesel-fuel detergents, the PIB with 'an Mn of 700 to 1300 is especially preferred'[7] [58].

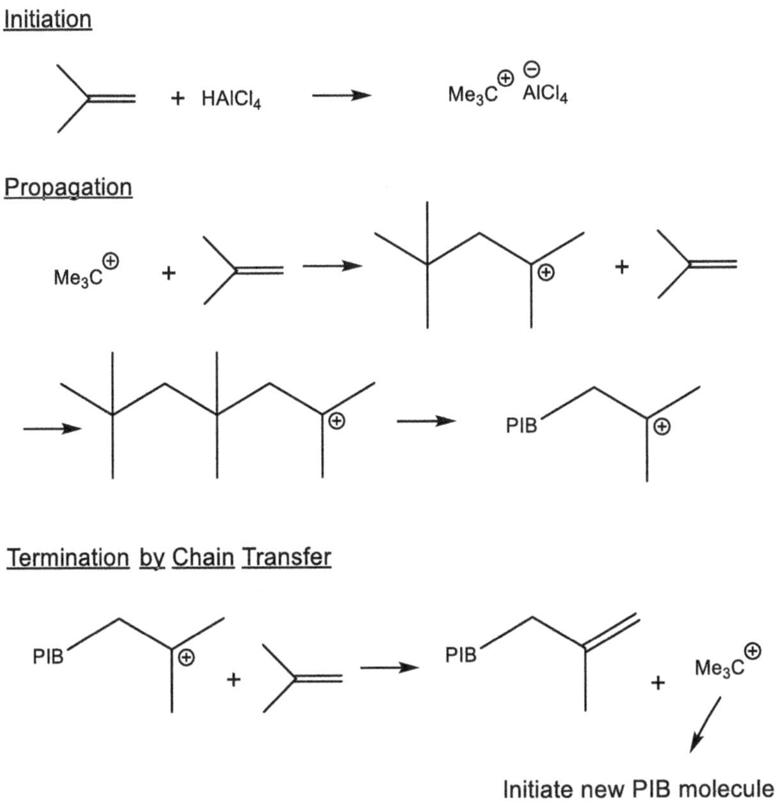

Figure 3.6 Cationic Polymerization of iso-Butylene.

3.4.2b PIBSA

Middle distillate detergents are most often modelled on lube oil dispersants, which are based upon PIBSA-PAM chemistry. The question is: how to attach a suitable polar end group to PIB to turn it into a detergent? Note that the only reactive group on the PIB molecule is a carbon–carbon double bond. When the polymerisation catalyst is boron trifluoride, the PIB molecule is mostly linear with a terminal methylene ($C = CH_2$). This is a relatively reactive (electron rich) and accessible double bond that is easily functionalised to a reactive group by heating with maleic anhydride. Maleic anhydride has an electron deficient carbon–carbon double bond so undergoes a thermal 'ene' reaction with electron rich alkenes, leaving no by-products. The product is PIB bearing a succinic anhydride end group, 'PIBSA' (Figure 3.7).

However, while the boron trifluoride derived PIB – known as reactive or highly reactive PIB – is generally preferred in this process, boron trifluoride is

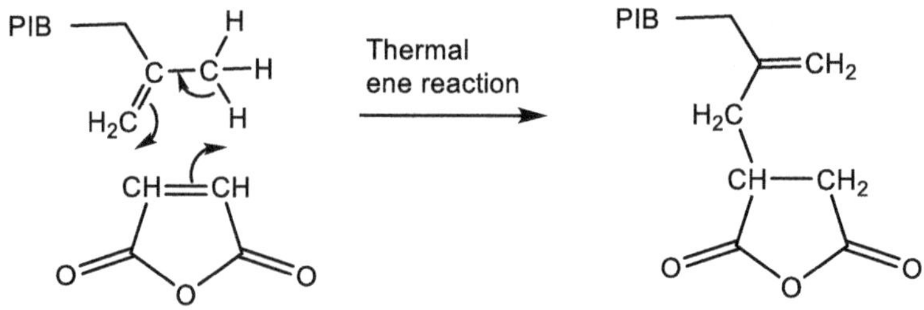

Figure 3.7 Ene Reaction of Maleic Anhydride with PIB.

toxic and requires an expensive, stringent recycling processes. The oil industry prefers to use aluminium chloride or di-ethyl aluminium chloride as the polymerisation catalyst; these are less expensive and less toxic than boron trifluoride, and can be hydrolysed by sodium hydroxide, to give relatively innocuous, water-soluble by-products. Unfortunately, aluminium chloride tends to scramble (isomerise) the carbon–carbon double bonds and the cation position so that some of them end up as the thermodynamically more stable tri- ($R_2C = CHR$) and tetra- ($R_2C = CR_2$) substituted double bonds, internal to the PIB molecule. Such structures are the result of alkyl group rearrangements – even the terminal carbon–carbon double bond is mostly PIB-C(Me) = CHMe [57]. Such sterically hindered double bonds are less reactive towards maleic anhydride and this PIB has to be coaxed into reacting by either chlorination then de-hydro-chlorination (at about 150°C), which gives more accessible double bonds, or by heating PIB with maleic anhydride to higher temperatures (about 200°C or higher), giving the product known as 'thermal PIBSA' [59]. Nowadays, fuel additives should not contain any chlorine (section 3.4.2d), so the thermal PIBSA process is preferred for fuel detergent applications, when made PIB from the aluminium chloride process.

3.4.2c PIBSA-PAM

The succinic anhydride group, now attached to the PIB, is reactive towards amine groups. Heating moderately, starting at 50°C to 100°C, then up to about 150°C, to remove water of reaction [60,61], primary amine groups react with the anhydride in a two-stage process: first to an amide then to give thermally stable imide links. It has been found that free amine groups are necessary for good fuel detergency so, if one amine group is needed to link to the succinic anhydride, others are needed to provide the detergency. Fortunately, polyamines, such as poly-ethylene-amines (PAM), are readily available. These are made by way of the reactive intermediate aziridine which polymerises as it is produced, under the conditions of the reaction, to give a range of poly-ethylene-diamines, $H_2N\text{-}(CH_2CH_2\text{-}NH)_nH$, such as 1,2-ethylene-diamine (EDA), di-ethylene-tri-amine (DETA), tri-ethylene-tetra-amine (TETA), tetra-ethylene-pentamine (TEPA), and so on up to PAM with about 12 EDA units. The mixture of polyamines is separated by fractional distillation to provide each of the lower members (*n* from 1 to 6), an intermediate mixed PAM and a distillate residue mixture of the highest PAM oligomers. The reaction of the PIBSA with the PAM may tie one PIB molecule to one polyamine (DETA in Figure 3.8) by a succinimide link [62].

Preferred proportions of PAM and PIBSA cover a range of 0.5 to 2.0 mole of PAM to 1.0 mole of PIBSA [63]. There are differing claims to the preferred structure. One description starts with the poly-ethylenediamine with one primary amine end group reacted with PIBSA, leaving a free amine at the other end (as Figure 3.8); a further molar equivalent of a PIBSA is then added to react with the other primary amine end-group – the succinic anhydride group, preferentially, gives imide links with the primary amine, thus tying one PIB group to each end of the PAM molecule (Figure 3.9) [63].

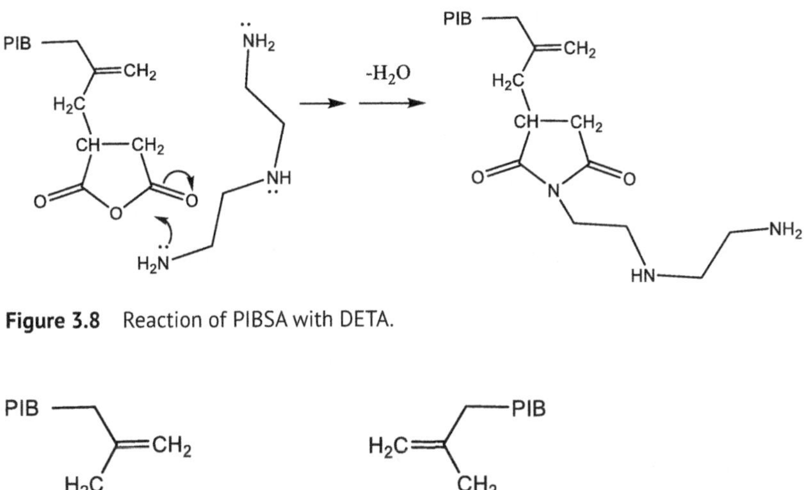

Figure 3.8 Reaction of PIBSA with DETA.

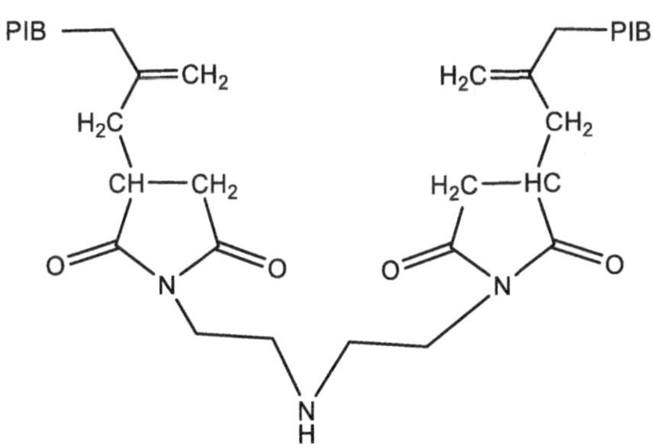

Figure 3.9 DETA with Two Molecules of PIBSA.

Some of the succinic anhydride groups may also react to give two amide links, instead of one imide, providing a branched structure [63]. Also, the first ene-reaction product of Figure 3.7 can react with more maleic anhydride to give some PIBSA molecules containing two succinic anhydride groups. As a result, there is potential for the formation of oligomers from the reaction di-functional PAM with difunctional PIBSA. Such processes would lead to higher molecular weight products, which is beneficial as a lubricating oil additive [64]. However, the higher viscosity and the potential for the formation of deposits from higher molecular weight materials themselves are not welcome in fuel applications. Thus lower molecular weight (900 to 1100) PIB-derived 'PIB succinimide' detergents are used for middle distillate detergents; also, their preparations use conditions that avoid oligomerization – the preferred molar ratio of PIBSA to PAM is claimed to be 1.0/0.9 for fuel applications [62].

The reaction of PIBSA with even a simple polyamine, for example TEPA, gives a mixture of imide, amide, imidazoline, diamide and alkyl-ammonium carboxylate groups [47]. Often, the less expensive PAM distillation residue is used, instead of one of the distilled poly-ethylene-amines (such as TEPA). Such PAM, having about 8 to 12 ethyl diamine segments, contains molecules that are branched at tertiary amine nitrogen atoms. This is because, in the PAM manufacture, the aziridine intermediate can add to an internal secondary NH, as well as to the terminal primary NH_2. So, even a 'simple' PIBSA-PAM dispersant, having one PIBSA (with one succinic anhydride group per PIBSA molecule) per primary amine group, could have two, three or four PIB chains attached by succinimide links to the amine core (Figure 3.10). For lubricant applications of PIBSA-PAM, it is claimed that highly branched PAM is, in fact, preferred [64]. As a result, PIBSA-PAM has a complex, cross-linked structure,

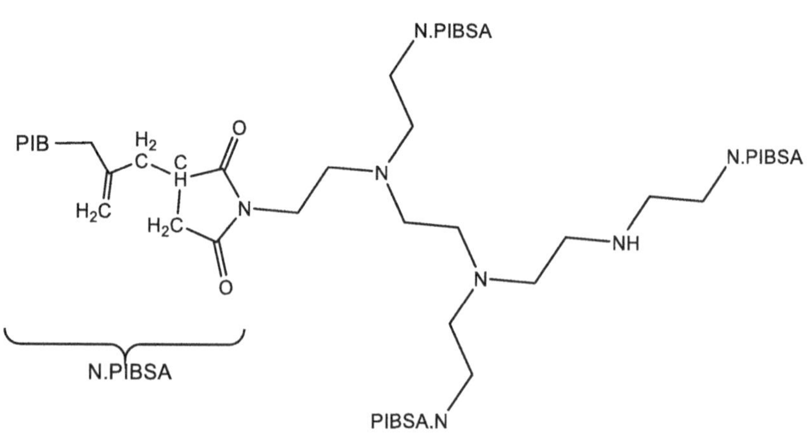

Figure 3.10 PIBSA Reacted with Branched PAM.

having three sources of branching: more than one succinic anhydride group per PIBSA, some succinic anhydride reacting to form diamides, and branching in the PAM.

In fuel, the polar parts of PIBSA-PAM – amide, imide, amine salt and, polyamine chains – are most likely to aggregate as cores of micelles (dipole–dipole and hydrogen bond interactions are, energetically, much more favourable than dispersion force interactions with hydrocarbons). The cores are surrounded by multiple PIB chains, that are highly fuel soluble, stabilising inverse micelles (as Figure 1.8). Any water in the fuel is attracted to the polar core, adding further stability[8].

This complex PIBSA-PAM (also known as PIB-succinimide) detergent has been shown to reduce inlet valve deposits (IVDs) [65], in gasoline engines, and to keep injectors clean in diesel engines [62,66]. PIBSA-PAM detergents are preferred over PIB-Amine and PIB-phenol Mannich detergents in diesel engines [4] (sections 3.4.2d and 3.4.2e). For example, when discussing diesel fuel additives, one quote states 'Recommended treat rates are around 200 ppm for the succinimides and up to three times that amount for the hydrocarbyl amines' [50]. PIBSA-PAM detergents used in diesel and gasoline not only keep injectors clean but may also have general engine cleanliness credits by adding to the dispersant in the lube oil. However, recent developments are mostly concerned with keeping the injectors of modern diesel engines clean [5,21,58,59], interactions with the lube oil are sometimes considered as potentially harmful.

Marine diesel fuels are generally higher boiling than those for on-road diesel. One problem is that the cylinder liner of a marine engine using fuels with 90% distillation point preferably above 420°C (cf. for on-road diesel in the Euro spec EN590, the 95% distillation this is limited to 360°C) becomes fouled by a lacquer. The detergent claimed to prevent this is a PIBSA-PAM detergent [67].

The reaction product of dodecyl-succinic anhydride (DDSA[9]) with hydrazine has been claimed to be effective in modern[10], common rail diesel engines [68]. Test data shows that while a PIBSA-PAM detergent is most effective in the older diesel technology ('XUD9' test), the DDSA-hydrazine material is most effective in the engine test run under modern injector conditions (CECF-98-08 DW10)[11] in which the fuel contains a few ppm of dissolved zinc – found to be a particular problem for modern injectors (section 3.5.1b). The reaction of excess hydrazine with DDSA (best products are from reactions using 3 mole hydrazine per mole DDSA) might be thought to give an N-amino succinimide (A. of Figure 3.11) and, possibly the dimer (B.) but it is now thought to be a complex mixed product containing various oligomers (such as C. in Figure 3.11 and as described in [68]). Interestingly, a PIBSA-hydrazine product was described earlier as a component for use in a lubricating oil additive, though without any structural details [61].

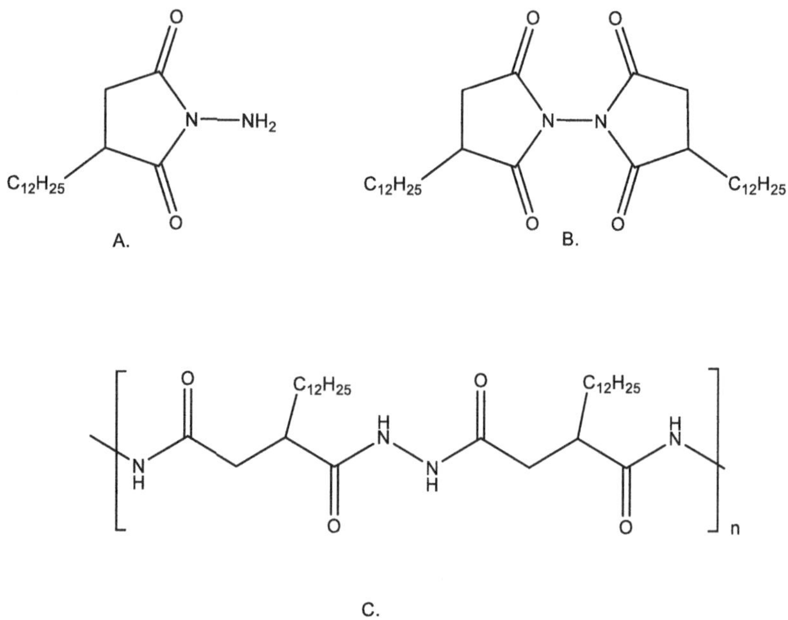

Figure 3.11 Possible PIBSA-Hydrazine Reaction Products.

Heating oils are usually middle distillates, the exceptions being heavy fuels that are used in some large heating facilities. Not surprisingly, then, the detergents in use in heating oils are the same as those used in diesel fuel. For example, in middle distillate heating oils containing FAME (a B10 fuel was used in testing), additive formulations that were shown to reduce deposit formation on storage – the source of problems for home heating systems with modern burner technology – needed to contain a PIBSA-PAM detergent derived, preferably, from reactive PIB [39].

3.4.2d PIB-Amine

Gasoline detergents are often PIB-amines, rather than PIBSA-PAMs, especially for IVDs [4]. Simple, primary amines were first claimed in the 1950s as carburettor cleanliness detergents [69], when the problem with carburettor fouling was understood. PIB amine detergents are in general use as gasoline detergents to control combustion chamber and injector deposits as well as IVDs [70] – low molecular weight PIB ('Mn of ...preferably from 500 to 1500' [71]) is the preferred oil soluble group.

Converting PIB to PIB-polyamine, via chlorination, was first claimed by Shell [72] for use in lubricating oil, but simple amines do not now figure in lubricating oil formulations [73]. Several companies then found that this route, applied to PIB with EDA, gave an effective gasoline detergent [23,74,75]. The chlorination of PIB was carried out under free radical conditions, in which chlorine atoms were expected to subsitute allylic C–H hydrogen atoms, preferentially [23,72] (Figure 3.12[12]). Reacting the PIB chloride with EDA gave a N-(PIB)-1,2-ethane diamine as the amine hydrochloride (Figure 3.12). Here, demonstrated using 'reactive' PIB from the boron trifluoride process but the allylic chlorination also occurs easily at methyls adjacent to the di- and tri-substituted alkene bonds, present in PIB from the aluminium chloride process. The hydrochloride was removed by a wash with aqueous sodium hydroxide. The nucleophilic substitution by the amine nitrogen was probably S_{N2} in such a low polarity medium.

Under the reaction conditions, some PIB was chlorinated by the addition of chlorine, producing chlorine on tertiary positions. Free radical chlorination can also introduce chlorine at tertiary positions. In addition, when the PIB was from the aluminium chloride process, there was some addition of HCl, resulting in more chlorine at tertiary positions. Chlorine at tertiary positions is difficult to displace (amines do not substitute tertiary alkyl chlorides, though they may cause elimination of hydrogen chloride [76]). In fact, it was admitted that the product does contain about 1% chlorine, but that it was not deleterious to the PIB amine detergency [77]. However, this created another problem. In the early 1990s it was realised that burning polymers containing chlorine can produce chlorinated dibenzodioxins; one of these, 2,3,7,8-tetrachloro-dibenzodioxin, is widely believed to be one of the most toxic man-made chemicals and is found as a contaminant in agent orange [78]. An explosion that released a toxic cloud containing this material in an industrial accident in Seveso (Italy), in 1976, alerted Europe to these materials. As a result, oil companies banned the use of compounds containing chlorine in fuels (amongst many other actions). As a result, the use of PIB-amine made by the chlorine route was displaced by non-chlorine PIB-amine.

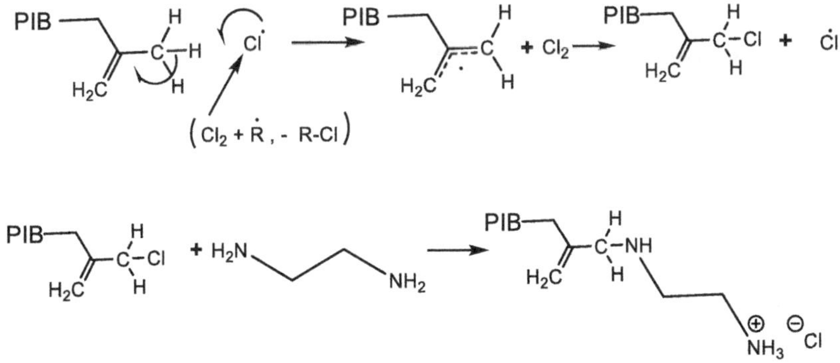

Figure 3.12 Formation of Chloro-PIB and Its Reaction with EDA.

Given the low yield of the simple, straight PIB chains with amine groups at the ends by the aluminium chloride route (PIB formed by $AlCl_3$ catalysis contains only 10% terminal unsaturation [71]), it might be expected that a product containing a high proportion of terminally functionalised PIB would be a more potent gasoline detergent. Reactive-PIB has a high proportion of terminal unsaturation, i.e. C = C at the end of the molecule. The Oxo process is a means of producing terminal functionality, by the addition of carbon monoxide and hydrogen, over a rhodium or cobalt catalyst, to a terminal C = C bond to give an aldehyde [76] (Figure 3.13) Carbon monoxide adds to the tertiary carbon because the Oxo process has cationic character. Oxo conditions are also reductive so, usually, the aldehyde is accompanied with some alcohol. The aldehyde may be hydrogenated completely to the primary Oxo-alcohol, as used in the production of many Oxo alcohols such as 'iso-decanol'; or the aldehyde may dimerised by an aldol condensation, then hydrogenated, as in the process used for 2-ethyl-hexanol from butylene via butanal [79].

Using the Oxo process to maximise aldehyde (1/1 molar = CO/H_2), then using reductive amination (ammonia or di-methylamino-propylamine and hydrogen over a Raney nickel or cobalt catalyst) converts the aldehyde to an amine [71] (Figure 3.13). In another patent [80], the process is described as a reductive Oxo on PIB, to give the alcohol, then reductive amination to the amine. Gasoline engine testing, measuring valve deposits, showed that these PIB amine products are more effective than products made from less reactive PIB, with more internal C = C, or by the PIB chloride route [71].

This PIB amine is both chlorine free and more effective than other PIB amine products as a gasoline detergent and is currently one of the few preferred gasoline detergents.

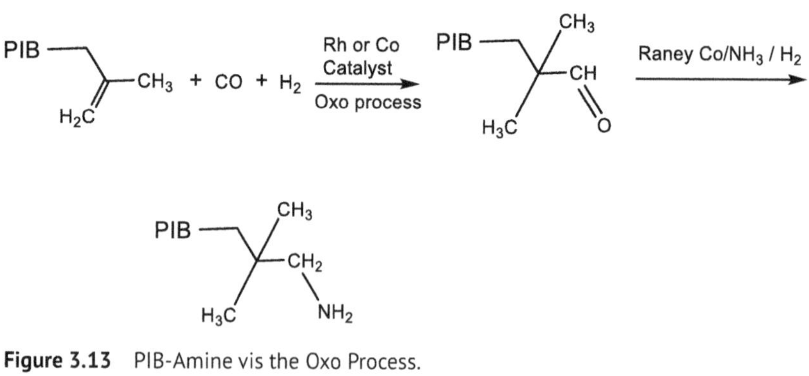

Figure 3.13 PIB-Amine vis the Oxo Process.

3.4.2e Mannich Detergent

Another fuel detergent that is of quite long standing, and is still in use, is one generally called a Mannich detergent. This is also chlorine free. For one version of this product, a PIB-phenol, from 1000 molecular weight reactive PIB, is reacted with an amine, such as N,N-dimethyl 1,3-propanediamine (3-DimethylAminoPropylAmine, DMAPA) or ethylene diamine, and formaldehyde, removing water of reaction [81]. In the Mannich reaction, an aldehyde reacts with an amine (R-NH_2) to give an imine (R-N = CH_2), the protonated imine (R-NH^+ = CH_2) gives an electrophilic substitution on an alkyl phenol to give a phenol-methylene-amine link (Figure 3.14). If a polyamine is used, the process can link two (or more) phenols together (various structures are described [82]).

The products were shown in engine testing [81] (CEC F-05 A-93 procedure in a Mercedes Benz engine) to reduce IVDs; the Mannich detergents were combined with a carrier fluid (section 3.4.2j) and added to an unleaded gasoline at 136 ppm of the detergent. The deposits on the inlet valves were measured after the engine test (Table 3.1).

Attention was drawn to the better performance of the EDA-derived detergent – bearing a primary amine group – over that derived from DMAPA or TETA. This detergency advantage was further demonstrated in the Peugeot XUD-9 indirect injection diesel engine test, in which the blockage of injectors is measured by the loss of fuel flow through the injector when the injector needle is lifted 0.1 mm (Table 3.2) [81].

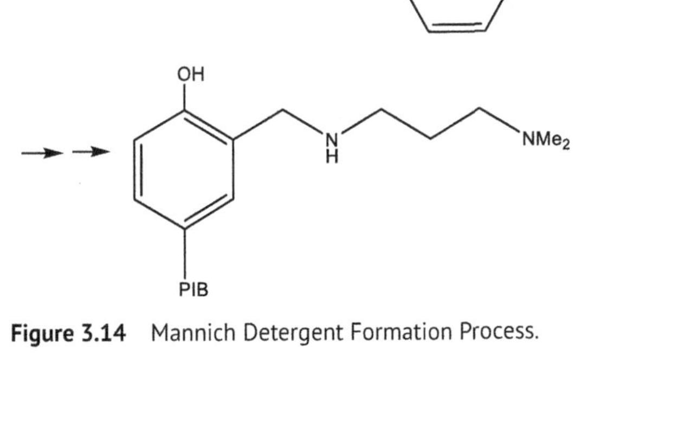

Figure 3.14 Mannich Detergent Formation Process.

Table 3.1 Inlet Valve Deposits at End of Mercedes Benz Engine Test.

Mannich Detergent by Amine on PIB-Phenol at 136 ppm	Deposits (mg)
None	278
EDA	16
DMAPA	30
TETA	89

Such PIB phenol Mannich compounds are described as 'traditional fuel detergents' [83] and are also used as lubricant dispersants [84]. Mannich detergents have been shown to be more effective than other amine-based gasoline detergents in keeping clean the injectors of direct injection gasoline engines [70].

3.4.2f Imidazoline

Imidazolines are used as corrosion inhibitors, particularly in refineries [85,86]. Simple imidazolines, such as those formed when a 1,2-diamine is heated with a long-chain carboxylic acid (Figure 3.15), such as oleic or stearic acid, were found to be effective carburettor detergents [87]. As already pointed out, such structures are present in PIBSA-PAM detergents.

A dodecylbenzene sulphonate of an imidazoline has been claimed as a carburettor detergent [88]. In the current environment of near zero ppm sulphur fuels, it is unlikely that sulphonates are in use as fuel detergents. However, alone, the strongly basic imidazolines would form a salt with the fatty acid lubricity improvers (Chapter 5). Imidazolines themselves are listed in older references as possible fuel detergents [24,50] but not in more recent articles [4,53].

Table 3.2 Peugeot XUD-9 Loss of Fuel Flow Reduced by the PIB-Phenol-EDA Mannich Detergent.

Mannich Detergent, Amine on PIB-Phenol	% Fuel Flow Loss at 0.1 mm Needle Lift
No detergent	89.6
EDA	75.6
DMAPA	86.4
TETA	81.7

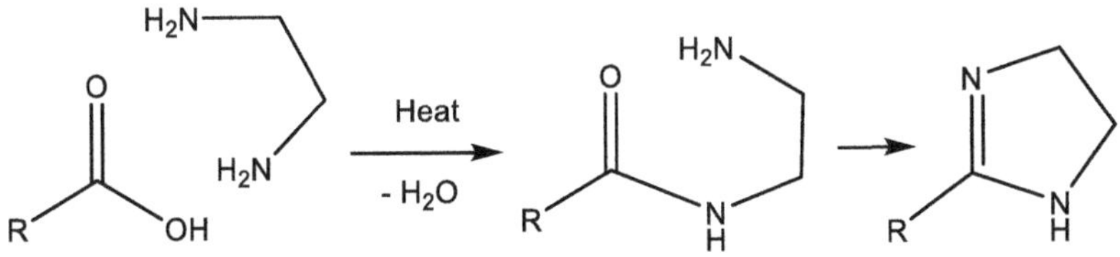

Figure 3.15 2-Imidazoline Formation.

3.4.2g PIBSA/Polyols

Amongst the PIBSA-derived detergents, most rely upon having basic, hence nucleophilic, nitrogen, which can attack engine seals causing them to swell and degrade. An alternative is to use nitrogen-free systems based on polyols, such as PIB-succinic acid reacted with penta-erythritol [89,90] to give an ester-linked polyol. The structure of such a product is unclear for, while one mole of penta-erythritol is charged per two equivalents of acid (judged from the acid number [89]) which should mean per mole of PIB-succinic acid, there are many uncertainties to the material balance: not all the PIB reacts with the maleic anhydride to form PIBSA (hydrolysed to PIB-succinic acid) and not all the penta-erythritol reacts (much is filtered off). The reaction of two hydroxyls from the penta-erythritol with the succinic acid group could give a nine membered ring, but such rings are strained [91], so, the structure is likely to be a mixture of oligomeric ester (diol/diacid), with some acid groups bearing a single penta-erythritol group. While such polyol esters have been used as lube oil additives, they do not appear in lists of fuel detergents, for example in ATC documents [53,92]. However, it is worth a reminder, here, that the jet fuel detergent is based upon a penta-erythritol ester of a thio-phosphonic acid (section 3.4.2k).

One recent fuel detergent approach is to react PIB-alkylated ortho-cresol (2-methyl-phenol) with formaldehyde to add further $-CH_2OH$ groups to the cresol ring. This was shown to be more effective at reducing inlet valve deposits, in a CEC SG-F-0202 engine test, than a 'typical alkylphenol amine detergent' – which was described as 'a commercially available nitrogen containing additive' (from the described method of preparation, this was a Mannich detergent [82]).

3.4.2h Polyether Amines

Polyether amines were introduced as carburettor cleanliness additives and variations are also effective as detergents, especially for gasoline fuel injectors and valves [93]. The balance of polarity for gasoline detergent solubility and activity is obtained by using the appropriate mixture of the polar Poly-Ethylene-Oxide

(PEO) with the non-polar Poly-Propylene-Oxide (PPO) or Poly-Butylene-Oxide (PBO)[13]. End groups can also be varied, such as hydroxyl (polar) or alkyl phenol (non-polar). Poly-ethers are made by Ring Opening Polymerization (ROP) of epoxides, which can be acid or, more usually, base catalysed. ROP is a form of living polymerisation, since the end group of the polymer may react with more monomer, as long as there is excess monomer [94]. After the first epoxide has all reacted, a second epoxide can be added that will react with the still 'live', anionic end group. This enables a relatively easy manufacture of block copolymers. An amine end group may be added after polymerisation by replacement of the secondary alcohol end group on the PPO block by ammonia, with the aid of a Raney nickel catalyst [93] (Figure 3.16).

Alternatively, the amine group may be present in the initiating monomer, for example from 2-(2-amino-ethoxy) ethanol [95] (Figure 3.17).

3.4.2i Quaternised Detergents

A further development of the Mannich and PIBSA-DMAPA detergent systems have been the quaternised amine versions. Quaternization relies upon there being one or more tertiary amines on the detergent, while any primary or secondary amines (as from PAM) would also react with a quaternizing

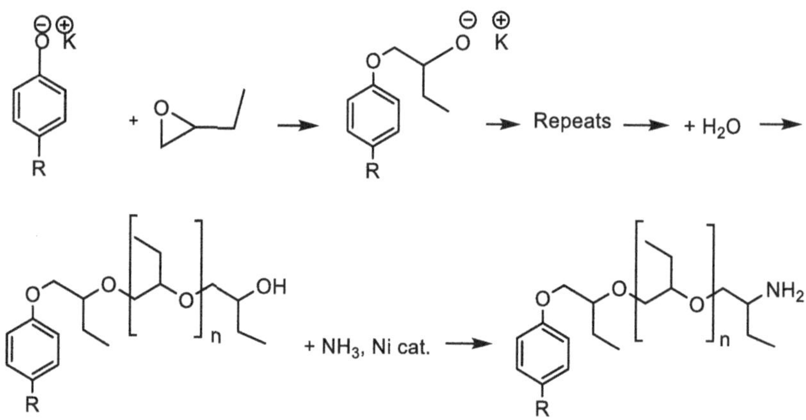

Figure 3.16 Formation of a Polyether Amine.

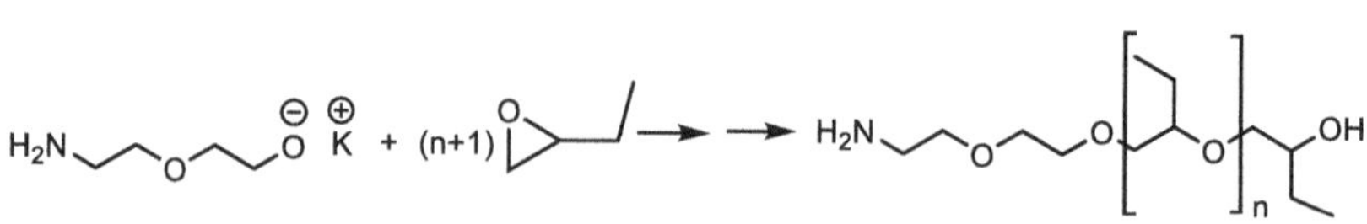

Figure 3.17 Alternative Route to a Polyether Amine.

agent – probably with no gain in activity. The Mannich detergent derived from the Mannich reaction of dimethylamine and formaldehyde with a PIB alkylated phenol gives one of the precursors that may be effectively quaternised [96] (Figure 3.18 A to B). Quaternization was demonstrated using reagents such as dimethyl sulphate or benzyl chloride but adding sulphur or a halogen to an additive for use in fuel is not acceptable. The process used to give a CHON only product is the reaction with styrene oxide and acetic acid (Figure 3.18 B to C).

PIBSA reacted with DMAPA produces a 'PIBSA-PAM' detergent which also bears a free tertiary amine that may be quaternized [97]: the quaternization method of choice is the reaction with acetic acid and propylene oxide [98]. Such quaternization was shown to provide a detergency advantage over the precursor, non-quaternised PIBSA-DMAPA, when tested in a John Deere engine, in ULSD, run for 8 hours [99]. At the end of this test, the loss of torque with 100 ppm of PIBSA-DMAPA was 8.16% while only 0.04% with the quaternized

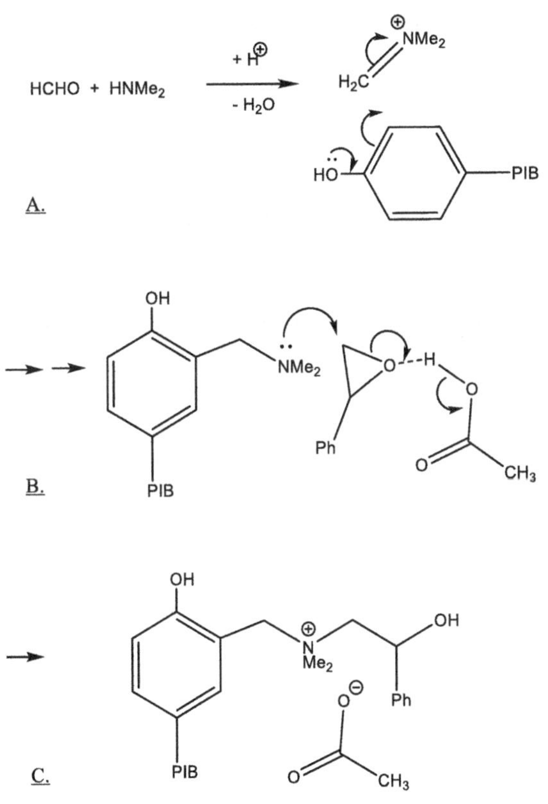

Figure 3.18 Mannich Detergent Quaternized with Styrene Oxide/Acetic Acid.

PIBSA-DMAPA; also, internal injector deposits were present using the former while no such deposits using the quaternized version. In a deposit clean-up test, the engine was run on ULSD with a fouling agent over various conditions for 32 hours, in which time clear signs of deposit formation were observed as their effect on the torque supplied by the engine: a loss of 11%. The quaternized detergent was then added to the fuel (100 ppm) and the test continued for a further 17 hours after which time there was a gain in torque of 7%, indicating deposit clean up activity.

Further testing showed that these quaternized detergents assist the polyetheramine in the removal of existing deposits in gasoline engines, caused by running automobiles (e.g. Toyota Corolla) for some thousands of miles on regular gasoline. Earlier work had indicated that the DMA-PIB-Phenol Mannich detergent improved the deposit removal by a polyetheramine[14] [100]. However, while the best of these results show that the inlet valve and combustion chamber deposits removal may be doubled by combinations of 2400 ppm of a polyether amine and 390 ppm of the Mannich detergent, increasing the proportion of Mannich and further decreasing the polyether amine treat gave very poor deposit removals (Table 6 of the patent [100]).

The latest patents are most concerned with the PIBSA-DMAP quaternised by reaction with propylene oxide /AcOH [98] (US 10479950) (Figure 3.19). It is recognized that the PIBSA DMAP reaction gives varying proportions of amide and imide linking. While the quaternised products of both of these (PO/AcOH) seemed effective at reducing power loss in the DW10 engine test, the version with a high amide content (low imide) gave a pass in the lube oil compatibility test[15] while the low amide/high imide product failed this test. Propylene oxide may have process advantages over the earlier version in which styrene oxide was used as the quaternizing agent [97].

3.4.2j Carrier Fluid

As with most fuel and lube additives, the main component, here the ashless detergent, is usually used in combination with other components for improved

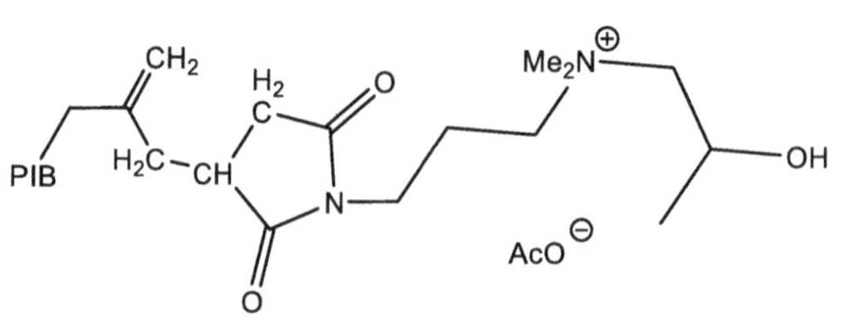

Figure 3.19 Quaternized PIBSA-DMAP Detergent.

activity. With fuel detergents, it was found that a non-volatile fluid, such as a lubricating oil, aided the action of the fuel detergent by providing a fluid that, being always present as a thin film, could help wash away deposits that had been solubilised (in the oil) by the detergent. Various proportions were recommended, such as 100 ppm of detergent with 400 ppm of oil [101]. This effect is referred to, and described, in a more recent patent that claims an improvement upon lube oil as the co-additive [80]. Here, the claim is that lube oil is not stable enough and that a polyether/ester is more effective. Data was presented that shows a clear advantage for combining the detergent (PIB-amine or PIBSA-PAM) with a poly-ether (e.g. tri-decanol-poly-butylene oxide or TDA-poly-propylene oxide) and a poly-ester (e.g. di-isotridecyl adipate or di-C_9/C_{10}-phthalate) in reducing IVDs and valve sticking to near zero [80]. Fuel treat concentrations were in the region of 100 to 300 ppm of each of the three components. A non-volatile polyether is likely to be an excellent solvent – such as the cellosolves or glymes (mono- or di-[ethylene-glycol]-mono or di-alkyl ethers). Detergent carrier fluids have also been described as non-volatile aromatic solvents, mineral oil or synthetic lubes (poly-1-alkenes) and poly-ether synthetic oils [102]. These are in some of the mixtures that are supplied as commercial fuel detergents [92].

3.4.2k Jet Fuel Detergent

Military jet fuel, JP-8, is made to stringent specifications in batches which could be stored for long periods of time, as the stock is run down. In use, in high-performance fighter jets, it suffers greater thermal stresses than does jet A-1 in commercial jet liners – fuel in the jet burner nozzle may reach temperatures above 200°C. If deposits form in the heat exchanger tubes, they could block fuel lines and close tolerance valves [15]. To prevent these problems, a combination of an anti-oxidant, a metal deactivator (previously known technology) and a new, phosphorus containing fuel detergent, SPEC-AID 8Q405, was developed to prevent the harmful deposits in Military JP-8 jet fuel – the more stable fuel grade was named 'JP-8 + 100' [17,18] (see section 2.4.3). SPEC-AID 8Q405 is a Betz product that is probably that described in a BetzDearborn patent published in 1998, using similar language about the need for a stabiliser for military jet fuels [103]. Testing of JP-4 and JP-8 fuels in a 'dirty F100-PW-200 engine' and long-term operation of an F-16 are described in this patent. The jet fuel additive is described as poly-isobutylene reacted with phosphorus pentasulphide to give a PIB-thio-phosphonic acid, which is then used to esterify penta-erythritol to give (acronym) PBTPA – from this description, the chemistry is likely to give mixed products at both stages. PBTPA was shown to clean up dirty jet engines and give better operation in long-term testing of a jet plane. SPEC-AID 8Q405 is specified as one that may be required in JP-8 fuel [16].

Jet fuel, the kerosene fraction of middle distillates, produces deposits that are much the same as for diesel (the formation of their oxidation products has been treated together [104,105]). SPEC-AID 8Q405, however, being phosphorus- and sulphur-containing, is unlike usual gasoline and diesel detergents; it has more in common with a lube additive: zinc dialkyl-dithio-phosphate (ZDDP). As such, its action is likely to provide stability against oxidative degradation – it appears to be used in combination with a metal deactivator and an antioxidant (such as BHT) in the fully formulated fuel [106]. In addition, such an additive forms an adsorbed, protective film on aluminium [18], which is widely used in airplane metal parts. By comparison, the ZDDP additive in lubricants is known to provide both anti-oxidancy and protective film formation [47].

3.5 Mechanism of Detergency Action

3.5.1 Chemical Identities of Deposits

3.5.1a Oxygenated Hydrocarbons

Chemical analyses of deposits have identified fuel oxidation products, carbon, sodium salts (possibly from refineries that use salt driers) and polymeric materials, some of which have been related to fuel and lubricating oil additives, such as detergents and detergents combined with acids [5]. Perhaps it is not surprising to find additives in deposits as they are designed to interact with deposit forming compounds to inhibit their continued formation and the formation of fuel oxidation product deposits. Work on unadditized fuels and hydrocarbons [107] supports a process of oxygen absorption to form hydroperoxides, slowly at lower (<80°C) temperatures. At higher temperatures, above 120°C, hydroperoxide decomposition aids rapid oxygen uptake. This work indicates a stepwise process of the formation of dimers and trimers – molecules with significant oxygen content, up to five oxygens per molecule – then higher polymers (up to 1,000 to 1,500 molecular weight) to form a soluble gum. Gum and deposit formation are slower than the rate of oxygen absorption, in support of a two-stage process – peroxide formation then thermal decomposition of the peroxides, at 100°C or more [104]. Hydroperoxides, alcohols, ketones, esters, and aldehydes were identified in the gums. Pyrolysis of the gum at higher temperatures, in the region of the engine (180°C and higher), converts the gums to solid deposits. A stepwise progression in molecular weight is consistent with a condensation polymerisation mechanism.

Absorption of oxygen into fuels at moderate temperatures (below 100°C), to form hydroperoxides, occurs in the refinery, during distribution and in long-term storage, such as with heating oil, and military jet and diesel fuels. Motor fuels get well mixed with air during pumping from large tank to road-tanker to smaller tank; and also, during the recycle of warm fuel, from the injector pump and injectors to the vehicle fuel tank. Oxygen absorption to form peroxides is a free radical process [47], dealt with in more detail in the discussions of autoxidation (Chapter 2):

$R\cdot + O_2 \longrightarrow ROO\cdot$, $R\cdot - H \longrightarrow R\cdot + ROOH$, where R–H is a fuel hydrocarbon molecule.

Preventing the formation of peroxy-compounds prevents the subsequent processes that lead to deposit formation. This is the role of antioxidants (section 2.5.6) but, while peroxide radicals are mostly trapped by anti-oxidants, some escape to propagate a number of steps before being trapped. Those that are trapped are trapped as alkyl[16] hydroperoxides or dialkyl peroxides – peroxy acids and esters may also be formed. These are quite stable below 100°C but decompose readily at above 120°C [105] to give a range of products, depending on the structure of the hydrocarbon group, and more oxygen absorbing alkyl radicals. In general, this would produce a range of aldehydes, ketones, alcohols, ethers, and carboxylic acids (see Chapter 2 for details of hydrocarbon oxidation).

Such temperatures as are needed for peroxide decomposition are quite relevant to fuel in contact with engines. For example, in a modern diesel engine, the initial pressurising of the fuel by the injection pump heats it up to about 100°C; it then travels to the fuel injector to reach temperatures, at the nozzle tip, of 250 to 350°C. Some of the fuel that has experienced these temperatures is returned to the fuel tank[17], where it may further oxygenate and pass through another injection cycle [58,82]. Ease of decomposition of peroxide compounds at these temperatures, is represented by their half-lives – the time taken for half of their quantities present to decompose. Half-lives of examples of the peroxy compound types that represent those that may be found in oxygenated fuel, demonstrate how readily they are decomposed at such temperatures (Table 3.3) [108].

In other words, while fuel is sitting in the injector body, waiting to be injected into the engine, at the temperatures of about 200°C, much of its peroxide content will have decomposed (the per-esters will have gone almost instantly and the dialkyl peroxides in a few seconds). In addition, the fuel sitting above the hot engine in the common rail, may be at about 100°C or more for a longer time period; some of this fuel will not be injected but will be circulated back to the fuel tank, possibly a number of times; so, while it is at a lower temperature than in the injector body, it is at this temperature for a longer time period.

Hydrogen atom abstraction by the peroxy radical, during the formation or decomposition of hydroperoxide, is more likely to be intra-molecular than inter-molecular, providing that a suitable hydrogen atom is available (section 2.5.5 and Figure 2.8). As a result, multi-functional oxygenated molecules are

Table 3.3 Temperatures (in °C) at which Relevant Peroxy Compounds Have Half-Lives of 6 Minutes or 36 Seconds.

Compound	Peroxy Compound Type	Temperature at Which Half-Life is 6 Minutes	Temperature at Which Half-Life is 36 Seconds
t-Butyl peroxy-2-ethylhexanoate	Peroxy ester	113	125
di-t-Butylperoxide	Dialkyl peroxide	164	190
t-Butyl hydroperoxide	Alkyl hydroperoxide	207	230

produced, hence the high oxygen content found in the gum intermediates [107], which interact at high temperatures to form cross-linked, polymeric materials. At first, the polymeric oxygenated gums are still soluble in fuel but are left behind on evaporation of the fuel – soluble gum that is left behind, when the fuel is removed by evaporation, is specified for gasoline at a maximum of 5 mg per 100 ml (EN 228).

Polymers result from inter-reactions of the multi-functional alcohols, carboxylic acids, ketones and aldehydes, such as poly-esterification of hydroxy-acids (example Figure 3.20) and multiple carbonyl condensation reactions, such as the aldol (Figure 3.21) and Claisen condensations. Note that the aldol and Claisen condensations can lead to 2,3-unsaturated carbonyl compounds that may polymerise by free radical mechanisms, or the aldehyde may be oxidised to produce another hydroxy-carboxylic acid.

Peroxide decomposition product ketones and acetals can form poly-ketals and poly-acetals (Figure 3.22).

These condensations are acid catalysed [76,109], the acidity being provided by the carboxylic acid lubricity additives and decomposition products of peroxides. Amines in the detergent may neutralise this acidity by forming salts with the acids, which should reduce this mechanism of gum and deposit formation; however, amine salts can still provide protons, though less efficiently – it is difficult to predict the acidity of compounds in a micellar environment, at high temperatures, in a hydrocarbon solvent.

The same series of reactions, with more emphasis on the role of unsaturation, apply to biodiesel fuels which are particularly susceptible to oxidative deposit formation [110]. Fatty acid methyl esters used for biodiesel are rich in unsaturated fatty acid esters, derived from acids such as oleic and linoleic, which are used because they have lower melting points and are more soluble

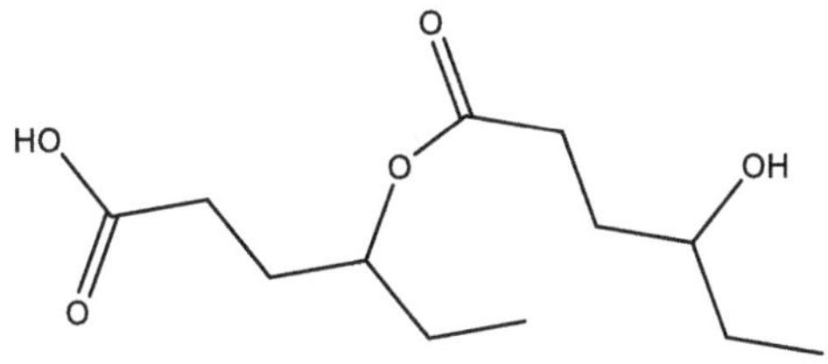

Figure 3.20 Polyester, Dimer-Intermediate.

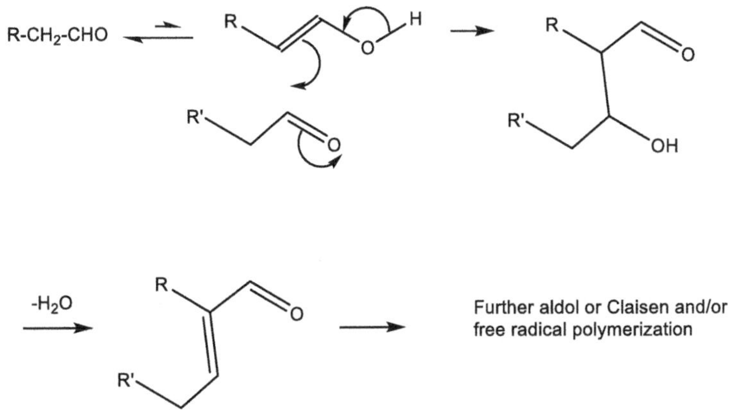

Figure 3.21 Aldol Condensation.

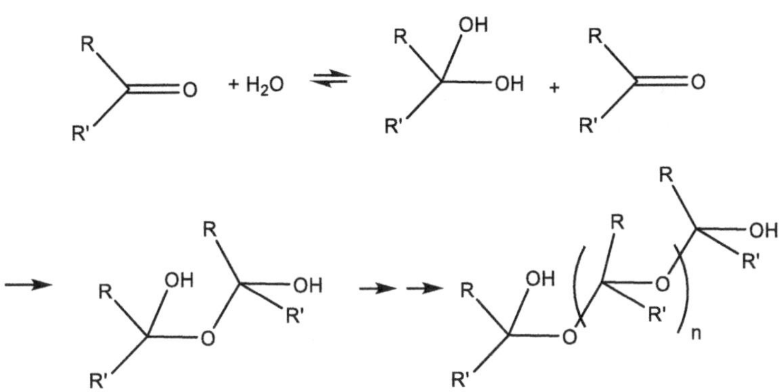

Figure 3.22 Formation of Poly Ketals (both Rs are alkyl) or Acetals (one R group is hydrogen).

in petroleum diesel at low temperatures. Hydrogen atoms on carbon atoms adjacent to carbon–carbon double bonds (allylic positions) are particularly susceptible to being removed, leading to the formation of free radicals that react with oxygen (see Chapter 2).

In the past, another source of gasoline deposits was identified as the oxidation of phenalene and its reaction with indoles – then, both were present in gasoline. However, this was preventable by hydrogenation and by removal of acid species; now, gasoline and diesel are routinely hydrogenated to remove sulphur.

3.5.1b Zinc Deposits

In modern diesel engines, operating on the Euro V EN590 specification (from Jan 2011), with finer injector nozzles, it has been reported that zinc soaps – produced by the interaction of acidic lubricity additives with brass pipe fittings – contribute to deposits inside injector bodies and nozzles [5,27–29]. It has been shown that low levels of zinc (1 to 3 ppm [27]) in the fuel contribute significantly to the deterioration in injector performance (power loss) over time. Standard diesel engine tests, in the Ford Puma and in the Peugeot DW10, show an increased loss of power in the presence of zinc, both for ULSD and for B10 diesel fuels; these tests also demonstrated the increase in power loss with increasing zinc concentration [25] (Table 3.4).

Such deposits appear to be more difficult to remove by standard detergents, which were sufficiently effective on fuel oxidation products alone. So, these zinc containing deposits require new detergent technology or higher treat-rates of existing detergents [25,27]. Available injector deposit analyses from these systems, using fuel doped with 1 ppm Zn (as the neodecanoate), show the presence of substantial amounts of zinc but fuel with doped zinc may not be representative of normal operation [25]. However, it is pointed out that as diesel injectors evolve (to Euro V and beyond) they will become more sensitive to the zinc deposit problem [27].

Zinc coordinates well with sulphur-containing ligands, but naturally occurring sulphur compounds are almost completely removed in current ultra-low-sulphur fuels. This may be one reason for the recent appearance of the problem

Table 3.4 Loss of Engine Power in Presence of Zinc in the Fuel.

Engine Test	Fuel	Zinc, ppm in Fuel	% Power Loss
Puma	ULSD	0	3
Puma	ULSD	0.6	9
Puma	ULSD	1.0	11
DW10	ULSD	0	0
DW10	ULSD	1.0	12
DW10	B10	0	3.6
DW10	B10	1.0	12

of zinc in deposits. Zinc, as carboxylates or oxides, would be expected to associate with the polar deposits and their precursors, but also with groups on the detergent – zinc appears to be able to coordinate well with all of the chemical groups discussed here [111]. It is, then, a matter of competition between the attraction of zinc to the deposit-forming compounds and to the detergent. Increasing the concentration of standard detergent can reduce or prevent zinc deposits but changing the detergent chemistry has been found to prevent zinc deposits at usual concentrations [30]. What are these new chemistries? Some recent patents point to a dodecyl-succinic anhydride/hydrazine detergent for protecting against power loss caused by fuel containing 3ppm of zinc [68]. Another claim is for the quaternised detergent PIBSA-3-trimethylammonium-propylamine salicylate (section 3.4.2i), which was tested in fuel containing 1 ppm Zn [112]. Currently there is insufficient chemical understanding of this problem for detailed discussion.

3.5.2 The Action of Detergents

If the engine only sees fuel that contains an adequate concentration of detergent, there should be no deposits to clean up as the deposit precursors will interact with the amine-based detergent in solution, preferentially, and be dispersed. Reactions with the detergent amine groups can tie up acid, ester, ketone, and aldehyde groups necessary for polymerisation. In addition, detergents form adsorbed layers on the metal surfaces, helping to inhibit the adsorption of polar deposits. While prevention is preferable, oil companies also claim that in their fully additized fuels, detergents can remove inlet valve, injector and combustion chamber deposits (CCDs), after their formation [31,32].

While removal of existing deposits is often claimed by various companies for the action of detergents, most patent test data is for *prevention* of deposit formation, such as clean injectors as measured by fuel flow at particular needle positions, the appearance of inlet valves or reduction in ORI. However, Lubrizol have published deposit removal data for some of their detergents, using a total concentration of around 3000 ppm, most of which is a polyether (primary) amine [100]. It is understandable that a treat of about twenty times the more usual treat is needed to demonstrate the effect in a reasonable test time, but it does demonstrate that it is much more difficult to remove existing deposits than to prevent their formation in the first place.

On the removal of existing deposits, there is interesting data in two patents, which are concerned with products that would be used to remove engine deposits during a service. One looked at CCDs that were scraped off the piston crowns of a gasoline engine that had been run for 300 hours. The presence of a poly-ester/poly-aldol structure in these deposits is consistent with their break-up and partial dissolution by reaction, at 70°C, with aqueous sodium hydroxide (50% dissolved) or aqueous 1,2-diaminoethane (38% dissolved) [113]. The purpose of reducing octane requirement increase was further demonstrated by spraying aqueous ethylene diamine into the combustion chamber of a Chevrolet engine that had been run long enough to have an octane requirement increase of 4 to 5 octane units; the EDA treatment, after 11 hours 'soak', removed all of the ORI.

In the second of these patents for engine deposit clean-up during a service, a wide range of simple solvents were considered for spraying into the air intake of a warm engine, followed by a soak before restarting. The purpose of this was to remove all engine deposits (CCDs), valve and spark plug deposits, even ring deposits [114]. A wide range of candidates were put through a screening test, the Cold Spark Plug Immersion Test (CSPIT): spark plugs, bearing heavy deposits but cleaned of oil by a solvent rinse, were immersed in 50 ml of the test fluid, at 21°C, for 5 minutes, then shaken and removed. The fluid, stained by the removed deposit, is diluted to 250 ml with water and its colour determined (the 'Orbeco-Hellige test') – the darkest colour was related to most or all deposit having been removed (100% removed) and no colour to no removal (0%) (Table 3.5). Given these mild conditions, this test is a measure of solvation activity and alkaline conditions: amines were found necessary to provide a high (alkaline) pH (aqueous ammonia was also effective). It does not, however, involve the high-temperature chemical interactions discussed below. Simple amines combined with water were effective while the dipolar aprotic solvents, with or without water, were not. However, the hydrogen-bond accepting dipolar aprotics, especially methyl formamide, are good co-solvents to isopropylamine and less likely than water to evaporate in a warm engine. The preferred formulations were methyl formamide and isopropylamine, with either water or isopropyl alcohol.

Amide formation from esters is not likely at 21°C, but alkylammonium salt formation with carboxylic acids in the deposit is a likely role of the amine. Also, amine/proton-source (water) reaction with aldol links is possible. Good solvency of the dipolar aprotics, such as acetonitrile (DMSO, DMF, nitromethane and acetone also), is an advantage for this application, when combined with the effect of an amine. Similar results were obtained for the different amines, hindered or not (n-, sec- and tert-butylamine) and primary, secondary or tertiary (isopropyl-, diethyl- and triethyl-amines), suggesting that the amine is acting as a base and hydrogen bond acceptor in the deposit solvation – but not reacting. The quaternised dispersants (section 3.4.2i) carry a range of solvating groups in their chemistries: phenolic and aliphatic hydroxyls as H-bond donors, the quaternised amine would be a good phase transfer agent (pulling any anionic species into the hydrocarbon medium) and some half-amide (part

Table 3.5 CSPIT Results of the Removal of Deposits by Various Solvents.

Non-Amine (%)	Water (%)	Isopropylamine (%)	% Deposit Removed
0	0	100	3
0	34	66	80
0	100	0	10
Acetonitrile	0	0	0
Acetonitrile	34	0	0
Acetonitrile	0	50	80
Methyl formamide	0	0	20
Methyl formamide	34	0	5
Methyl formamide	0	80	100
Formamide	0	0	10
Formamide	34	0	10
Formamide	0	80	95
Ethylene glycol	0	0	0
Ethylene glycol	34	0	0
Ethylene glycol	0	50	80
Methyl alcohol	0	0	1
Methyl alcohol	34	0	1
Methyl alcohol	0	70	60

of the PIBSA-DMAP product) that would provide an electronic push-pull solvation and chemical interaction with ester groups.

The surface-cleaning action of detergents – the removal of dirt from a surface and keeping it from re-adsorbing – is usually considered in the context of the removal of oily dirt from fabrics or crockery, into water [45]. Baked-on domestic deposits also need the chemical action of enzymes, bleaches, alkali, or some combination of these[18]. Much of this general process probably applies to the action of fuel detergent on fuel deposits. Adapting the general process to the action of fuel detergents, and inverting the roles of the polar and non-polar parts of the detergent molecules from those in water, we have the following:

Detergent molecules adsorb at interfaces: those between deposit and metal, deposit and fuel, and metal and fuel. Detergent molecules can adsorb onto both metal and the deposit by their polar segments (i.e. the amines and carboxylic derivatives). Initially, these interactions are dipole–dipole and hydrogen bonding but, under the high temperature conditions around fuel injectors, inlet ports and burners, they are probably followed by chemical reactions between the deposit and the detergent (section 3.5.4). Chemical reactions enable the detergent polar groups break up the deposits into smaller particles, by binding to and breaking up polyester and poly-acetal segments, then carry them away as dispersions in the fuel. Where deposits or varnish are adsorbed onto a metal surface, the polar groups of detergent molecules may adsorb onto and interact with both the metal surface and the deposit; when this is at the edge of the deposit, the large PIB groups increase the contact angle between the deposit and metal and the detergent gradually rolls back the deposit until it is detached, wholly or in part. Clearly, it is difficult to say whether the detergent chemically only breaks up any adsorbed deposit and carries it away, or does it also take part in a roll-back mechanism.

3.5.3 Stabilisation of Dispersed Deposit or Particulate Material by Fuel Detergents in Gasoline and Middle Distillates

Detached deposit material, non-adsorbed particulates and gums must be kept in a stable dispersed phase. Being in fuel or oil, the mechanism of stabilisation of the dispersed phase by the ashless detergent is steric [73]. Conditions for effective steric stabilisation of dispersions are [55,56]:

a) The particle should be entirely covered by detergent/polymer.
b) The detergent molecules must be strongly adsorbed or anchored, onto the particle.
c) The stabilising part of the detergent (PIB) must be soluble in the medium (fuel).
d) The adsorbed layer must be thick enough to prevent flocculation.

In steric stabilisation, as two particles come together, the adsorbed polymers on the two particles, here PIB, must occupy the same space, in increasing concentration, which is entropically unfavourable. This action keeps the particles far enough apart that attractive interactions, which are shorter range, do not come into play (Figure 3.23). Detergent conditions c) and d) are well met by the PIB – being a highly branched alkyl group, of low crystallinity, it is very soluble in hydrocarbon fuels. The molecular weight is chosen to be great enough to provide an adsorbed layer thickness sufficient to give stable dispersions of the deposit/gum material, while not being too viscous in solution. Also, increasing the size of the non-polar PIB becomes less efficient, on a weight (and cost) basis, beyond the PIB molecular weight at which effective, stable dispersion has been achieved.

Complete coverage of the particle will be achieved by a detergent that can bind strongly to the particle and which has a roughly cone-shaped molecular structure (conditions a) and b)). For the PIB dispersants described here, a cone-like geometry results from a PIB chain that is linked by a position along the molecule chain, which is the case for most PIB (from the $AlCl_3$ catalysed route) used in the PIBSA-PAM detergents. In Figure 3.23, it is assumed that the imide link to the PAM is near the middle of a perfect PIB chain, though the position of attachment varies molecule to molecule; also, such PIB molecules usually contain branched structures. PIBSA-PAM and Mannich PIB-phenol detergents have large, multi-polar groups to give strong adsorption on polar deposits. These are balanced, for solubility, by the PIB molecules attached to the polar groups (see description of PIBSA-PAM, section 3.4.2c).

The end functionalised PIB chain, from the BF_3 route, is used in a popular PIB-amine detergent for gasoline. This PIB is linear – PIB has a rigid structure because of the methyl groups attached to every other backbone carbon

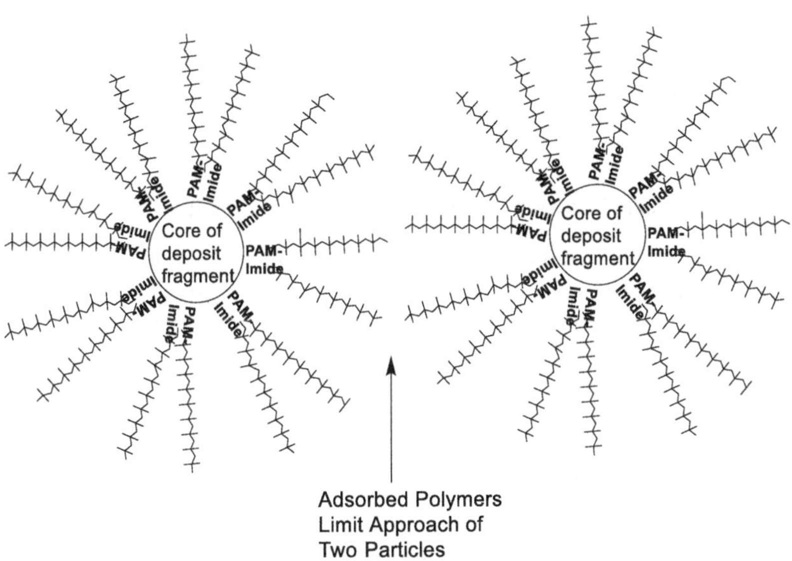

Figure 3.23 Particle Stabilisation by Adsorbed PIBSA-PAM.

[115]. Unlike the PIBSA-PAM, the PIB-amine has a single polar group, the primary amine, so full coverage of particles must come from more molecules becoming attached to the particles. Perhaps the primary amine on the PIB-amine is more efficient at trapping errant, oxidised molecules before particle formation.

That the detergent molecule should be strongly adsorbed onto the particle may seem obvious, but the question is how strongly adsorbed does it need to be. Initially, adsorption interactions will be dipole–dipole and hydrogen bond interactions but, given the make-up of the detergents (containing many N–H and carbonyl groups) and of the deposits and their precursors (with carbonyl and C–O bonds of aldehydes, ketones, alcohols, carboxylic acids and esters), chemical reactions probably occur at the temperatures experienced.

In the aqueous systems, charge provided by adsorbed ionic species plays an important part in the stabilisation of dispersions of particles or emulsions. Water has a high dielectric constant, as well as the ability to provide hydrogen bonding, so is easily able to support charge separation. Hydrocarbons have low dielectric constants and can provide no hydrogen bond stabilisation so were thought to be unable to support this method of particle stabilisation. However, detailed work has shown that particle charges can provide electrostatic stabilisation in hydrocarbons – a favoured example is the finding that carbon black dispersions in aromatic hydrocarbons are only stable when the particles are charged [116]. Charges are stabilised in the inverse micelles formed by the dispersants, which contain the water always present in fuels. This mechanism could, therefore, be providing an extra boost to the dispersing activity of the quaternized dispersants (section 3.4.2i)

3.5.4 Chemical Reactions of Dispersants with Deposits

It seems unlikely that such physical, solvency interactions alone would be enough to remove baked-on-deposits from a metal surface that would, otherwise, require strong abrasion – as found in the clean-up experience described above (section 3.5.2). Also, as just discussed, to be effective at stably dispersing deposit particles, the detergent needs to strongly anchor onto them. So, further chemical reactions are involved to explain the ability of some fuel detergents to remove deposits, as well as halting the polymerisation reactions that lead to their formation and effectively dispersing deposit forming and deposit sourced particles – analyses of deposits have identified detergent additives [5,117].

A chemical linkage that has been proposed is of imine formation (Figure 3.24), which is the first-stage reaction of an aldehyde or ketone with a primary amine, with removal of water (the temperatures in the regions of deposit formation are usually well above 100°C).

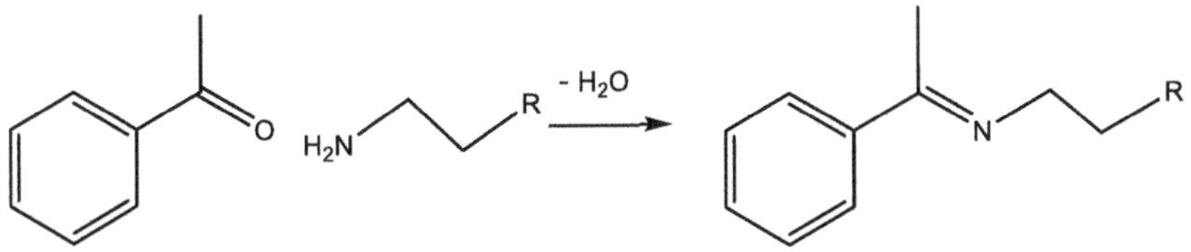

Figure 3.24 Schiff Base Formation.

Imines from aliphatic ketones and aldehydes are generally unstable, unless one or both ends of the imine group is attached to a benzene ring [76]. Gasoline, the lightest fuel having the lowest molecular weight hydrocarbons (C4 to C12), is the one in which the simple PIB-amine is the most preferred detergent type. Gasoline is also the fuel in which alkyl aromatics are maximised for their high octane, though limited by regulation (currently up to 35% in Europe and up to 45% elsewhere [118]). Alkyl aromatic compounds will preferentially form peroxides at the benzyl position (the alkyl carbon alpha to the phenyl group) and decomposition of the peroxide can lead to a benzoyl carbonyl which, in turn, can react with primary amine to produce a relatively stable imine. Imines and their further condensation reactions with other carbonyl deposits, and their precursors, can chemically attach oxidation products to the dispersing power of the PIB amine molecule.

Heating oil may also contain high levels of aromatics – the molecules more susceptible to oxidation. While deposits are less critical in heating systems, they can build up in storage tanks, in which heating oil may spend several months. These fuels too need to rely upon fuel detergents and antioxidants to prevent the formation of such deposits. From the discussion above, it would be unlikely that peroxide decomposition takes place quickly during storage at normal ambient temperatures. However, sludges do form on long-term storage of heating oils that contain fractions of aromatic cracked stocks. Such deposits have been described as containing 'sediment', rust/dirt and water emulsions [119] so should be susceptible to prevention and dispersion by the fuel detergents used in diesel fuel.

Aromatics, with low cetane numbers, are minimised in diesel fuels – for example, typical European fuels are reported as having 14% aromatics [120]. PIBSA-PAM detergents are preferred in diesel, gas oil and other high boiling fuel; PIBSA-PAM contains many secondary amine groups but few or no primary amine groups. Individually, secondary amine groups can react with aldehydes and ketones to give reactive enamines that are likely to form further condensation products with other carbonyl compounds [104]. In PIBSA-PAM, the secondary amines are separated by two carbons, allowing for further chemical reactions. Given the temperatures that a fuel might experience (around 200°C), as it flows near to and around the injectors of a diesel engine, or heat exchange pipes of a jet engine, and taking into account the reactions of poly-amines with carbonyl compounds in the syntheses of heterocyclic compounds [109,121], a number of reactions are likely:

1) Above 100°C, primary and secondary amines react readily with carboxylic acids and esters to produce the more stable amides, which would cause the break-up of any poly-esters and would link a detergent strongly to deposits or gums or precursors.
2) 1,2-diamines heated with carboxylic acids or esters give thermally stable imidazolines or, with ketones and aldehydes, give imidazolidines [121]. 1-Primary-2-secondary diamines, found in some gasoline detergents, can give 2-imidazolines (Figure 3.15) while 1,2-secondary diamines, found in diesel detergents, can give imidazolidines (Figure 3.25) and 4-imidazolines (Figure 3.26).
3) The oxidation of longer chain hydrocarbons, found in diesel, can lead to poly-carbonyl compounds (above and Chapter 2) which can form a range of heterocyclic compounds with amines (e.g. pyridines and pyrroles).

Poly-acetals (or poly-ketals), formed with the help of carboxylic acid catalysis, are formed reversibly so are a latent source of carbonyls that may be unlocked by reactions with the amines of the detergent, catalysed by amine salts. Such a process would unravel deposit or gum poly-acetal/ketal structures.

The result of such reactions is that deposit-forming polymeric material can get broken down. In the process, the fragments become chemically attached to the detergent molecules which, bearing multiple, large, fuel soluble PIB chains, can carry them away into stable dispersions in the carrier fluid and fuel. These same reactions enable detergents to form chemical links, more easily, with

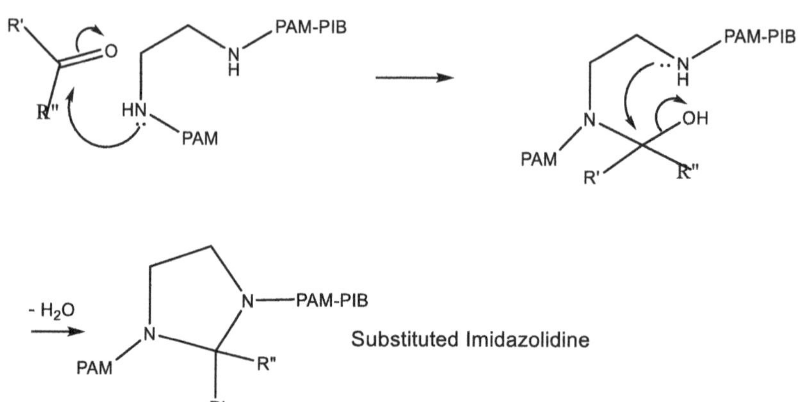

Figure 3.25 Condensation of sec-Diamine with a Ketone or Aldehyde to Form an Imidazolidine.

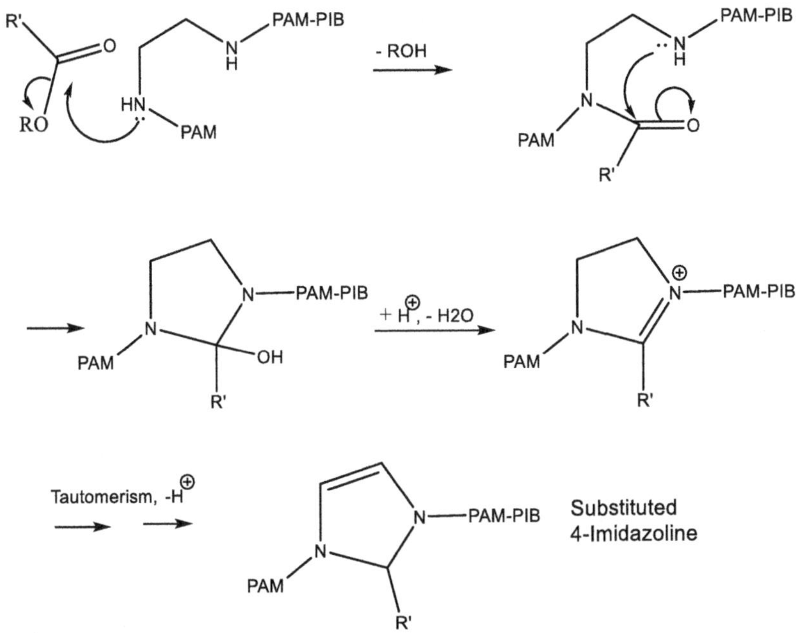

Figure 3.26 Formation of 4-Imidazolines.

soluble deposit precursors, breaking them up and dispersing them as stable, unreactive species.

In heavy fuel, the need is to keep (mainly) asphaltenes in stable dispersion. Since asphaltenes are highly polarisable (being poly-aromatics) and contain heteroatoms (nitrogen, oxygen, and sulphur) [122], they should bind well with the polar end groups of a PIBSA-PAM fuel detergent. Then the fuel detergent can stabilise the asphaltene in dispersion, with the solubilising effect of the PIB chains (as described above).

Notes

1 In a pintle fuel injector, the nozzle of the injector is closed by a pintle that is shaped along with the details of the nozzle; the pintle, at the end of a metal rod, is moved up and down to control the flow of fuel through the nozzle and its spray into the cylinder. Diagrams are abundant on the web and engineering literature.

2 Such water boilers may be for heating large housing and industrial buildings or to raise steam for use in power generation or in multifarious applications in refineries.

3 In January 2019, CEIC reported production of gasoline, jet fuel, diesel and residual fuel was split as 22%, 6%, 44% and 27% respectively [12].

4 IUPAC nomenclature poly-2-methylpropene does not explain the widely known acronym 'PIB'; hence the industry's continued use of pre-IUPAC nomenclature.

5 Names such as ethylene glycol and poly-ethylene amines have such widespread industrial use that I have used these terms rather than those derived from di-hydroxy-ethane or di-amino-ethane.

6 Aluminium chloride is used as a slurry in the butane/butylenes mixture, in which it is not soluble.

7 In this patent [58], the preferred Mn refers to a 'hydrocarbyl group' attached to a succinic acylating group and 'in especially preferred embodiments the hydrocarbyl substituted acylating agent is a polyisobutenyl substituted succinic anhydride'.

8 The 0.05 to 0.1 wt.% of water, expected to be in fuel, is comparable with the treat rate of detergent at 0.02 to 0.05 wt.%. However, the molecular weight of a detergent is about 100× that of water, so this means that there may be 100 molecules of water per molecule of detergent, stabilising the inverse micelle.

9 DDSA is the ene reaction product of 1-dodecene with maleic anhydride.

10 Euro 5 and 6 from 2011 and 2015, respectively, for all new cars.

11 Diesel detergency tests have developed from Peugeot XUD-9 to Peugeot DW10 (Chapter 18 of reference [70] – a practical description of the detergency tests is given [62].

12 While there is a methylene allylic position also, this position is very hindered, compared with the allylic methyl.

13 While the IUPAC names for these are oxiranes, they are widely known by these pre-IUPAC names.

14 The polyetheramine used was tridecyl alcohol reacted with butene oxide (1/20 mole ratio) then cyanoethylated and hydrogenated.

15 The lube compatibility test looks for adverse interaction, causing gelling, between the lube oil and the fuel additive, some of which inevitably ends up in the lube oil.

16 Alkyl or aralkyl.

17 To maintain the high pressure needed for fuel injection and to ensure constant availability, excess of fuel is pumped into a common rail. Escape systems after the pump, and in the injector, send the excess fuel back to the fuel tank.

18 As alkali, washing soda and many other components are present in dishwasher formulations – Some patents are eye openers to the complexities of such formulations [123].

References

1 Henley, T.J., Kulinowsky, A.M., and Roos, J.W. (1999). The role of fuel additives in formulating clean transportation fuels chapter 6, pp. 61–74. In *Designing transportation fuels for a cleaner environment* (ed. M. Khan and J.G. Reynolds), s.l. London: Taylor and Francis Group. ISBN101560328134.

2 Chevron corp (2009). Motor gasolines technical review. chevronwithtechron.com. http://www.chevronwithtechron.com/products/documents/69083_MotorGas_Tech_Review.pdf (Cited: 24 September 2013).

3 Kerr, J. (2011). auto-tech-direct-versus-port-fuel-injection/. autos.ca. 19 October http://www.autos.ca/auto-tech/auto-tech-direct-versus-port-fuel-injection (Cited: 24 September 2013).

4 Tze, C.J. and Passut, C.A. (2009). Applications of surfactants in lubricants and fuels. In: *Handbook of Detergents, Part E: Applications* (ed. U. Zoller), 331–344. Taylor & Francis group, CRC Press. Ch.13.

5 Ullmann, J., Geduldig, M., Stutzenberger, H. et al. (2009). *Effect of Fuel Impurities & Additive Interactions on the Formation of Internal Diesel Injector Deposits.* Esslingen: Fuels Technische Akademie Esslingen (TAE) 7th Annual Colloquium.

6 Caprotti, R., Breakspear, A., Graupner, O. et al. (2008). Beyond The challenges for diesel detergency. Infineum.com. http://www.infineum.com/Pages/technicalpapers.aspx (Cited: 24 September 2013).

7 Wartsila (2013). Wartsila.com. http://www.wartsila.com/en/engines/low-speed-engines/RT-flex96C.

8 Infineum International Ltd (2016). Infineum marine fuel survey 2016. December http://www.infineuminsight.com/media/205989/Infineum-Marine-Fuel-Survey-2016.pdf (Cited: 14 December 2016).

9 Benwell, S. and Corradi, M. (2019). Heading for IMO 2020. Inevitable change for marine industry stakeholders ahead. infineuminsight.com. 3 December https://www.infineuminsight.com/en-gb/articles/fuels/heading-for-imo-2020 (Cited: 24 April 2020).

10 Witzel, L., Moszkowicz, P., and Claus, G. (1995). Mechanism of particulate reduction in heavy fuel oil combustion. s.l.: Butterworth-Heinemann, *Fuel* 74 (12): 1881–1886.

11 Danilov, A.M., Okina, N.G., Mitusova, T.N. et al. (2003). *Increase in the Efficiency of Use of Furnace Residual Fuel Oil.* Plenum Publishing corp., Vol. 39. 6–9. Chemistry & Technology of Fuels & Oils.

12 Russia Oil Consumption (2019). CEIC. 6 March https://www.ceicdata.com/en/indicator/russia/oil-consumption (Cited: 24 April 2020).

13 Baker Hughes (2013). Fuel-additive-technology. bakerhughes.com. http://www.bakerhughes.com/products-and-services/downstream-chemicals/fuel-additive-technology?gclid=CP3R_7SthrQCFaTMtAodACAAlw (Cited: 24 September 2013).

14 Chevron (2006). Aviation fuel technology review. cgabusinessdesk.com. https://www.cgabusinessdesk.com/document/aviation_tech_review.pdf (Cited: 24 September 2013).

15 Grinstead, B. and Zabarnick, S. (1999). Studies of jet fuel thermal stability, oxidation and additives using an isothermal oxidation apparatus equipped with an oxygen sensor. American Chemical Society, *Energy & Fuels* 13: 756–760.

16 ExxonMobil Aviation: World Jet Fuel Specifications (2005). ExxonMobil web site. http://www.exxonmobil.com/AviationGlobal/Files/WorldJetFuelSpecifications2005.pdf (Cited: 21 October 2014).

17 Heneghan, S.P. and Harrison, W.E. (1997). *JP-8+100: The Development of High Thermal Stability Jet Fuel*. Vol. 1. 271–283. Vancouver: 6th International Conference on Stability and Handling of Liquid Fuels.

18 Vangsness, M.D. and Heneghan, S.P. (1997). *Additive Stability in JP8 – Metal Alloy Systems at Elevated Temperatures*. Vol. 1. 285–293. Vancouver: 6th International Conference on Stability and Handling of Liquid Fuels.

19 Fawley Refinery (2019). exxxonmobil.com. 9 May https://www.exxonmobil.co.uk/Company/Overview/UK-operations/Fawley-refinery (Cited: 4 April 2020).

20 Bauldreay, J.M., Blundell, W.G., Heins, R.J. et al. (1997). *Testing JP8-100 Packages in European Jet Fuels – Evaluating and Understanding Benefits and Drawbacks*. Vol. 1. 333–352. Vancouver: 6th International Conference on Stability and Handling of Liquid Fuels.

21 Grabarse, W., Bohnke, H., Tok, C. et al. (2011). Acid-free quaternized compounds and use thereof as additives in fuels and lubricants. US Patent 2012/0,010,112. Filed 6 July.

22 Lange, A., Rath, H.P., Posselt, D. et al. (2011.). Method for producing mannich adducts that contain polyisobutylene phenol. US Patent 2011/0,251,114. Filed 24 June.

23 Honnen, L.R. (1972). Fuel additives and use thereof. GB Patent 1,398,067. Filed 26 December.

24 Russel, T.J. (1989). Additives influencing diesel fuel combustion. [book auth.] In: *Gasoline and Diesel Fuel Additives* Vol. 3. (ed. K. Owen), 65–104. John Wiley and Sons for the Society of Chemical Industry. Ch 3.

25 Barbour, R.H., Arters, D.C., Dietz, J.G. et al. (2007). Injector fouling effects in modern direct injection diesels, lubrizol ltd. Bangkok: The 13th Annual Fuels & Lubes Asia Conference. Vol. 13th Paper.

26 Chevron (2007). Diesel fuels technical review. chevron.com. https://www.chevron.com/-/media/chevron/operations/documents/diesel-fuel-tech-review.pdf (Cited: 26 November 2012).

27 Caprotti, R. and Breakspear, A. (2006). *Diesel Injector Deposits Potential in Future Fueling Systems*. Toronto: Society of Automotive Engineers, Warrendale, P.A. 2006-01-3359, Powertrain and Fluid Systems Conference and Exhibition. ISSN 0148-7191.

28 Caprotti, R., Breakspear, A., Graupner, O. et al. (2005). *Detergency Requirements of Future Diesel Injection Systems*. San Antonio: Society of Automotive Engineers, Warrendale, P.A. Powertrain & Fluid Systems Conference and Exhibition. ISSN 0148-7191.

29 Caprotti, R., Breakspear, A., Graupner, O. et al. (2004). *Impact of Fuel Additives on Diesel Injector Deposits*. Tampa: Society of Automotive Engineers, Warrendale, P.A. 2004-01-2935, Powertrain & Fluid Systems Conference & Exhibition. ISSN 01487191.

30 Caprotti, R., Bhatti, N., and Balfour, G. (2010). Deposit control in modern diesel fuel injection systems. Infineum.com. http://www.infineum.com/Documents/FuelsTechnicalPapers/SAE/2010-01-2250.pdf. ISSN 0148-7191 (Cited: 27 September 2013).

31 Shell V-Power Diesel (2020). Shell.co.uk. https://www.shell.co.uk/motorist/shell-fuels/shell-v-power/shell-v-power-diesel.html (Cited: 26 April 2020).

32 BP ultimate diesel with ACTIVE technology. (2020). BP p.l.c., https://www.bp.com/en_gb/united-kingdom/home/products-and-services/bp-ultimate-fuels/bp-ultimate-diesel-with-active-technology.html.

33 Hundere, A. (1970). Apparatus for determining the thermal stability of fluids. US Patent 3,670,561. Filed 30 January.

34 Anderson, D.G. and Yang, G. (2010). Detecting a short in an apparatus and method for determining the thermal stability of fluids. US Patent 2012/0,014,408. Filed 16 November.

35 Bahn, O.K., Tang, S.Y., Brinkman, D.W. et al. (1988). *Causes of Poor Filterability in Jet Fuel*. s.l.: Vol. 67. 227–237. Butterworth & Co Publishers Ltd. February Fuel.

36 Armour and Company (1955). Improvements in and relating to fuel oil compositions. gb patent 802,345. Filed 3 October.

37 Armour and Company (1955). Improvements in or relating to fuel compositions. GB Patent 817,262. Filed 2 February.

38 Claydon, D.J. (2008). Fuel compositions. US Patent 8,709,108. Filed US, 24 September.

39 Henry, C.P. and Spinelli, H.J. (1989). Branched polymers as fuel oil additives. US Patent 4,985,160. Filed 8 February.

40 Siegel, J.R. (1957). Stabilized hydrocarbon fuel oil composition. US Patent 2,958,590. Filed 15 October.

41 Benwell, S. and Ashton, R. (2020). *Post IMO 2020 Marine Fuels, Combatting the Stability Issues Arising with Very Low Sulphur Marine Fuels*. Infineum International Ltd. Insight. 23 June https://www.infineuminsight.com/en-gb/articles/fuels/post-imo-2020-marine-fuels (Cited: 30 June 2020).

42 Hertel, G., Feustel, M., Bystrom, H. et al. (1997). Heavy oils having improved properties and an additive therefor. US Patent 6,488,724. Filed USA, 27 October.

43 Bernasconi, C., Faure, A., and Thibonnet, B. (1986). Homogenous and stable composition of asphaltenic liquid hydrocarbons and additive useful in industrial fuel. US Patent 4,622,047. Filed 6 January.

44 Porter, M.R. (1994). *Handbook of Surfactants*, 2nd edition. Chapman and Hall. ISBN 751401706.

45 Reynolds, P. (200). Wetting of surfaces. In: *Colloid Science, Principles, Methods and Applications*, Chapter 10, 2nd e (ed. T. Cosgrove), 159–179. Wiley-Blackwell.

46 Strethsa, L.K. and Aramaki, K. (2011). Structures of nonionic surfactant micelles in organic solvents, a SAXS study. In: *Self Organised Surfactant Structures*, Chapter 2 (ed. T.F. Tadros), 17–57. Wiley-VCH.

47 Mortier, R.M. and Orzulik, S.T. (1997). *Chemistry and Technology of Lubricants*. Blackie Academic & Professional. ISBN 9780751402469.

48 Meffert, M.W. and Quinn, T.G. (2006). Additizing heavy fuel oil at terminals. US Patent 2008,021,713. Filed 18 July.

49 Sanders, H.R., Morway, A.J., and Clark, N.J. (1958). Residual fuels containing insoluble additives. US Patent 3,036,901. Filed US, 24 November.

50 Coley, T., Owen, K., and Weaver, C.S. (1995). Automotive fuels reference book. 2nd. Society of Automotive Engineers, Warrendale, P.A. ISBN 1560915897.

51 ATC Europe (2008). Lube additives and the environment. atc-europe.org. http://www.atc-europe.org/publications.asp (Cited: 1 October 2013).

52 Marshall, E.L. and Motor Gasoline, O.K. (1995). Society of chemical industry. ISBN 0854044094.

53 ATC. Fuel Additives and the Environment (2004). Technical committee of petroleum additive manufacturers in europe. 1 January http://www.atc-europe.org/public/doc52.pdf (Cited: 1 October 2013).

54 BP Statistical Review of World Energy (2019). Oil. 2019, 68th edition, bp.com/statisticalreview. June https://www.bp.com/content/dam/bp/business-sites/en/global/corporate/pdfs/energy-economics/statistical-review/bp-stats-review-2019-oil.pdf (Cited: 30 March 2020).

55 Tadros, T.F. (2007). General principles of colloid stability and the role of surface forces, Chapter 1. In: *Colloid Stability and Application*, 1 (ed. T.F. Tadros), 1–22. Wiley VCH.

56 Tadros, T.F. (2013). *Emulsion Formation and Stability*. Weinheim: Wiley-VCH. ISBN 9783527319916.

57 Kennedy, J.P. and Marachel, E. (1982). *Carbocationic Polymerisation*. New York: J.Wiley. ISBN 9780471017875.

58 Reid, J. and Burgess, V. (2011). Fuel composition comprising detergent and ammonium salt additive. EP 2,966,151. Filed US, 10 March.

59 Blackborow, J.R. and Clarke, M.J. (1995). Fuel compositions containing a polyisobutene succinimide detergent. US Patent 5,588,973. Filed 7 June.
60 Norman, G.R. and LeSuer, W.M. (1965). Compositions containing substituted succinic acid derivatives. GB Patent 1,121,168. Filed GB, 17 August.
61 LeSuer, W.M. (1962). Oil-soluble nitrogen-containing products and process for preparing the same. GB Patent 1,021,183. Filed GB, 15 August.
62 DeCanio, E.C., Herbsman, S., Kaufman, B.J. et al. (1996). Detergent additive compositions for diesel fuels. US Patent 5,925,151. Filed 19 September.
63 Branch, S.J., Coupland, K., and Richardson, D.A. (1968). Ashless dispersants. GB Patent 1,162,436. Filed 8 March.
64 Meinhardt, N.A. and Davis, K.E. (1979). Novel carboxylic acid acylating agents, derivatives thereof, concentrate and lubricant compositions containing the same. US Patent 4,234,235. Filed 18 February.
65 Abramo, G.P. and Trewella, J.C. (1989). Polyalkylene succinic deposit control additives and fuel compositions containing the same. EP 0,376,578. Filed EU, 18 December.
66 Andress, H.J., Ashjian, H.H., and Hills, F.J. (1990.). Reaction products of alkenyl succinimides with ethylenediamine carboxy acids as fuel detergents. US Patent 4,971,598. Filed 22 March
67 Allen, R.W. and Attfield, M.J. (1997). Marine diesel process and fuel therefor. WO Patent 9,744,414. Filed WO, 19 May.
68 Breakspear, A.P., Caprotti, R., and Thompson, R.M. (2007). Method and Use For The Prevention Of Fuel Injector Deposits. EP 2,088,185. Filed 3 July.
69 Buckmann, J.P., Linda, Y., Liggett, F.S. et al. (1958). Detergent automotive fuel. US Patent 3,011,879. Filed 20 January.
70 Richards, P. (2014). *Automotive Fuels Reference Book*, Third edition. Warrendale, P.A: Society of Automotive Engineers. ISBN 9780768006384.
71 Kummer, R., Franz, D., and Rath, H.P. (1987). Polybutyl- and polyisobutylamines, their preparation, and fuel compositions containing these. US Patent 4,832,702. Filed US, 25 March.
72 Wagenaar, A.H. (1964). Aliphatic amines suitable as ashless lubricant additives. GB Patent 1,004,411. Filed 4 August.
73 Tze, C.J. and Passut, C.A. (2006). Detergent formulation in lubricants. In: *Handbook of Detergents, Part D: Formulation* (ed. U. Zoller), 437–471. CRC Press, Taylor & Francis group. Ch.15.
74 Alquist, H.E. and Ebersole, G.D. (1974). Fuel additive package containing polybutene amine and lubricating oil. US Patent 4,022,589. Filed 10 October.
75 Stuebe, C.W. (1973). Nitrogen-containing organic compositions, processes for making them and fuels and lubricants. GB Patent 1,487,860. Filed US, 26 December.

76 Smith, M.B. and March, J. (2001). *March's Advanced Organic Chemistry, Reactions, Mechanisms & Structure*, 5th edition. John Wiley & Sons. ISBN 0471585890.

77 Nakaguchi, G.M. (1973). Detergent automotive fuel composition. US Patent 3,884,647. Filed 9 August.

78 DeMarchi, B., Funtowicz, S., and Ravetz, J. (1996). Seveso: A paradoxical classic disaster. In: *The Long Road to Recovery, the Community Responses to Industrial Disaster* (ed. J.K. Mitchell), United Nations University Press. 29 May ISBN 9789280809268. https://unu.edu/search/the+long+road+to+recovery (Cited: 28 April 2020).

79 Wikipedia_contributors (2020). 2-ethylhexanol. Wikipedia. 1 April https://en.wikipedia.org/w/index.php?title=2-Ethylhexanol&oldid=948522914 (Cited: 28 April 2020).

80 Vogel, H., Rath, P., Jacob, C.P. et al. (1989). Motor fuel for internal combustion engines. US Patent 5,004,478. Filed 31 October.

81 Moreton, D.J. (1997). Detergents for hydrocarbon fuels. US Patent 5,876,468. Filed 5 September.

82 Reid, J. (2008). Fuel compositions. US Patent 2010,258,070. Filed 25 September.

83 Stevenson, P.R., Moreton, D.J., Vilardo, J.S. et al. (2011). Nitrogen free deposit control fuel additives. US Patent 2012,317,874. Filed 15 June.

84 Karll, R.E. and Lee, R.J. (1977). Mannich reaction products made with alkylphenol substituted aliphatic unsaturated carboxylic acids. US Patent 4,142,980. Filed 19 September.

85 Dajani, M. (1962). Inhibiting corrosion in petroleum refining processes. US Patent 3,247,094. Filed 23 November.

86 Seffens, W.E. (1972). Corrosion inhibitors for refining & petrochemical processing equipment. US Patent 3,766,053. Filed 29 June.

87 Zumstein, F., Assmann, E., Koenigsberger, R. et al. (1967). Gasoline fuel. GB Patent 1,162,919. Filed 31 August.

88 Bonazza, B.R. and Holtz, H.D. (1978). Imidazoline fuel detergents. US Patent 4,247,300. Filed 27 April.

89 van_der_voort, H.G.P. (1974). Process for the preparation of an ester of an alkyl- or alkenyl succinic acid and a polyvalent alcohol. GB Patent 1,483,728. Filed 11 September.

90 LeSuer, W.M. (1969). Lubricating oil or fuel containing sludge-dispersing additive. US Patent 3,639,242. Filed 29 December.

91 Casadei, M.A., Galli, C., and Mandolini, L. (1984). Ring closure reactions. 22 kinetics of cyclization of diethyl (omega-bromoalkyl) malonates in the range of 4- to 21-membered rings. The role of ring strain. *Journal of the American Chemical Society* 106: 1051–1056.

92 Fuel Additives: Use and Benefits (2013). ATC Doc. 113. atc-europe.org. http://www.atc-europe.org/public/Doc1132013-11-20.pdf (Cited: 2 February 2014).

93 Rath, H.P., Mach, H., Oppenlaender, K. et al. (1991). Gasoline-engine fuels containing polyetheramines or polyetheramine derivatives. US Patent 5,112,364. Filed USA, 29 April.

94 Nuyken, O. and Pask, S.D. (2013). Ring-opening polymerization – An introductory review. *Polymers – Open Access* 5: 361–403. ISSN 2073-4360.

95 Campbell, C.B. and Buchanan, J.A. (1984). Deposit control additives – Hydroxy polyether polyamines. WO Patent 8,501,956. Filed USA, 30 October.

96 Moreton, D.J., Stevenson, P.R., and Thetford, D. (2011). Quaternary ammonium salt of a polyalkene-substituted amine compound. EP 2,455,406. Filed 6 November.

97 Barton, W., Davies, M.C., Moreton, D.J. et al. (2011). Quaternary ammonium salt detergents for use in fuels. US Patent 2011,185,626. Filed 13 April.

98 Stevenson, P.R., Ray, J.C., Moreton, D.J. et al. (2017). Quaternary ammonium amide and/or ester salts. US Patent 10,479,950. Filed 26 July.

99 Bush, J.H., Barbour, R.H., and Moreton, D.J. (2013). Diesel detergent without a low molecular weight penalty. EP 3,489,332. Filed 22 October.

100 Garrett, D.L. and Jackson, M.M. (2005). Additive and fuel compositions containing detergent and fluidizer and method thereof. EP 1,812,534. Filed 19 October.

101 Chloupek, F.J. and Chao, T.S. (1974). Gasoline fuel composition and method of using. US Patent 4,039,300. Filed 3 June.

102 Cherpeck, R.E. (1994). Fuel compositions containing aryl succinimides. US Patent 5,620,486. Filed 30 December.

103 Wright, B.E. and Witzig, W.L. (1996). Methods for reducing fouling deposit formation in jet engines. US Patent 5,725,611. Filed 12 August.

104 Mayo, F.R. and Lan, B.Y. (1987). Gum and deposit formation from jet turbine and diesel fuels at 100°C, 26. In: *American Chemical Society*, 215–220. Industrial &. Engineering Chemistry Product Research and Development.

105 Mayo, F.R. and Lan, B.Y. (1986). Gum and deposit formation from jet turbine and diesel fuels at 130°C. In: *American Chemical Society*, 25, 333–348. Industrial &. Engineering Chemistry Product Research and Development.

106 Dacre, B. and Hetherington, J.I. (1997). *Electrical Conductivity of HITTS Additive Packages for the JP8+100 Programme*, 315–332. Vancouver: 6th International Conference on Stability and Handling of Liquid Fuels.

107 Zinbo, M. and Siegl, W.O. (1985). *Chemical Composition and Origin of Engine Deposits* (ed. L.B. Ebert), Plenum Press. ISBN 030641936X.

108 AkzoNobel Functional Chemicals (2010). Initiators and reactor additives for thermoplastics. AkzoNobel.com. http://www.akzonobel.com/polymer/system/images/AkzoNobel_Initiators_and_Reactor_Additives_for_Thermoplastics_Low-res_protected_July2010_tcm96-39468.pdf (Cited: 22 August 2013).

109 Clayden, J., Greeves, N., Warren, S. et al. (2001). *Organic Chemistry*. s.l.: Oxford University Press. ISBN 9780198503460.

110 Waynick, J.A. (2005). Characterization of biodiesel oxidation and oxidation products CRC project no AVFL-2B. November https://www.nrel.gov/docs/fy06osti/39096.pdf. NREL/TP-540-39096 (Cited: 30 April 2020).

111 Archibald, S.J. (2004). Zinc. In: *Comprehensive Coordination Chemistry II* Vol. 6. (eds. J.A. McCleverty and T.J. Meyer), 9–12. Transition Metal Groups. s.l.: Elsevier, Chapter 6.

112 Roger-Gopfert, C., Bohnke, H., Grabarse, W. et al. (2012). Quaternized nitrogen compounds and use thereof as additives in fuels and lubricants. US Patent 2013,133,243. Filed 28 June.

113 Robbins, J.L. and Lunt, III R.S. (1993). Method for carbonaceous deposit removal and for reducing engine octane requirement. EP 580,413. Filed 20 July.

114 Hughett, P. (2008). An upper internal combustion engine cleaning composition. EP 2,138,557. Filed 27 August.

115 Olah, G.A. and Molnar, A. (1995). *Hydrocarbon Chemistry*. s.l.: John Wiley & Sons/Wiley Interscience,. ISBN 047111359X.

116 Poovarodom, S. and Berg, J.C. (2010). Effect of particles and surfactant acid-base properties on charging of colloids in apolar media. Elsevier Science B.V., *Journal of Colloid and Interface Science* 346: 370–377.

117 Ullmann, J., Geduldig, M., Stutzenberger, H. et al. (2008.). *Investigation into the Formation and Prevention of Internal Diesel Injector Deposits*. Detroit,_ Michigan: Society of Automotive Engineers, Warrendale, P.A. 2008 World Congress. pp. SAE Technical Paper 2008-01-0926. ISBN 9780768016390.

118 Fahim, M.A., Al-Sahhaf, T.A., and Elkilani, A.S. (2010). Fundamentals of petroleum refining. Elsevier B.V.

119 Marsh, B.E., DuBrow, B., DuBrow, P.L. et al. (1955). Fuel oil compositions. US Patent 2,851,344. Filed 2 February.

120 Caprotti, R., Takaharu, S., and Masahiro, D. (2008). *Impact of Diesel Fuel Additives on Vehicle Performance*. Shanghai: Society of Automotive Engineers, Warrendale, P.A. 2008 SAE International Powertrains, Fuels and Lubricants Congress. pp. Technical Paper 2008-01-1600. ISSN 01487191.

121 Norman, R.O.C. (1968). *Principles of Organic Synthesis*. Methuen & Co. Ltd. SBN 4164719502.

122 Holder, G.D., Enick, R.M., and Mohamid, R.S. (1996). Solids formation in hydrocarbon systems. s.l.: Elsevier Science B.V., *Fluid Phase Equilibria* 117: 126–137.

123 Weber, H., Ettl, R., and Tropsch, J. (2008). Dishwasher detergent formulations comprising a mixture of hydrophobically modified polycarboxylates and hydrophilically modified polycarboxylates. US Patent 2010,234,265. Filed USA, 10 October Detergent. BASF SE.

4

Cold Flow Improvers

4.1 Introduction

Cold flow problems arise in the flow and pumping of various fuels at the low temperatures that occur in winter, compared with those in summer. These flow problems are caused by the separation of crystalline wax. Fuels most at risk are diesel fuels, heating oils and other gas oils. Heavy fuels used for marine, large-scale heating and electricity generating plants have viscosity problems at ambient temperatures year-round, resulting from a combination of wax and asphaltene separations; such fuels are blended with less valuable gas oils to control their viscosities and are usually kept in heated storage.

The waxes that crystallise from middle distillate fuels are mostly, or entirely, composed of n-alkanes and are known as paraffinic waxes. Crude oils contain various proportions, usually in the region of 15% to 30%[1], of wax-forming n-alkanes. At the higher molecular weight end of the molecular distribution, long-chain alkanes, bearing short sidechains, and cyclo-alkanes also contribute to the wax separating from vacuum gas oils, lubricating oils, heavy fuels and crude oil. In gasoline, carbon numbers are low ($<C_{12}$) and the components that are used to blend gasoline usually contain low levels of n-alkanes to maintain high octane numbers (n-alkanes have low octane numbers). C_4 to C_{12} iso-alkanes and alkyl aromatics, which have low melting points, are preferred for gasoline because they have high octane numbers. So, gasoline does not have a wax crystallisation problem. Jet fuel (refined kerosene) is distilled to have no wax separation above the temperature that it is likely to experience, −60°C to −40°C, depending upon grade.

Middle Distillates (MD) – diesel fuel, gas oil, jet fuel and kerosene – are major refinery products around the world, especially so in Europe (Chapter 1).

Fuel Additives: Chemistry and Technology, First Edition. Robert D. Tack.

Often, refinery economics revolve around the production of these middle distillate fuels, so their production from crude oil needs to be maximised. MDs contain high proportions of n-alkanes. While these have favourably high cetane numbers (good for diesel), the higher n-alkanes in an middle distillate, usually n-$C_{17}H_{36}$ to about n-$C_{30}H_{62}$, crystallise out in the cold, below the fuel cloud point[2], as thin plates (photo-micrograph Figure 4.1). Crystals such as these block narrow pipes and filters and, eventually, gel the fuel when the temperature is below its pour point. In the past, wax problems were avoided by specifying regional maximum cloud points for winter middle distillate fuels at the lowest temperature expected 90% of the time; cloud points in summer were, and are, relaxed to higher values but are not usually limiting.

To meet the cloud point limitation in winter, a refinery would be running sub-optimally, compared with its summer operations. Additives were developed first to depress the pour point, so alleviating the risk of a fuel gelling in a storage tank. Then, more effective cold flow improvers were developed to aid the operation of diesel engines and the flow of fuels, between storage and tankers or burners, at temperatures below their cloud points. Over the years, further developments have been made to meet various further cold flow problems resulting from variations in fuel refining, such as the use of waxy crudes and biodiesel, and to meet more stringent fuel specifications. The routine use of cold flow improvers by refineries has enabled them to produce more diesel fuel from each barrel of crude more economically than is possible without them.

In this chapter, the applications, effects and chemistries of the Wax Crystal Modifying (WCM) cold flow improvers will be discussed. The diesel application

Figure 4.1 Wax Crystals (about 0.3 mm) Separating from Diesel Fuel below Its Cloud Point.

(on- or off-road and marine) is the most important, and the most exacting, so takes up the lion's share of the discussion. When heating oil is a similar middle distillate product to diesel, the same additives are used; otherwise, specific cold flow improver variations have been developed. Such additives, which are blends of WCM compounds, are generally known as Middle Distillate Flow Improvers (MDFIs). Jet fuel, however, is refined to not produce wax at low temperatures, though some consideration has been paid to the possibility of jet fuel cold flow improvers.

Organic synthesis has been widely covered for this application and a wide range of compounds have been found effective as middle distillate cold flow improvers: depressing the cloud points and pour points and, more importantly, lowering the temperatures at which almost-summer-grade diesel and heating fuels could be used in winter.

Understanding the crystallisation of n-alkanes has attracted much academic interest, perhaps because of the ease with which the process can be studied by optical microscopy. n-Alkane crystallisation was also extensively studied by electron microscopy [1–3] in confirmation of the Burton, Cabrera and Franck (BCF) dislocation theory of crystal growth [4,5], around 1950. Optical micrographs of n-paraffin crystals, showing growth spirals, were published in 1937, before the BCF theory was developed [6]. More recently, theoretical calculations of crystal morphology, and the effects of surface roughening on crystal growth, have been applied to the crystallisation of n-alkanes [7,8].

Various cold flow tests have been developed at various times for the various applications of middle distillate fuels. Descriptions and details of such tests have been given in many places so a list of tests with short descriptions and references will suffice here.

4.2 Detailed Problems and What Cold Flow Improvers Do

Cold flow problems can exist for any middle distillate, especially those that are higher boiling than kerosene: diesel, gas oil (for heating and off-road vehicles) and distillate and residual-fuel-containing marine fuels. Cold flow problems are caused by the separation of paraffin (n-alkane) wax from middle distillate fuels at low temperatures. In untreated fuels, the most obvious problem is that the fuel gels into a solid mass as the fuel cools below its pour point, typically 3–6°C below the cloud point, the temperature at which wax is first visible during cooling. This gel prevents any movement of the fuel: from storage tank to road tanker, or from storage tank through fuel lines to diesel engines or heating appliances. Between the cloud point and the pour point, fuel will still be mobile and may be moved between storage tanks and road tankers, which have coarse filter screens. However, the wax crystals in middle distillates are thin plates that rapidly coat any fine filter with a thin, continuous glaze of wax that blocks the flow of fuel, preventing the operation of any diesel engine or heating appliance.

4.2.1 Diesel Vehicle Fuel Systems and Operability

The automotive diesel engine fuel system is particularly sensitive to cold flow problems. To understand these problems and their solutions, a general explanation of the diesel engine and its fuel system is necessary as it will be referred to many times in this chapter.

A diesel engine fuel system is made up of a fuel tank, fuel line leading to the fuel filter, then to a high-pressure pump and injection of the fuel, under pressure, into the engine cylinder (Figure 4.2) [9]. Fuel lines are designed to eliminate bends and low spots at which water might accumulate and cause blockage by ice. Fuel is drawn from the tank under the vacuum supplied by the fuel injection pump; trucks often also have a lift pump before the fuel filter, but passenger cars and vans usually do not. The fuel filter is designed to capture small particles of grit or rust that are one to two microns across, and larger, to prevent them from causing wear to the valuable, high-pressure injection pump.

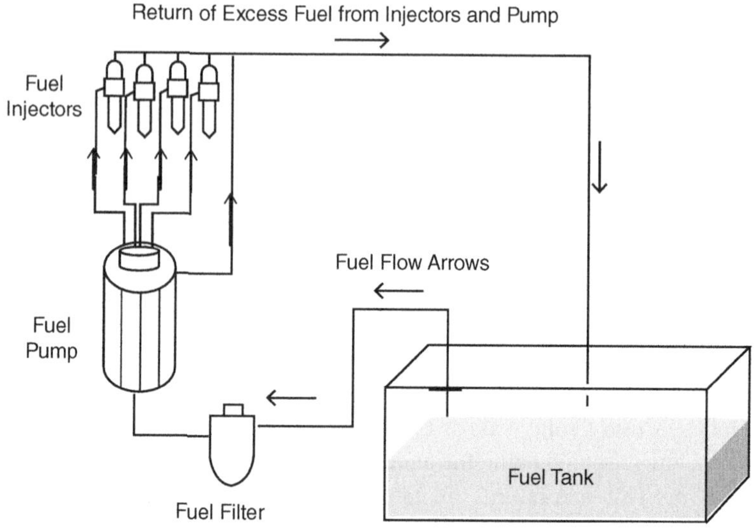

Figure 4.2 Diesel Engine Fuel System.

When ambient temperatures fall below the fuel cloud point, wax in the fuel collects on the fuel filter. If the wax crystals are in their untreated form of thin plates, they quickly form an impermeable layer on the filter, preventing the flow of fuel to the injection pump and the injectors, so the engine stalls for lack of fuel. MDFI-treated fuel produces wax crystals with a compact shape which form a wax cake on the filter that allows the fuel to pass through. If the wax cake were to build up continuously, the fuel filter container would become full of wax and the narrow fuel line would then become blocked. However, the fuel system has return lines from the pump and the injectors for excess fuel. This fuel is heated, initially, by the energy of the electric fuel pump and, later, by the heat of the engine (fuel filters are located near the engine block). When the warm fuel returns to the fuel tank, it does so at a point close to the take-off point where it warms the fuel, melting wax along the line and, eventually, in the filter[3].

4.2.2 Cloud Point Limitation

The earliest answer to problems caused by wax separation in middle distillate fuels, in the first half of the twentieth century, was to ensure that no wax would form under normal winter conditions. This was done by setting a maximum cloud point specification at the lowest temperature expected in normal winter conditions. Cloud point specifications were, and are, varied to match conditions in different climatic regions around the world. Cloud point depends upon the highest boiling fractions in the fuel because these contain the highest molecular weight, least soluble, wax-forming n-alkanes. A fuel with a higher upper boiling point (referred to as the Final Boiling Point, FBP) contains more of the less soluble, high C-number n-alkanes, so has a higher cloud point. As a result, the upper boiling range, hence yield, of middle distillate fuel, is limited by the necessity of meeting the cloud point specification. Summer temperatures enable higher maximum fuel cloud points, allowing a refinery to optimise yields with less constraint on fuel FBP.

In winter, however, a refinery had to reduce the FBP significantly, to reduce the cloud point to the expected winter minimum weather temperature – set as the winter cloud point specification maximum. This reduced the yield of middle distillate from each barrel of crude. Alternatively, for a given fuel, more kerosene (a diesel blend already contains kerosene) could be added to lower its cloud point for winter operation. However, it takes a large amount of kerosene to give the required reduction of cloud point. For example, kerosene needs to be added at 50% of the fuel to lower its cloud point by 10°C [9]. A refinery would usually be unable to provide such a large extra amount of kerosene, so would have to reduce its production of middle distillates by lowering the FBP of the gas oil part of the diesel blend. The unused, higher boiling gas oil would not be needed for the fluid catalytic cracker (FCC) as refinery operations would already be producing enough gasoline to meet demand. So, this high boiling gas oil would end up in the much lower value heavy fuel pool. Cloud point depressants (CPDs) were considered but the fuel treatment quantities were prohibitively expensive; one patent [10] claims that 25°C of cloud point depression were obtained using 7% of poly lauryl methacrylate)[4].

The limitation in winter diesel production, resulting from the cloud point limit, was severe in Europe in the 1950s. At that time, middle distillates were more widely used for heating than now (the widespread provision and use of natural gas had yet to be developed) so this put further demand on winter middle distillate production. In North America, gasoline was the major transportation fuel, much more so than in Europe; refiners optimising for gasoline production also produced more kerosene (FCC by-product) which could be used to lower the cloud point of gas oil and, to a lesser extent, of diesel (aromatics in FCC kerosene have low cetane, but North America had a lower cetane diesel specification than had Europe).

Jet fuel, given that any fuel-derived failure is not acceptable, still uses the most fool proof approach of specifying that it should not form any wax at temperatures that it is likely to experience. Such temperatures may be in the region of −50°C, as may have been observed by anyone who has travelled in a passenger jet that shows the outside temperature on an information screen. Because there is a certain amount of super-cooling when measuring a fuel cloud point on cooling, jet fuels are specified by the temperature at which the wax redissolves on reheating, the Freeze Point. The main international fuel, Jet A-1, has a freeze point[5] of -47°C. This temperature is chosen to be below the lowest expected jet fuel temperature in passenger jets, which is higher than the low air temperature at high altitudes, a result of air friction heating the wing tanks.

4.2.3 Pour Point Limitation

Pour points, also dependent upon the FBP, gave further limitations to the production of middle distillates in winter. Pour point problems were critical to lubricating oils and effective Pour Point Depressants (PPDs) had been discovered for lubricating oils as far back as 1930 [11]. Unfortunately, lubricating oil PPDs either do not work, or are needed in ≥ 1% concentrations for the required pour point depressions in middle distillate fuels. It was not until the late 1950s that low molecular weight polymer PPDs were discovered that are effective in middle distillate fuels (heating oils, gas oils, and diesel and jet fuels) at economically low treat rates. These were, initially, branched polyethylene and ethylene-propylene copolymers – the effective pour point depression by one branched, low molecular weight polyethylene was compared with the much less effective pour point depression provided by a lubricating-oil-PPD in a gas oil (Table 4.1) [12]. Ethylene/vinyl ester copolymers were then found to be yet more effective [13], especially ethylene/vinyl-acetate copolymers, EVACs [14]. EVAC PPDs may depress a fuel pour point to as much as 20°C below the cloud point, at low treat rates (concentrations) of around 100 ppm [9].

Initially, EVAC PPDs were used for fuel storage as an insurance against extended periods of low temperatures and for domestic heating oil storage facilities. In North America, where these PPDs were discovered, that was as far as their use progressed in the 1960s (pour point and cloud point were still the main gas oil specifications in North America at the millennium). In Europe, however, the fuel economy of the diesel engine and the need to optimise all motor fuels from each barrel of crude oil were more important – Europe having no indigenous source of oil, at that time. Fortunately, it so happened that middle distillate fuels in Europe were particularly responsive to pour depression by the EVACs, raising the possibility of below-cloud-point operability of diesel fuels that contained EVACs.

4.2.4 Diesel Vehicle Operability and the Cold Filter Plugging Point

An important result of their greater responsiveness to the effects of the EVAC PPDs was that EVAC-treated European diesel fuel enabled vehicles to operate below the cloud point of the fuel. An extensive field trial was carried out under T. Coley, of Esso Petroleum, in Sweden (1965), to determine the limits of low temperature operability for a range of vehicles, using a range of untreated and EVAC-treated fuels. This data and the same fuels were used to develop a small

Table 4.1 Depressions (°C) of the Pour Point of a Gas Oil by a Lubricating Oil PPD ('PARAFLOW™'), a Low Molecular Weight Branched Polyethylene, and EVAC[6].

Additive ID	% Additive	Gas Oil Pour Point, °C	Pour Point Depression, °C
None	0.0	−3.9	0.0
'PARAFLOW™'	0.2	−6.7	2.8
'PARAFLOW™'	1.0	−15.0	11.1
Low MW PE	0.2	−15.0	11.1
Low MW PE	1.0	−34.4	30.5
Low MW EVA	0.05	−40.0	36.1
Low MW EVA	0.1	<−56	>52

scale, rapid test that would correlate with the field data and, thus, predict the lowest temperature at which the treated, wax-containing fuel could be used. The test was named the Cold Filter Plugging Point (CFPP) test. CFPP is now widely used for setting middle distillate fuel cold flow specifications in most countries, world-wide [15]. The compact, rapid CFPP test is used in many refineries to provide a timely measure of a diesel fuel's cold flow limitation, so that it may meet local CFPP specifications and can be relied upon to have no wax related problems in normal winters. The details of the CFPP test and its development are well described by Coley himself [9,16].

Field trials are difficult and expensive means of proving the worth of a diesel fuel treated with a cold flow improver, but were needed in the early years to convince the oil companies, and specification-setting bodies, of the effectiveness of both the additives and the CFPP test. Eventually, field trials were replaced by the more predictable environment of the Cold Climate Chassis Dynamometer (CCCD) [9]: Vehicles with the test fuels in their tanks are positioned on rollers in a large refrigerator and cooled down slowly, overnight, to a night time minimum temperature that would be expected for the region from which the fuel is sourced. The next morning, the vehicle is started and run through a sequence to imitate start-up and running on various road conditions. It would either pass or fail, depending upon the lowest temperature reached and cold flow improver treatment. If it failed, the process would be repeated, with fresh fuel, at a higher temperature; if passed, the repeat would be at a lower temperature. The objectives of the test sequence are to find the lowest temperature at which the test fuel permits the vehicle to pass, and the highest temperature at which it fails. CCCD testing has been used to confirm the validity of the CFPP test [17,18] and is still often used to confirm the effectiveness of any new development.

The use of cold flow improvers and the CFPP test provide the refiner with greater flexibility and a large economic advantage. To illustrate: the summer CFPP specification for the UK is −5°C, which the refiner may meet by setting the cloud point at −5°C (Figure 4.3). At this cloud point, the refinery can optimise its operations to meet the 'demand barrel' – the distribution of products needed in the marketplace – and maximise its profitability. However, when relying upon cloud point (and no wax separation) to ensure the diesel fuel gives reliable operability in winter, the cloud point must be set at the lowest temperature expected in most winters – usually for 90% of winters, from records. For example, this is −15°C in the UK. Very occasionally, there will be a risk of an even lower temperature, such as January 14, 1982, when the temperature dropped to −16·6°C in Oxfordshire, while the long-term average for January is 3·6°C. Minus 16.6°C was the lowest temperature recorded in Oxfordshire since 1815, so the risks of such occurrences are low. When using

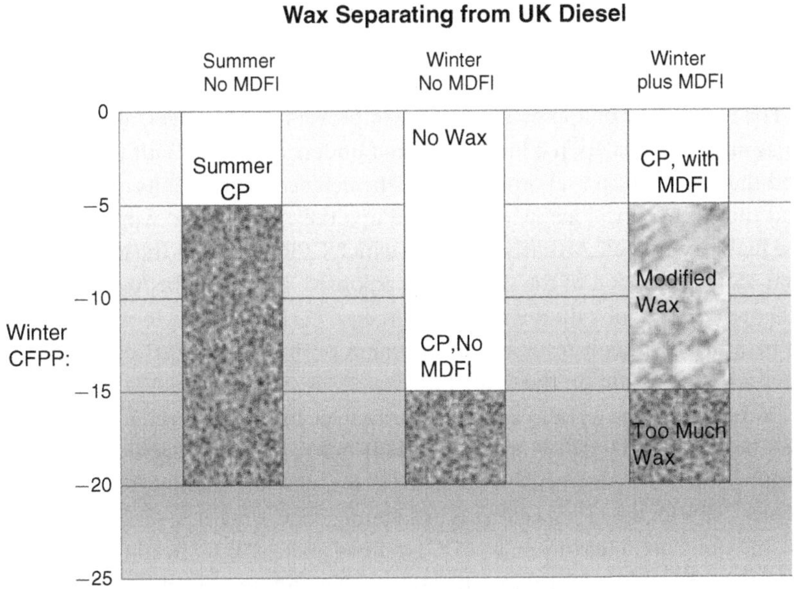

Figure 4.3 Use of Cloud Point and CFPP in Summer- and Winter-Diesel, in the UK.

cold flow improver additives (usually referred to as middle distillate flow improvers, MDFIs), coupled with CFPP [9] and a CFPP specification of −15°C, the winter cloud point may be raised by 10°C to −5°C with the same level of vehicle operability risk as having a cloud point of −15°C (Figure 4.3). As a result, the refinery can be operated at, or close to, its summer efficiency. The cloud point, with no wax separating, is replaced by CFPP, with wax separating, as the test of reliable low temperature operability.

Depending upon the fuel and the additive, 10°C of CFPP depression may be obtained by 150 to 800 ppm of MDFI [9]. The costs of applying the MDFI are much less than the extra profit made possible to the refiner who is then able to produce more diesel fuel [19]. One study (report in 2001) was quoted as saying that $4.4 worth of extra diesel was produced for every $1 worth of cold flow improver used [20]. However, there is a limit to the amount that a refiner will pay as an external cost, compared with internal economics of an alternative refinery solution. An additive supplier must balance the treat costs of his MDFI, per Tonne of fuel (treat rate multiplied by MDFI price) to satisfy the refiner and to compete with other additive suppliers. There are also two other limitations to the amounts of MDFI used: one is that the CFPP test is considered to provide a reliable prediction of safe diesel operation down to 10°C below the cloud point but such predictions may be less reliable at lower temperatures [9]. The other is that it has long been advised to use a maximum of 500 ppm of EVAC in a fuel, to avoid the risk of polymer separating out of solution in cold fuel.

The technology of applying cold flow improvers to best effect, at minimum treat rate, depends upon a knowledge and understanding of both the additive and the fuel. Critical fuel properties are the detailed distributions of n-alkanes and their concentrations, which depend upon distillation and wax content of the fuel to be treated. An early patent, in which CFPP is used as the test, showed that 15°C depression of the CFPP of a responsive fuel could be obtained with 200 ppm (0.02%) of a dual EVAC combination [21]. However, in what proved to be a more difficult-to-treat-fuel, 500 ppm of the same EVAC combination gave a 1°C elevation[7] in the CFPP [22].

Domestic heating oil (also known as furnace oil, fuel oil or gas oil [23, 24]) may also be specified by CFPP and cloud point but to a less stringent level, since heating systems are less sensitive than diesel engines. For example, in France the winter CFPP for diesel is specified as maximum −15°C while it is −5°C for heating oil; the cloud point maximum is −5°C for diesel and + 2°C for heating oil.

In the USA, fuel oil for domestic and small community heating equipment rely upon pour point specifications (−18°C for grade 1 and −6°C for grade 2) [24]. In Japan, domestic heating is mainly kerosene and larger heating facilities use A-fuel oil, which has a 5°C pour point specification [25]. There is a wide range of heating facilities using a wide range of fuels, often heavier than diesel or gas oil, including a number of grades containing residual fuel, that are specified by viscosity at 40 or 100°C (USA) or 50°C (Japan). Furnace residual fuels have been widely used in the past, especially in Russia [20] though recent reports say that Russian demand for heavy fuel oil has been declining since 2015 but their refiners have been slow to change and now have to export the surplus [26].

4.2.5 Cold Flow Improvement and Fuel Variations

At this point in the development of MDFIs, in the early 1970s, it seemed as though the general problems of cold flow in middle distillates were under control. So, CFPP and MDFI technology spread more widely throughout and outside of Europe. Then several problems appeared. The wider the net was spread, the wider the variations in middle distillate fuels and their cold flow responses to the standard EVAC MDFI. Even within Europe, winter fuels in Scandinavia had considerably lower FBPs, lower cloud points and sharper distillation cuts than those around the Mediterranean. Both fuel types needed different MDFIs, which were slowly developed. For example, in high FBP fuels the basic EVAC additive had to be supplemented by a lubricating oil PPD (section 4.3.3a). As the use of MDFIs and the CFPP spread around the world, a confusing array of fuels was uncovered that needed a range of cold flow improver variations to treat them effectively.

To deal with the diesel fuel variations that affected MDFI additive response, a useful, empirical classification of fuels was devised (Figure 16–20 of a fuel handbook [9], published 1995). Based upon the standard fuel distillation plot[8] (Figure 4.4), this classification (Figure 4.5) arranged fuels in order of FBP, distillation breadth (boiling temperature range between T90% and T20%) and sharpness of distillation cut-off (efficiency of fractionation, °C between FBP and T90%[9]). Generally, a relatively low FBP was accompanied by narrower distillation ranges and cut-offs (T90% – T20% and FBP – T90%). Because of the limited amount of kerosene being available for reducing the cloud point of diesel, the FBP is decreased in winter in order to reduce the cloud point. The loss in yield, resulting from a lower FBP, can be offset by increasing fractionation efficiency (smaller FBP – T90%). This works because an inefficient fractionation leaves a 'tail'[10] of high boiling, low solubility, high molecular weight n-alkanes, while an efficient fractionation reduces the tail. Small amounts of high boiling distillate increase the cloud point disproportionately because the solubilities of wax n-alkanes fall exponentially with increasing molecular weight and boiling point. Thus, a narrow boiling distillate (usually referred to as an NBD or, previously, 'NARROW' [9] had a FBP below 370°C, T90% – T20% less than 100°C and distillation cut-off (FBP – T90%) less than 25°C. In the 'Extreme Narrow' fuels, these figures were < 360°C, < 70°C and < 20°C, respectively (Figure 4.5).

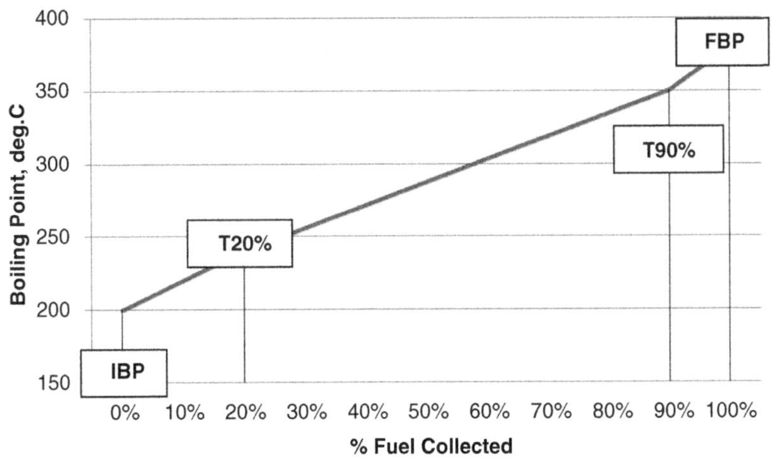

Figure 4.4 Standard Distillation Plot for a Middle Distillate (Schematic). IBP – Initial Boiling Point, FBP – Final Boiling Point

Distillate Class:	Extreme Narrow	Narrow	Broad		High FBP	Extreme High FBP
FBP:	< 360	360–370	370–380		>385	>395
		360–380 Normal				
FBP - T90%:	=<20	20 – 25			>30	>35
		>25 Normal				
T90% - T20%:	<70	70 – 100				>110
		100 – 120 Normal				

Figure 4.5 Middle Distillate Type by Distillation, 1995 (Temperatures, °C).

At the other end of the scale, with a very broad distillation, were the 'High FBP' distillates: FBP above 385°C, T90% – T20% more than 110°C and FBP – T90% more than 30°C. Also, 'Extreme High FBP' distillates: FBP > 395°C, T90% – T20% > 110°C and FBP – T90% > 36°C. Between these two extremes are the fuels then labelled as 'Normal' – also generally known as broad boiling distillates (BBD [9]), which were the fuels that responded most easily to CFPP depression by EVAC based MDFIs.

The classification of diesel fuels by boiling ranges and their related CFPP responses to standard, EVAC based MDFIs was, and still is, useful for selecting the MDFI best suited to a fuel. A variety of cold flow improver[11] additives have been developed to effectively improve the cold flow of all these fuels. However, the occurrence of fuels distributed across the range described must be updated. Over the last few decades, environmentally oriented specifications have moved all fuel distillations in the same direction – lower FBPs, lower densities, lower aromatic contents and near zero sulphur levels (the difficulty in de-hydro-sulphurising compounds that boil above about 330°C also leads to lower FBPs). As a result, the distribution of fuels found in the world-wide marketplace may be described afresh.

The 'Infineum Worldwide Winter Diesel Fuel Quality Survey' (WDFQS) in 2012 [15] and 2018 [27] covered 42 and 53 countries, respectively. These show that high FBP road diesel fuels are becoming scarce. This is because high boiling fractions are responsible for harmful particulate emissions, so such fuels are being ruled out by modern specifications. Fuel specifications around the world are moving into line with those coming out of Europe in its progress to improve the quality of diesel fuels. Such fuels enable modern vehicles to produce negligible harmful emissions. Brazil (Table 4.2) was the only country in 2012 to have an average FBP > 390°C, with four out of the five test fuels having FBPs > 390°C. However, change can occur rapidly: in the 2018 WDFQS, the highest FBP in Brazil was 376°C and the average FBP (of six test fuels) was 366°C. In only seven countries are the average FBPs in the region

Table 4.2 Winter Diesel Fuel Distillations and Cloud Points in 2012/2018 Compared (Lightest to the Heaviest[14], °C, Averages for Each Country).

Distillation, BP°C, at % collected	Fuel from: Finland	Sweden	UK	Kuwait	Brazil
IBP	168/180	187/187	169/170	204/193	161/169
20%	197/216	216/223	221/223	262/249	220/228
50%	225/247	240/251	270/268	298/290	277/272
90%	290/297	296/299	332/333	354/350	361/337
95%	314/311	313/315	349/354	370/368	386/356
FBP	332/326	326/327	358/361	376/378	394/366
CP, °C	−32/−29	−30/−29	−7/−9	+2/−1	+4/−4
CFPP, °C	−45/−44	−31/−31	−18/−21	+2/−4	−5/−10
T90-T20	93/83	110/76	111/110	92/101	141/109

of 370–379°C in 2018 and only seven fuels (out of 355) had FBPs above 380°C in 2018, compared with twenty in 2012. This illustrates the changes that have taken place since about 1990, when the approach described in Figure 4.5 was set out. Overall, the movement has been to lower boiling diesel fuels, taking out the high boiling fractions; the current Euro specification for T95% distillation is 360°C[12].

Initial boiling point (IBP) and the temperature at which 20% is recovered (T20%) have not varied very much, because all diesel fuels contain quantities of kerosene (except for a Kuwait fuel in the survey). However, the boiling temperatures for the last 10% to distil (90%, 95% recovered and FBP) can be ordered as a steady progression upwards, in line with their cloud points (Table 4.2). UK and central north European countries, such as Germany, France and Benelux, have diesel fuels that are in the middle of the range of fuels and use mainstream MDFIs; all winter diesel fuels in these countries are MDFI treated.

Finland and Sweden are both examples of having low boiling, low cloud point fuels that are more difficult to treat and need MDFI formulations to deal with their low FBPs and low temperatures of operation. Comparing cloud points with CFPPs of fuels from these two countries, it appears that while Finland does use MDFIs, Sweden avoids the problem by not using MDFI and relying on low cloud points[13]. Canadian winter diesels have similar properties and appear to contain no MDFI (cloud points and CFPPs < 4°C apart), though many contain pour point depressant (PP 10°C or more below cloud point).

Kuwait and Brazil, in 2012, are examples of high FBP fuels that would also require cold flow improver reformulation. However, Kuwait appeared not to use MDFI while Brazil did, in both its high FBP fuels, and, in 2018, in its lower FBP fuels (cloud point −4 and CFPP -10°C).

The distribution of diesel fuels with respect to FBP (2018 WDSQF [27]) (Figure 4.6), compared with the previous classification (Figure 4.5), suggests that current day 'Normal' FBP has an average of 359°C, compared with an average of 370°C in the previous classification – in fact, the FBP of 2018 would have been classified as 'Extreme Narrow' previously. Previous categories 'Extremely High' and 'High FBP' have all but disappeared from the road diesel market; however, these fuels remain in use in for marine diesel, heating oils, off-road gas oils and power generation – the larger of these applications are often supplied by the various heavy fuel grades.

On distillation cut-off ('tail', FBP-T90%, Figure 4.7), no country average has FBP-T90% of more than 30°C, previously found in those with high FBP. For most individual fuels, this measure is between 25°C and 30°C; more tightly limited than the previous statement for normal of more than 25°C. This is a result of refiners balancing the upper boiling range and a narrow cut-off (to maintain yield) with MDFI responsiveness.

And on distillation breadth, T90%–T20% (Figure 4.8), about half of current fuels are now narrower than the previous 'Normal' (100 to 120°C); those that are broader are so because of more low boiling fractions, since their FBPs are low.

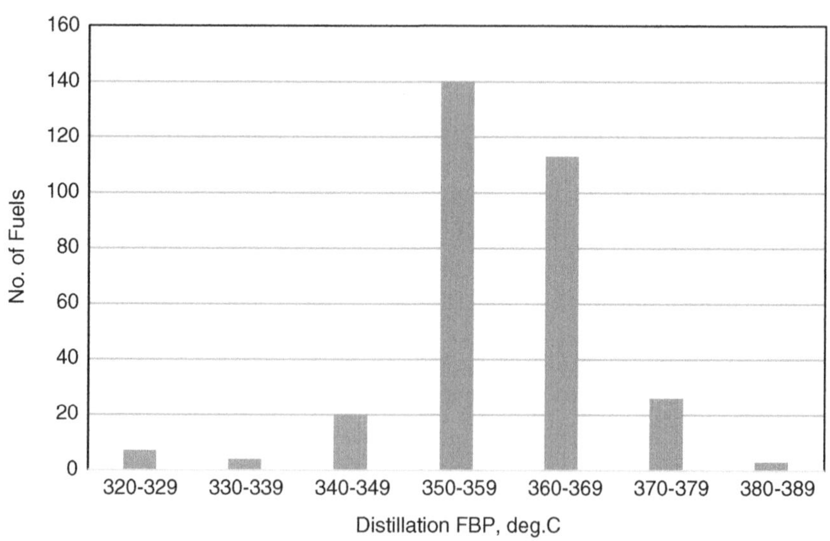

Figure 4.6 2018 WDFQS Fuels by FBP[15].

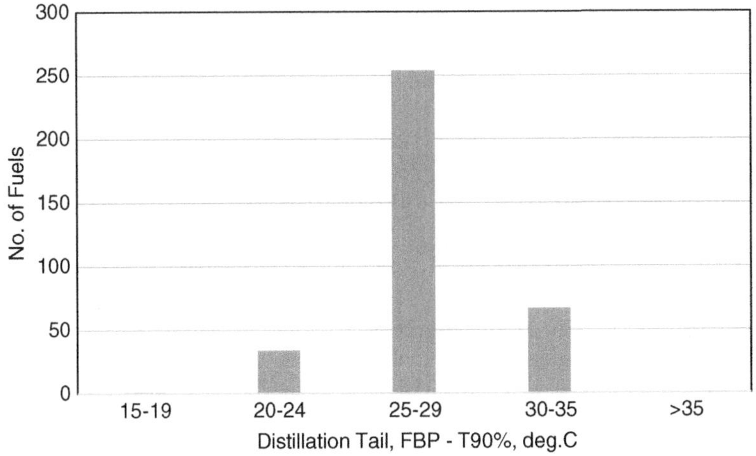

Figure 4.7 2018 WDFQS Fuels by FBP-T90%.

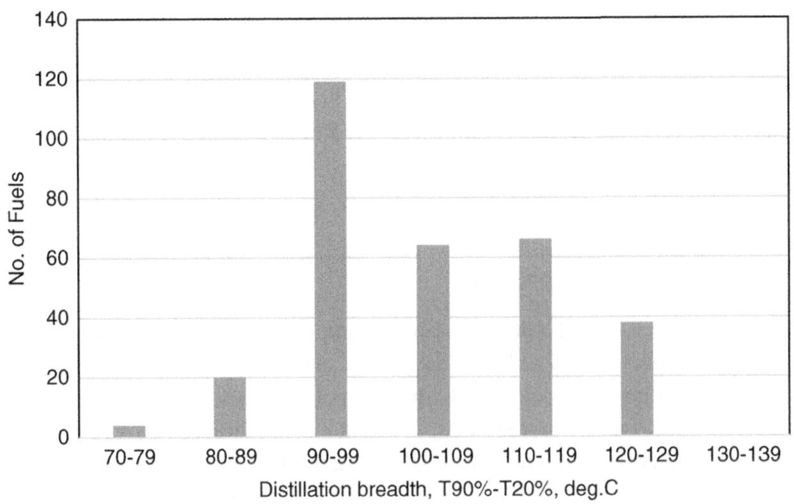

Figure 4.8 2018 WDFQS Fuels by T90%–T20%.

In summary, for automotive diesel fuels, the 'new Normal' is like the previous Narrow classification and MDFIs have gradually changed to take this into account. Compared with the previous 'Normal', the 'new Normal' has:

- Lower FBPs, 350–370°C.
- Narrower range of distillation tails (FBP-T90%) of 25–30°C.
- Distillation breadth (T90%–T20%) of 90 to 120°C, with a lower T20%.

In keeping with the previous order, fuels that are higher boiling than the new normal would be the new broad – a higher boiling group that is lower boiling than the previous high FBP version:

- FBP 370–380°C.
- Narrower distillation tail, 20–35°C, now high FBP fuels often narrower tail.
- Distillation breadth of 100–110°C, usually.

The 'Extreme Narrow' group is now lower boiling also, though not as narrow as previously:

- FBP less 350°C (320–350°C observed).
- Distillation tail also 20–30°C.
- Distillation breadth mostly ≤ 100°C.

That is to say that the FBP is the major differentiator, followed by breadth of distillation, while distillation tails are now similar for most fuels.

In the recent, environmental movement to narrower fuels, the approach has been to balance the fuel distillation narrowness with polymeric MDFIs developed for such fuels. To differentiate such narrow boiling fuels more selectively, distillation narrowness may be replaced by descriptions of fuels n-alkane distributions[16] [28]. Such descriptions had been used earlier to show that a waxy NBD fuel may be made responsive to MDFI by adding particular n-alkane ('paraffinic') waxes [29] or small amounts of high boiling fractions, such as Heavy Atmospheric Gas Oil (HAGO [30]).

4.2.6 Cloud Point Depression

Cloud point (CP) is still a defined specification – either alone or, more usually, in combination with a CFPP specification. Customers in North America continue to rely upon CP combined with pour point, to define a middle distillate fuel that will show no wax related problems in winter; this provides an economic incentives for the use of CP depressants [31]. There has also been interest in lowering the temperature at which wax separates from jet fuel; various tests were used which show that some of the claimed additives move the separation of wax to a lower temperature [32–34]. Where CFPP is used as the major specification, it is usually combined with a CP requirement that is derived from the advice that the CFPP should be no more than 10°C below the CP; for example normal winter diesel in France should have a maximum CP of −5°C and maximum CFPP of −15°C [27]. The reason for this advice is that the protective correlation of the CFPP with field or CCCD operability is less reliable at larger CP/CFPP differences [9]. In addition to, or in place of, national CP specifications, exchange agreements[17] may also specify a CP maximum. As a result of day to day (or hour to hour) variations, the middle distillate fuel may be a degree above this maximum, so there may be a need for a cloud point depressant to correct a temporarily off-spec CP.

CP depressants (CPD) for diesel fuel were known as far back as 1944 [10,35] – these and others that have been claimed as CPDs [36–38] are comb polymers. A number of materials have been developed as CPDs, but their general use has been limited by two problems. First, many interact adversely with MDFIs, reducing their CFPP depressing abilities [36]. Second, CPDs are needed at much higher treats rates than are MDFIs for the same degree of cold flow improvement, for example 5°C of cloud point depression required 0.2% (2,000 ppm) of polymer [37] compared with 200 ppm of a suitable MDFI for the same CFPP depression. In another patent, CFPP depressions of 10 to 12°C were obtained with 300 ppm of various MDFI combinations while only 3°C CPD was obtained, at best, using 500 to 625 ppm of CPD additives [38]. However, in North America, in which the cloud point is a major specification and CFPP not an official specification, there did appear to be an economic value for the use of CPDs [31]. Such additives are included in some supplier's additive lists [39].

4.2.7 Wax Anti-Settling

Heating oil may be stored in variously sized, unheated tanks. It is often blended from fractions of similar boiling ranges as those used for a refinery's diesel and has a similar or slightly higher CP. After the introduction of effective, EVAC PPDs, fuel CPs were raised while the pour points were controlled by the additive. As a result, fuel stored in unprotected tanks could experience extended periods of temperatures below its CP. During such conditions, PPD treated wax would separate and would settle to the bottom of the non-gelling fuel. In the 1970s, problems caused by the settling of the treated wax were recognised [40]. Wax crystals, treated with EVAC, are compact prisms that can move freely within the fuel – unlike the thin, plate-like wax crystals that gel untreated fuel. Being denser than the fuel, wax crystals slowly settle to the bottom of the fuel to form a wax rich layer, about 10%–25% of the fuel volume at 10°C below the CP.

Tanks at a refinery are very large and continuously receive new fuel, which is warm, being the rundown at the end of refinery processing, so these tanks do not get cold in winter. Tanks at a distributor are of an intermediate size which, combined with its turnover (fuel being sold, and the tank being replenished) may withstand a certain degree of extended cold weather. Filters on the outlet from such tanks, through which fuels pass that are transferring to delivery tankers, are quite coarse (60 mesh# or square opening 0.24 mm across) so a modest amount of wax might not cause a problem. However, over periods of prolonged cold, as wax settles in such tanks, it can build up to a thick, wax-rich layer having a high CP. Such a layer will exit the storage tank unevenly [9] – the first batch of fuel taken will carry an excess of wax as a funnel is formed in the wax layer (Figure 4.9).

Following batches may then have less wax than the original fuel but, if the storage tank is run low, batches of fuel near the bottom (the last 25%) will also contain excess wax. Any fuel containing wax more than the original, on-specification fuel is likely to be off-specification, having higher CP and CFPP. As such, it may well cause waxing problems in a customer's heating or off-road fuelling facility. Heating oil users are varied: German domestic-oil customers usually have their fuel stored in their cellars; community heating oil users may also have protected fuel storage. However, other domestic-oil customers and small-scale users, such as farmers, may have unprotected external fuel storage. Fuel from external storage is susceptible to irregular waxing problems of fuel line and filter blocking, because of wax-settling.

Automotive fuels are taken from the refinery, or pipeline-fed terminal, by tanker to underground storage at service stations, which usually protects diesel fuel from temperatures below the CP. In the 1980s, Europe was the main market for MDFIs, based upon EVACs. Oil companies were making full use of the benefits of using MDFIs in winter diesel fuels. During the unusually cold winters in the mid-1980s, European diesel vehicles were failing, apparently at temperatures above their CFPPs. Adding more EVAC MDFI did not alleviate the problem, though it did depress the CFPP further [9]. The

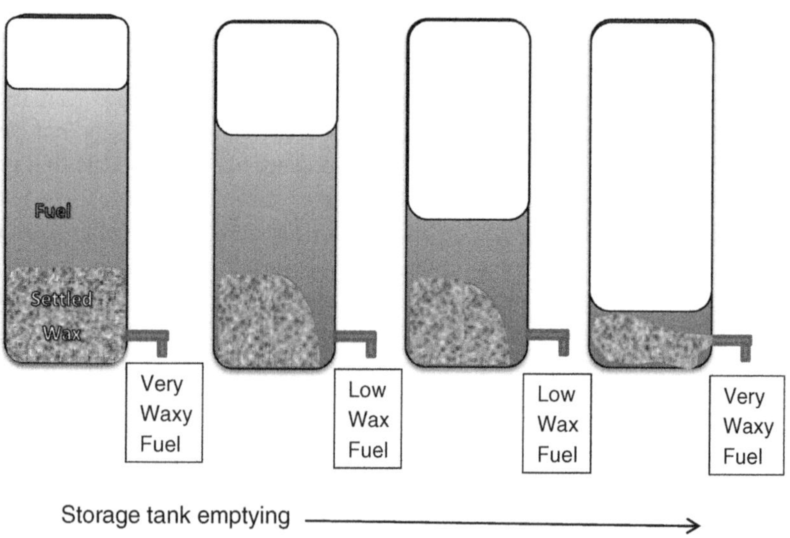

Figure 4.9 Fuel Take-off as Tank Empties after Wax Settling.

problem was assigned to wax-settling in the vehicle fuel tank, overnight, with the result that when the vehicle was started, wax-enriched fuel was drawn from the bottom of the tank. Furthermore, this lower layer now has a much higher cloud point than the original fuel, so may not have fully redissolved during repeated night/day temperature cycling, when the vehicle was not in use as over a weekend [41]. During progressively colder nights, in undisturbed fuel tanks, wax can build up at the bottom of the tank, producing progressively deeper wax-rich layers that can be drawn off on start-up. The excess wax may then fill and block fuel lines and the fuel filter, starving the engine of fuel, causing the engine to stop or, at least, operate irregularly, at a temperature above that specified by tests on the whole fuel before wax-settling.

To counter the problem, Wax Anti-Settling Additives (WASAs) were developed from long-chain alkylamine salts and amides. These were introduced commercially in the late 1980s [19]. When used in conjunction with the EVAC MDFI, the WASA produces much smaller wax crystals that do not settle within an overnight period, or longer. As a result, the small wax crystals remain fully dispersed throughout the fuel, so wax does not build up concentrated layers and the cloud point of the fuel remains the same throughout.

Wax anti-settling tests have been applied to measure the amount of wax settling during overnight cooling or for longer periods of storage. Addressing long storage periods, one procedure used a freezer with programmed cooling from above the cloud point to the test temperature. For example [41], one sample of a fuel was treated with MDFI (EVAC) and another sample of the same fuel was treated with MDFI plus WASA ('WAFI'). After cooling both samples (initial CP −7°C) from + 5 to −16°C, at 1°C /hour and holding at −16°C for 5 days, top and bottom samples (20% samples syphoned off) from the two fuels had cloud points that clearly indicated wax settling in the MDFI treated sample, but no wax settling in the WAFI treated sample (Table 4.3).

In general, in order to obtain results in a practical time frame, the settling period is usually over-night, for example sixteen hours, after a one-degree-per-hour cooling from above the cloud point to −15°C, as in a S. Korean fuel

Table 4.3 Wax Settling in MDFI Treated Fuel Compared with WAFI Treated Fuel.

Fuel Additive:	**MDFI**		**WAFI**	
Sample:	Top	Bottom	Top	Bottom
Cloud Point °C	−13	+5	−7	−7
ΔCP	na	8	na	0

(Starting cloud point −7°C, see below for ΔCP)

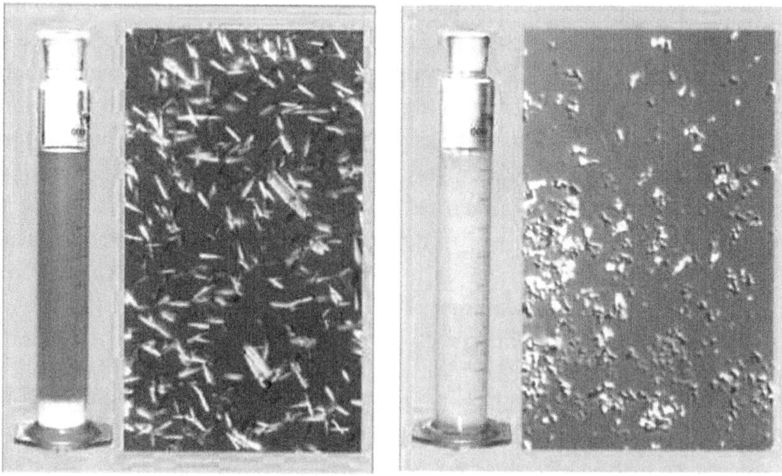

Figure 4.10 Wax Settling Test of Fuel Treated with MDFI (Left) Compared with Fuel Treated with WAFI (Right), after 1 °C /Hour Cooling and Hold at −15 °C for 16 hours.

(Figure 4.10). In this case, the fuel had CFPP reduced from + 8°C to −19°C by a MDFI/WASA mixture (50%/50%) and by the same total treat of MDFI alone. After the wax settling test, the CFPPs of the top and bottom 20% samples (after reheating) were obtained; the sample from the bottom of the 100% MDFI treated fuel was −8°C, while that from the 50% MDFI/50% WASA (i.e. WAFI) treated fuel was still −19°C (± 1°C).

The photographs (Figure 4.10) show the 500 ml cylinders of fuel, after 16 hours at −15°C. MDFI-treated fuel (left) shows a white layer of settled wax at the bottom with clear fuel above. The WAFI-treated fuel (right) shows the fuel with the wax in complete dispersion throughout the fuel. The pictures to the right of each fuel sample are micrographs of the wax crystals, the wax crystal sizes are: 50 microns across on the left, in the MDFI treated fuel, and about 10 microns across in the WAFI treated fuel.

Now, a more easily applied wax settling test is the widely used Aral Short Sediment Test [42]. In this test, 500 ml of treated fuel, in a measuring cylinder, is placed in a freezer, set at the target temperature. After overnight cooling (which includes a final, static temperature period) visual examination immediately indicates whether and how much wax settling has occurred. Then the top 20% of fuel is syphoned off as a top sample, the next 60% of fuel is syphoned off carefully and the remainder, containing any settled wax, is the bottom sample. After heating sufficiently to fully dissolve the wax, CP and CFPP are measured on the top and bottom samples. With no wax settling, CPs and CFPPs of the top and bottom samples should be the same; with extensive wax settling, the top and bottom sample CPs and CFPPs can vary widely, as in the longer tests. The results are often expressed as the ΔCP (or ΔCFPP), where ΔCP is the difference between the original fuel CP and that of the bottom 20% of the wax settling test - with no wax settling, ΔCP is zero, and the more wax settling there is, the greater the value ΔCP (as in the bottom row of Table 4.3).

In addition to preventing the formation of wax-rich layers, the compact, small crystals in a WAFI-treated fuel can form a wax cake on the fuel filter that is more compact than that formed by the larger crystals in fuel treated with EVAC alone, yet it is still porous. This allows the fuel filter to hold more wax before the wax starts to back down the narrow inlet pipe, which would result in a fuel line blockage. A major benefit from using the WASA-MDFI combination (WAFI) over MDFI alone was demonstrated by diesel vehicle CCCD testing: a diesel fuel with −1°C cloud point had its CFPP depressed to −24°C (much more than 10°C below the cloud point). One fuel sample was treated with MDFI alone, and the other with a WAFI combination (in both cases, 0.1% of additive). When the vehicle was running on the MDFI treated fuel, it failed at −23°C; when using the WAFI treated fuel, it was still operating at −26°C [41].

4.3 The Organic Chemistry of Wax Crystal Modifying Cold Flow Improvers

Anticipating the next section, cold flow improvers work by interacting with growing n-alkane-wax[18] crystals. They must adsorb onto the fastest growing surfaces, those leading to the formation of plates, and reduce their growth rates. As a result, the crystals then grow in a more compact habit (shape), less able to cause gelling and more able to form porous layers on filters. These fast-growing surfaces are those on thin edges that exhibit the long, polymethylene segments; the slow growing, expansive surfaces are those made up of the methyl, n-alkane end groups (Figure 4.11). To adsorb onto the polymethylene surfaces and into the crystal sites, the additive must also carry polymethylene segments. Thus, polymers such as polypropylene, polybutylene, polyisobutylene, polycyclopentene and polystyrene, that cannot fit into such crystal sites, are not effective as PPDs [12]. Cold flow improver additives in use, as described in this section, bear polymethylene segments or n-alkyl groups, apart from one special case.

The use of the plural in 'segments' is intentional because the additive has to compete with n-alkanes separating from solution onto the growing crystal surface. To adsorb more strongly than the n-alkanes, the additive molecule needs to carry more than one polymethylene segment: adsorption of multiple polymethylenes from one molecule provides an advantage of a lower

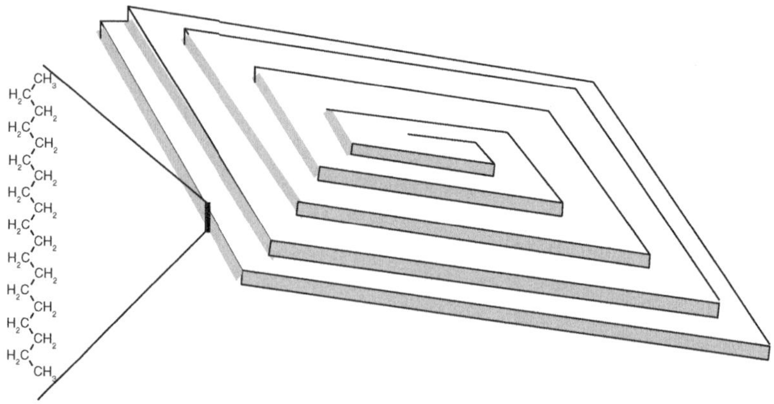

Figure 4.11 Schematic Wax Crystal Plate, Indicating n-Alkane Edges.

loss of multiple translational enthalpies of adsorption, compared with several n-alkane molecules. Translational entropy is, largely, a per-molecule event, so is similar for one additive molecule and one n-alkane molecule. The drivers for adsorption are the same as for detergents (section 3.5.3).

The special case that bears branched alkyl groups are the alkyl-phenol-formaldehyde resins that do not interact with the wax but with the WASA, enhancing its activity (section 4.3.6).

Most cold flow improver molecules are wax crystal growth inhibiting additives. Nucleation is largely controlled by the distribution of n-alkanes in the fuel and their interactions with these growth inhibitors. However, there are compounds that seem to provide more wax crystal nuclei, leading to more, and hence smaller, crystals – smaller crystals generally improve the flow of fuel laden with wax crystals. Again, these must carry n-alkyl groups or segments to attract n-alkanes coming out of solution. Many compounds can perform this activity, for example higher n-alkanes [38] [29] and polyol esters from linear fatty acids, such as sorbitan tri-stearate [43]. They do so because of the long n-alkyl chains that they bear, their crystal structures and their relatively low solubilities; but their low solubilities cause additive handling problems and unreliable activity[19].

Many cold flow improvers are combinations of two or more components. Blending a limited number of components allows a supplier to match the needs of many different fuels and environments. Two-component cold flow Improvers based on combinations of 'growth arrestor and nucleation polymers' [44] are widespread. WAFI additives may contain three or four components [45, 46] and fine-tuned MDFI additives may contain four or five components, for example: a combination of a dual EVAC, a dialkyl-Fumarate/Vinyl-Acetate

Copolymer (FVAC) and an ethylene/vinyl-neodecanoate copolymer [47]. One problem posed by multi-component additives is making the blend of components stable, as different polymers have an innate tendency to separate. Another problem to consider is that if a component has a low solubility, it may separate out and settle to the bottom of the storage tank. Refineries deal with liquids because they flow through pipes that can be monitored remotely, so they rarely deal with solids in everyday fuel blending operations. Additive blends need to be liquids that can be stored and added easily to a fuel stream as supplied, or after further dilution in a refinery stream.

Fuel additive suppliers refer to their cold flow additives by their registered trademarks plus an additive number. These can be seen on company websites and may be referred to in fuel additive literature, for example Dodiflow® and Dodiwax® are cold flow names registered to Clariant [48]; Keroflux® is the BASF MDFI trademark; Afton uses the name HiTEC® for its additives (for example HiTEC® 4566 is for cold flow [49]. Others use their company name plus a number such as Infineum R# for fuel additives, for example R591 and R309 are referred to in a Shell publication [50]. To determine what the commercial additive chemistries are is not straightforward but there are some clues from company websites and, more often, from papers and patents. Certain chemical descriptions of cold flow additives arise repeatedly in patents, often when the additive is being used as a co-additive in the patenting of a new component. It is assumed that such occurrences indicate commercial additive availability – sometimes such information is disclosed.

Most wax crystal modifying (WCM) compounds used as fuel cold flow improvers fit into one of the following descriptions. Details of each type are covered in the following sections. Lists such as these are often used to define potential co-additives for new WCM components described in patents [51–53].

- Linear, ethylene copolymers with a further unsaturated monomer, such as ethylene copolymers with vinyl esters, alkyl acrylates or 1-alkenes. In these, the polymer backbone provides the polymethylene (n-alkane-like) sequences (section 4.3.1). Such polymers, especially EVACs, are 'still the mainstay of the industry' [19].
- Comb polymers, for example poly-tetradecyl-methacrylate, in which a continuous array of n-alkyl groups is pendant from the polymer backbone, in a comb-like fashion (section 4.3.3a). Patents of such polymers usually describe their application as co-additives to other MDFIs, such as EVACs and WASAs.
- Long mono-, di- or tri-n-alkyl (such as n-octadecyl) amines and their derivatives such as amides and amine salts, which are often referred to as 'polar nitrogen compounds' (section 4.3.4). These are usually found in wax anti-settling additives.

- Wax crystal nucleators (section 4.3.5) may be long chain n-alkyl esters of poly-carboxylic acids or long chain linear carboxylic esters of polyols, both of which may be poly-ethoxylated. Sometimes, these are referred to as 'poly-oxyalkylenes' and are believed to be wax crystal nucleators. Certain block copolymers, prepared by anionic polymerisation followed by hydrogenation, are also used as wax crystal nucleators; one block is usually 1,4-poly-butadiene which, after hydrogenation, becomes a polyethylene block and the other block is a fuel soluble block, such as may be derived from isoprene.
- An Alkyl-Phenol-Formaldehyde Condensate (APFC), or similar (section 4.3.6).

4.3.1 Linear Ethylene Copolymers

The most important ethylene copolymer cold flow improvers are those made by free radical polymerisation with vinyl ester co-monomers. Ethylene/vinyl acetate copolymers (EVACs) with various structural details of vinyl acetate content, degree of alkyl side branching and low molecular weight, are widely used as MDFIs. The first of these was commercialised by Esso Chemical (now Exxon Chemical) around 1960, referred to as PARADYNE™ 20 [54] and the subject of several patents, one of which refers to an EVAC 'commercial fuel pour point depressant' [55]. From related patents, this polymer would have had 35 wt% vinyl acetate with 65 wt% ethylene, an Mn (number average molecular weight, given the method of measurement) of about 2,000 and was made in benzene at 150°C (using t-butyl peroxide as initiator) [30, 56, 57]. At this reaction temperature, the EVAC had a relatively high degree of alkyl side branching, mostly butyl groups formed by the backbiting mechanism found in the free radical polymerisation of ethylene (section 4.3.2). Branching is usually measured as the number of methyls (end of the butyl group) per 100 backbone methylenes, both of which can be determined by H-NMR spectroscopy. EVAC made at 150°C was shown to have a branching of 12 [58] which, combined with about 8 acetate groups from the vinyl acetate, gives a high total of 20 interruptions per 100 methylenes. Such a polymer is expected to have little crystallinity.

The cold flow activity of EVACs was improved by reducing the amount of side branching, achieved by running the polymerisation at 105°C, with a lauroyl peroxide initiator [58]. This second version of the EVAC pour point depressant had alkyl side branching of 3.5 [58] (still with the 8 acetates, so total 11.5 interruptions per 100 methylenes). The need for the improvement was that the first EVAC version, developed as a PPD, had poor effectiveness in the CFPP test, in many fuels, requiring high treats or not able to provide the necessary CFPP depression at all (see table III of the patent for the new EVAC [58]); the low branched EVAC, described as a 'growth arrestor' or 'growth inhibitor', was more effective at treating fuels for good cold flow, as measured in the CFPP test[20].

When the growth arrestor was combined with a second, higher molecular weight EVAC[21], described as a 'nucleator' (section 4.3.5), it was particularly effective as a cold flow improver in diesel and heating oils – as reflected by the CFPP test [21]. EVAC combinations such as a 3/1 mix of growth inhibitor/nucleator are found in many later patents in 1970s to 2000 [40, 59, 60] and appears, with small changes to the vinyl acetate content and Mn of the EVACs, in later patents by the original applicant company, and also in competitors' patents [47].

An average MDFI-EVAC molecule, with the 30–36 wt.% vinyl acetate, would have an average structure approximately as Figure 4.12.

Butyl and acetate side-groups are randomly distributed along the linear, polyethylene backbone, leaving uninterrupted polymethylene sequences of various lengths. The random, distribution of side-groups produces a normal distribution of such sequences with some molecules having long sequences, about twenty to twenty-five carbon atoms long. These polymethylene segments are the segments that can most effectively adsorb onto the A and B faces of the growing n-alkane crystals (section 4.4.5b). The segments attached to these, that carry more side groups, are more fuel soluble, so they remain extant in the fuel as loops and ends; their occupation of the space around the crystal surface inhibits the approach of n-alkane molecules.

Inexpensive EVAC-MDFI technology gained widespread usage and is still in use today. Companies have developed their own EVACs, often a single polymer with around 30 wt.% vinyl acetate (close to the average proportion of vinyl acetate in the 3/1 growth arrestor/nucleator mix just described). For example: Clariant describes the 'Classic Poly-E/VA-Technology' as its additive for various fuels in its website [48]; Innospec has a variety of 'EV' copolymers [44]; BASF has an MDFI Keroflux 5486™ or Keroflux H™, described as an EVAC [61]; in China the CFPP additive T1804 is an EVAC [62]; Mitsui of Japan [63] and Equistar of the USA [64] describe EVAC-MDFIs in their patents.

Other vinyl esters than vinyl acetate have also been considered, though such vinyl esters are more expensive. In the early days of ethylene/vinyl ester copolymer pour point depressants, a range of ethylene copolymers with vinyl esters from vinyl butyrate to vinyl laurate were made and examined for PPD potency [13]. Such monomers are not easy to come by, so the alternative

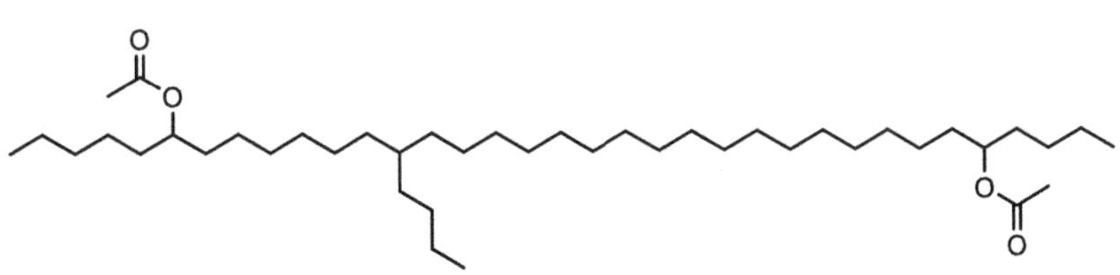

Figure 4.12 A Typical Section of an Average MDFI EVAC Molecule (Diagrammatic).

approach of trans-esterifying an EVAC MDFI, with various carboxylic acids, was carried out; replacing the acetate group by the octanoate group was shown to give significant improvements in CFPP depressing potency in moderately difficult to treat fuels [65]. Clariant started to combine a commercially available vinyl neodecanoate ('Vinyl Versatate™', then a Shell Chemicals product) with ethylene and vinyl acetate in its EVAC-MDFI, to improve its solubility in fuel and to avoid problems of polymer separation in cold fuel, above its cloud point [66]. Clariant refers to its MDFIs as 'ethylene/vinyl-acetate/vinyl-neodecanoate terpolymers', such as one with 13.7 mole% vinyl acetate and 1.4 mole% vinyl neodecanoate in a short list of 'commercial products' [67]. Others have been specific on the identity of Clariant's (then Hoechst) ethylene/vinyl-acetate/vinyl-Versatate terpolymers being the commercial Dodiflow™-V-4159 [68].

Exxon Chemical also introduced the use of terpolymers of ethylene/vinyl-acetate/vinyl-2-ethylhexanoate as cold flow improvers, judging from its wide appearance in patents [69, 70] and later used as reference components ('EVE') in the testing of another new additive [71]. BASF had two other alternatives to EVAC: one, the ethylene/2-ethylhexyl acrylate (50 wt.%) copolymer and the other an ethylene/vinyl-propionate (40 wt.%) copolymer, included in its list of 'conventional ethylene copolymer flow improvers' [72, 73]. The first of these is believed to have arisen in the 1970s [74] while the second is subject of an early 1980s patent [75]. Sanyo also describes terpolymers of ethylene, vinyl acetate and vinyl neodecanoate [76]. The structures of the most used vinyl esters are shown in Figure 4.13.

Given the widespread patent coverage of the ethylene/vinyl-acetate/higher vinyl-ester terpolymers, the advantage that they have over plain EVAC MDFIs compensates for the higher cost of low levels of higher vinyl esters. The advantages of a range of ethylene/vinyl-acetate /vinyl-2-ethylhexanoate terpolymers, over EVAC (at similar total mole ester content), in CFPP-depressing-potency, has been demonstrated in two quite narrow boiling middle distillates: having FBPs of 358 and 367°C, FBP-T90% of 23°C and 15°C and T90%–T20% of 91°C and 83°C, respectively [69]. 'Terpolymers' are sometimes referred to specifically in commercial literature on diesel cold flow improvers [44].

4.3.2 The Free Radical Polymerisation Process

If an effective polymer can be made by free radical polymerisation, then this is the preferred process because it is relatively easy to carry out. It is not sensitive to water or small amounts of oxygen and tolerates polar substituents, such as amide, ester and anhydride, unlike anionic, metallocene and Ziegler–Natta catalyses. The chemical process of free radical polymerisation is well understood, and the

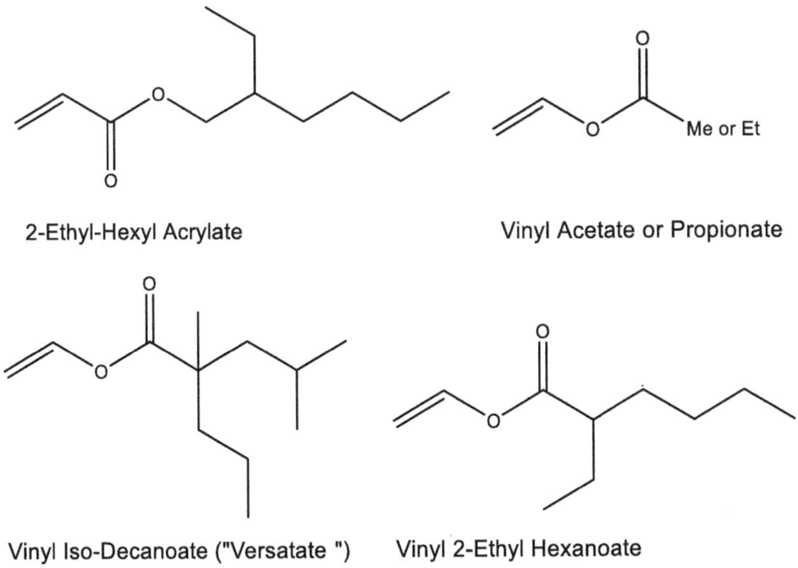

Figure 4.13 The Most Available Ethylene Co-Monomer Vinyl Esters.

details are covered in many books [77]. A general scheme for the polymerisation of vinyl acetate is:

1) Initiation: Initiator, usually a peroxide as dibenzoyl peroxide, is thermally decomposed to provide the initiating free radicals, such as benzoyl and phenyl radical (Figure 4.14).

These radicals initiate a free radical chain by adding to a monomer, here a molecule of vinyl acetate, producing a new radical on the first monomer molecule (Figure 4.15).

2) Propagation: A further monomer molecule adds to this radical and the radical is transferred to the second monomer (Figure 4.16). Such monomer additions are rapidly repeated, producing a growing polymer chain.
3) Termination and chain transfer: In competition with propagation is the termination process. More rapid termination, relative to the rate of propagation, leads to a lower molecular weight polymer. For some polymerisations in solution, such as polymerisation in cyclohexane (as described for EVAC [58]), termination is mostly by chain transfer to the solvent – the growing polymer chain picks up a hydrogen atom from a solvent molecule, transferring the free radical to the solvent molecule which then initiates another polymer chain (Figure 4.17).

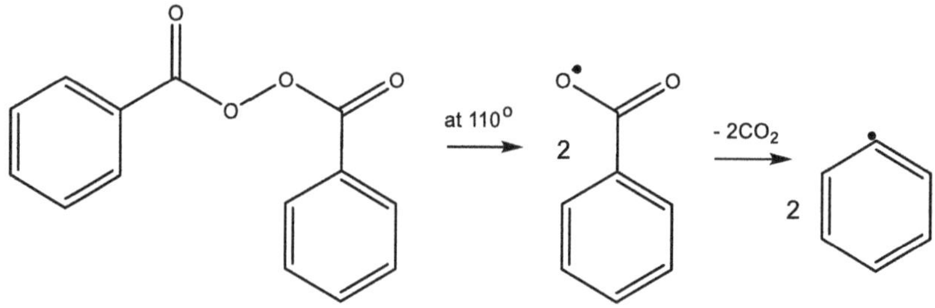

Figure 4.14 Dibenzoyl Peroxide Initiator.

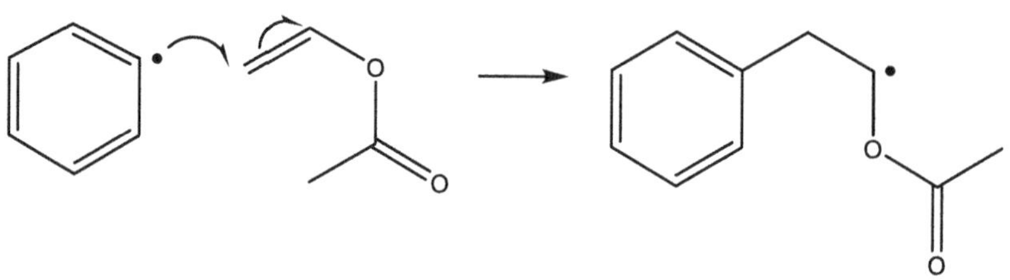

Figure 4.15 Initiator Radical Adds Monomer.

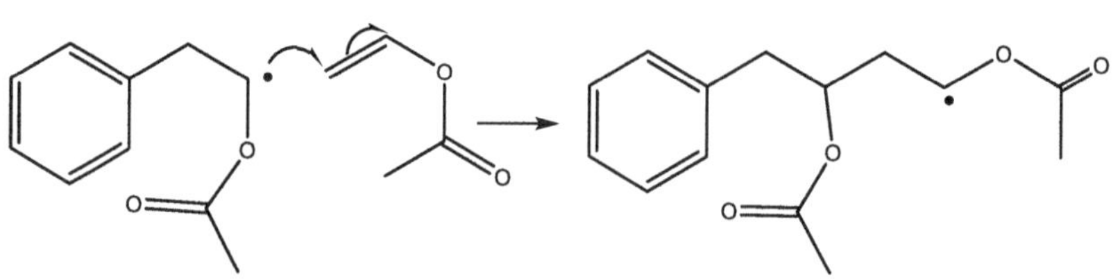

Figure 4.16 Propagation.

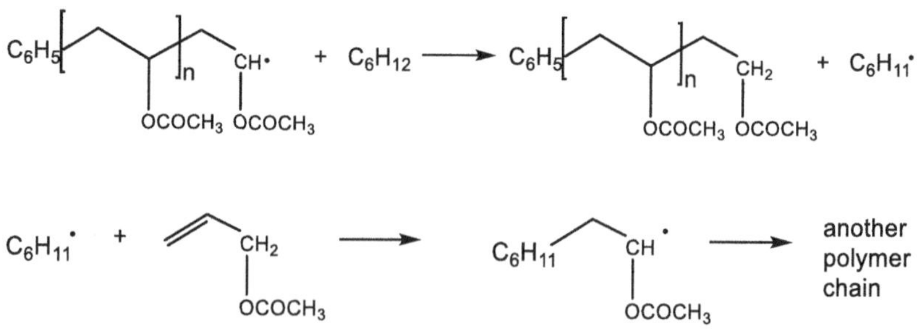

Figure 4.17 Chain Transfer.

Other termination mechanisms occur when two radicals come together and the two polymer chains either combine (for example polystyrene) or disproportionate, leaving one chain with a saturated end group and one with an unsaturated end group (such as methyl methacrylate above 60°C).

While the example given is for the polymerisation of vinyl acetate alone, the process works much the same for other substituted-ethylenes, ethylene itself and mixtures of monomers, except for substituents that provide allylic hydrogens, such as the higher 1-alkenes. Allylic hydrogen atoms present in most non-ethylene alkenes such as 1- or 3-hexene, are easily lost[22] causing rapid chain transfer, which prevents the formation of polymer with more than a small number of monomer units. With ethylene as the monomer, the substituent on the double bond would be a hydrogen in place of the acetoxy (in Figure 4.17). Ethylene/vinyl acetate mixtures tend to give random copolymers because these two monomers have similar reactivity ratios with each other in free radical polymerisation[23] – they are both 1.0 [77]. For cold flow improver applications, the proportions of vinyl acetate used are usually in the region of 10 to 20 mole%. The resulting copolymer is mainly polyethylene (i.e. n-alkyl-like) with vinyl acetate molecules randomly dispersed along the poly-methylene chains, leaving alkyl segments of two to about twenty methylenes (Figure 4.12).

EVAC is also made in high pressure reactors without any solvent. To produce low molecular weight polymers in such reactors, volatile chain transfer agents, such as propionaldehyde [78] or butan-2-one [79], are added.

Side branching, (section 4.3.1), is common in the free radical polymerisation of ethylene. This occurs as the free radical on the poly-ethylene chain end abstracts a hydrogen atom from the fifth carbon atom down the chain. Polymerisation then continues on the free radical left on the fifth carbon, leaving it with a butyl sidechain (Figure 4.18).

The poly-ethylene chain then continues to grow, as before, until terminated. Alternatively, some of these first branch positions add one mole of ethylene, abstract a hydrogen atom from a solvent molecule, or another backbiting of the polymer backbone, and become ethyl groups. The amount of branching produced

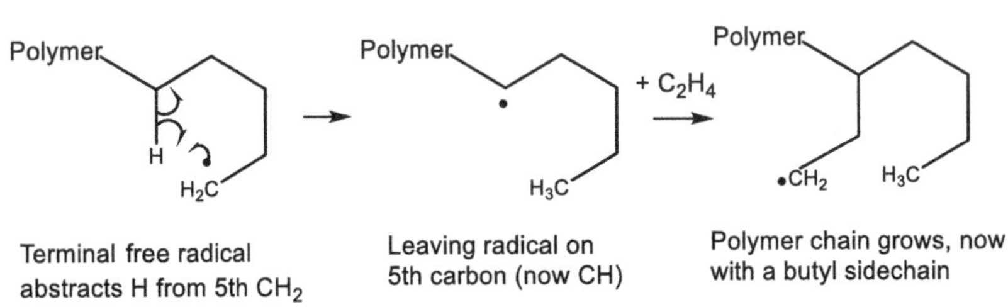

Figure 4.18 Backbiting Mechanism during Free Radical Ethylene Polymerisation.

in a cold flow improver EVAC molecule depends upon competition with propagation and temperature. In comparable solution co-polymerisations of ethylene with vinyl esters, even a small lowering of polymerisation temperature from 115°C to 90°C, make measurable reductions in the degree of side chain branching which, in turn, gives the improvement in cold flow potency of the polymer [69].

In the high pressure, solvent-free polymerisation process, the concentration of ethylene is increased with the result that propagation rate in the EVAC polymerisation is increased, while the unimolecular backbiting rate (independent of ethylene concentration) is unchanged, at the same temperature. Overall, the EVAC polymer from this process may have lower branching than that from the solution process, even though such processes (the tubular or autoclave reactor in continuous processes) are carried out at higher temperatures. For example, pressures of 150–275 MPa and temperatures of 200–280°C are preferred for cold flow improver ethylene/vinyl ester copolymers (EVECs) which have low degrees of branching [79]. In such processes, with rapid throughput, only 10%–20% of the monomers are converted to polymer and the excess monomers are quickly evaporated ('flashed off') and recycled. This appears to narrow the distribution of polymer properties. The resulting polymers have lower fuel solubilities than those from solvent based processes and this that can cause problems. When fuels treated with such polymers, in vehicle and small storage tanks, suffer periods of low temperatures, just above the fuel cloud point, low solubility cold flow improvers may separate and block fuel filters in the absence of wax.

It has been recognised that there are advantages to handling and solubility for the EVAC having a moderate degree of branching, which is provided by the conditions in the solution process. In the high-pressure process, extra branching may be supplied by, for example, the inclusion of a small proportion of propylene, as a third monomer [79]. Alternatively, a small molar proportion of vinyl neodecanoate, a highly branched ester, may be included to improve the EVEC solubility while also improving CFPP depressing efficiency, over EVAC, in some fuels [66]. Certain terpolymers, which contain vinyl-2-ethyl-hexanoate as well as the VA in the EVEC, can be more effective MDFI copolymers than EVAC itself [69]. These vinyl-2-ethyl-hexanoate containing EVECs are made at a lower temperature (90°C) than comparable EVACs (105°C), so also have low branching, quoted as 2.2 to 3.6 methyls per hundred methylenes [69]. Polymers with higher proportions of such vinyl esters (and less vinyl acetate) have been proposed for some fuels [80] but such vinyl esters are much more expensive than vinyl acetate – the refiner always has the options of adjusting the refining process to make the fuel more responsive to EVAC-MDFIs, or of lowering the FBP and adding more kerosene to give a lower cloud point and use less MDFI.

4.3.3 Comb Polymers

Comb polymers are so named because of their molecular structure. They comprise a polymer backbone bearing long alkyl chains (usually C_{12} to C_{22}) regularly spaced along the backbone. Such polymers may be poly-1-alkenes, poly(alkyl-methacrylates), olefin/maleic anhydride copolymer esters or dialkyl-fumarate/vinyl acetate (FVAC) copolymers. Such polymers are used mainly in combinations with EVACs or WASAs for treating some fuels. Cloud point depressants (CPDs) are usually comb polymers bearing tailored side chains.

4.3.3a Free Radical Comb Polymers

Poly(alkyl–methacrylates) or poly(alkyl–acrylates) ('PAMAs' or 'PAAs') had early use as PPDs for lubricating oil but were ineffective in middle distillates [12]. A PAMA lubricating oil PPD Acryloid™ 150[24] was found, in 1948, to be much more effective as a PPD in diesel fuel when combined with microcrystalline wax[25]: while 1,500 ppm of Acryloid 150 were needed to achieve a 10°C PPD in gas oil when used alone, only 100 ppm were needed when combined with 0.6% microcrystalline wax [81] Poly(lauryl-methacrylate), however, was reported to act as a CPD in 1946 [10].

Coming back to the present, PAMAs are still in use in lubricating oils as PPDs, viscosity index modifiers and dewaxing aids. PAMAs have been offered as MDFIs (for CFPP depression) but are often not cost competitive with EVACs. However, PAMAs, under the trade name 'Viscoplex™', a trade name of Evonik, are claimed for MDFI use in biodiesel [82]; the Viscoplexs are claimed to have the advantage of preventing filterability problems, caused by the sterol glycoside impurities in biodiesel, as well as providing good cold flow improvement. Also, a Chinese commercial CFPP additive, T1805, is described as a poly-acrylate, with 'alkyl chains ... perpendicular to the main chain' [62].

Alkyl acrylates and methacrylates polymerise readily under free radical conditions. The only problem is in controlling the polymer molecular weight, which tends to be high. High molecular weight PAMAs are viscous and difficult to handle as fuel additives, so their molecular weights must be reduced by including a chain transfer agent in the polymerisation. Dodecyl mercaptan has been used for this purpose [83] but this would leave too much sulphur in the additive for use in today's ultra-low sulphur diesel (ULSD) fuels. However, there are other methods for controlling the molecular weights of PAMAs that are, as particularly mentioned in the Evonik[26] Rhomax patent [83], ATRP, RAFT and NMP[27]; examples of ATRP block copolymers are also presented in another Rhomax patent, concerning PPD for lube oils and biodiesel [84]. So, it seems possible that Evonik is using one of these processes to produce the

Viscoplex™ MDFIs. Low MW poly-methacrylates may also be made using catalytic chain transfer (CCT), as described by Du Pont [85, 86].

A diagrammatic representation of part of a poly(hexadecyl-acrylate) molecule shows how the molecule is arranged – in solution or when it crystallises – as a double comb structure (left of Figure 4.19). On the right is an end on view of two units, showing how the hexadecyl groups can arrange themselves alongside an n-alkane molecule on its crystal surface; in this conformation, the polymer backbone projects into the liquid fuel, inhibiting the approach of n-alkanes to the wax crystal surface.

Dialkyl-fumarate/vinyl acetate co-polymers (FVACs) were introduced as lubricating oil PPDs in the 1950s; early patents date from 1956 [87, 88]. The free radical polymerisation of a dialkyl fumarate alone is much slower than its copolymerisation with vinyl acetate. Also, the fumarate C = C is electron poor, relative to ethylene, while that of vinyl acetate is electron rich, so these two monomers prefer to react with each other more quickly than with themselves. As a result, in the FVAC, the two monomers are alternating; in polymerisation kinetics, this is described by the reactivity ratios which are low for both the VA and the dialkyl fumarate when copolymerising [77], the product of these ratios being 0.005.

A wide range of FVACs, having varied ester alkyl groups, have been described for fuel applications. FVAC from di-dodecyl fumarate (C12 FVAC [89]), mixed dodecyl/tetradecyl ester (C12/14 FVAC) [90], C14 FVAC and C14/16 FVAC [38], have been described for use in middle distillate fuels of particular boiling ranges, especially specified by their FBPs. The performances of a range of individual alkyl chain length FVACs in low boiling [37] and in high boiling [38] fuels have been described, illustrating the relationship between the longer n-alkanes in the fuel and the length of the comb polymer alkyl groups, which are splayed, alternating at each side of the backbone (a double comb, as in Figure 4.19). The FVACs are usually used in combination with EVAC or ethylene/vinyl ester terpolymer and in WASAs [46]. C16/18 (hydrogenated tallow[28]) and 'behenyl' (a C18/20/22 alkyl mix) FVACs are used as PPDs in heavy fuels and crude oils [91].

Olefin/Maleic Anhydride copolymers (OMAs) are also prepared by free radical polymerisation. The double bond in maleic anhydride is electron poor and those in 1-alkenes are slightly electron rich, though not as much as in vinyl acetate. The resulting polymer is approximately alternating but can carry segments of two or more adjacent maleic anhydride groups. Esterification of the OMA with long chain n-alkanols gives a n-alkyl rich comb polymer – three n-alkyls per four backbone carbon atoms (one from the alkene and two from the maleic di-ester), though it is difficult to attain complete esterification of such polymers. This seems to allow quite a fine tuning of the cold flow activity

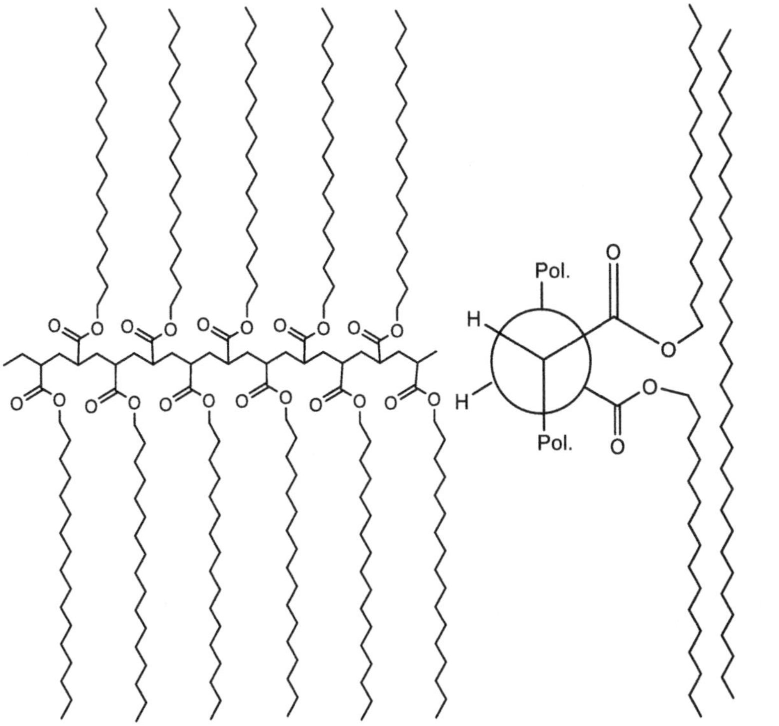

Figure 4.19 Poly (Hexadecyl Acrylate).

of combinations with EVAC, or ethylene/vinyl ester terpolymers, for improved effectiveness in FAME-containing biodiesel [92] and high boiling distillate and residual fuels [93].

4.3.3b Poly-1-Alkenes

The simplest comb polymer structure is that of the poly-1-alkenes (also known as Poly Alpha Olefins, PAOs, see Figure 7.1). However, these have gained only sporadic mention in fuel cold flow improver literature. Until the late 1980s, there were no reliable processes for preparing PAOs of a moderate molecular weight: free radical and cationic processes give extremely low molecular weights, because of rapid chain transfer, and Ziegler-Natta gives high molecular weight, high viscosity polymers. Metallocene catalysts allow easy control of molecular weight by temperature of reaction. Such polymers were the object of a patent for general cold flow applications [94], as a diesel fuel CPD [95] and as a jet fuel CPD [96]. However, there is no indication of commercial application.

Polymerisations on Ziegler-Natta or metallocene catalysts occur as the carbon-carbon double bonds of 1-alkenes (or ethylene) repeatedly insert themselves into the carbon-metal bond of the catalyst [77]. The Ziegler-Natta catalyst is usually based upon titanium or vanadium chloride reacted with tri-ethyl aluminium. Such a catalyst may be supported on small particles of alumina or silica for the ability to produce isotactic, stereo-regular poly-1-alkenes, in particular poly-propylene (the atactic version is soft and non-crystalline while the isotactic version is a strong, crystalline solid). However, it is difficult to produce low molecular weight polymers by such a catalyst – hydrogen may be used to terminate a growing polymer molecule, but the process is not easy to control. Metallocene catalysts are individual, soluble molecules of a metal atom (usually zirconium or hafnium) co-ordinated with various ligands, especially bridged cyclo-pentadiene structures. In the unactivated catalyst, one of the zirconium bonds is occupied by a chloride. To activate the metallocene catalyst, the chloride is displaced by a methyl group, with the aid of a methyl alumoxane[29] co-catalyst; preparative details are described in the patent claiming their potential use as fuel additives [94].

4.3.4 Polar Nitrogen Compounds – Long Chain Alkyl-Amine Derivatives

Long chain, linear alkyl-amines are derived from the acids that are found as glycerol tri-esters in fats and oils. Saturated palmitic (C_{16}) and stearic (C_{18}) acids and their unsaturated versions are the most plentiful in many vegetable oils and in animal fats. Next most plentiful, and in high demand for cosmetic products, are the lauric (C_{12}) and myristic (C_{14}) mixed acids found in coconut oil. Such fatty acids, made by saponifying (hydrolysing with alkali) the natural oils, are mostly commodities. They are hydrogenated to the saturated acids or to saturated linear alcohols or converted to amines. The fatty acid mix from animal fats are known as tallow acids and hydrogenated products derived from them bear the name Hydrogenated Tallow (HT), which is composed of mainly hexadecyl and octadecyl (a ratio of about 30/70) alkyls.

Reductive-amination of the acids yields a mixture of saturated primary, secondary and tertiary alkyl-amines via the amides, which are hydrogenated in the process. Process conditions are controlled to produce mainly the primary and secondary amines. For example, olive oil, mostly glycerol tri-oleate, would be processed by saponification, and glycerol separation, to give, mainly, oleic acid (top Figure 4.20). Reaction of oleic acid with ammonia and hydrogenation gives octadecyl-amine (middle Figure 4.20). A repeat of the same reaction between octadecyl-amine, in place of ammonia, and oleic acid gives di-octadecyl-amine (bottom Figure 4.20). By the same process, tallow would give HT-amine (primary) and 2HT-amine (secondary). Coconut-oil-derived acids are mostly dodecanoic (lauric) and tetradecanoic (myristic) acids, which provide the

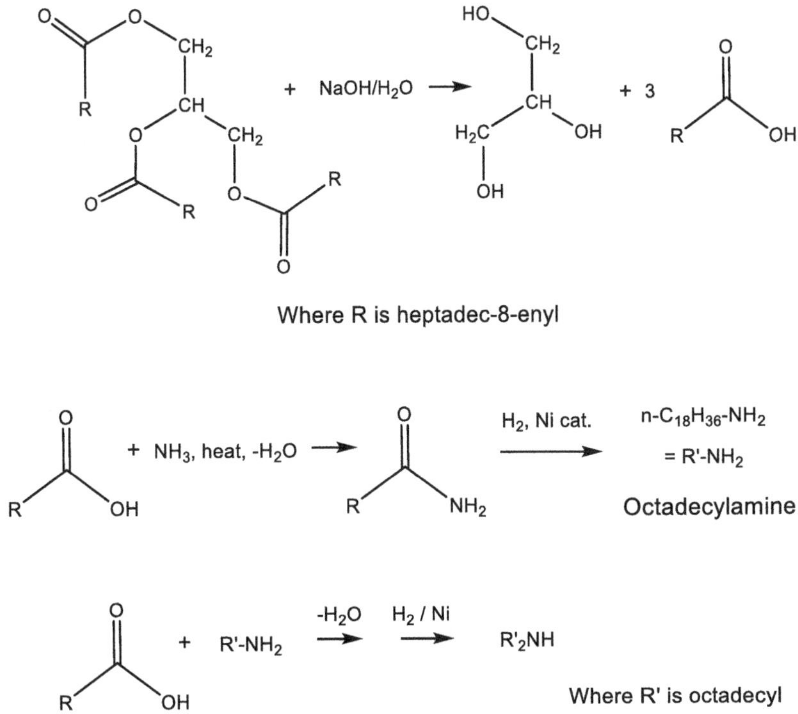

Figure 4.20 Conversion of Olive Oil to Octadecyl-Amine.

so-called C alkyls. So, the reductive amination process yields C (coco) and 2 C (di-coco)-amines. 2HT- and 2 C-amines are the amines usually used in the WASAs described below.

These amines are n-alkane like, since the amine nitrogen is not far from the tetrahedral structure of the carbon atom in a saturated alkane (the nitrogen lone pair occupies one position). So a 2HT amine behaves like a mixture of C_{33}, C_{35} and C_{37} n-alkanes in a paraffin wax crystal. Such amines have been found to act as wax crystal nucleators and as wax crystal growth inhibitors [97]. However, these amines are, perhaps, a little too much like the n-alkanes, fitting into the same crystal structure, and are not effective WCMs. However, various derivatives have been found to be commercially effective WCMs used as WASAs. The most widespread amine derivatives in use as WASAs are the reaction products of 2HT or 2 C amines with poly-carboxylic acids such as EDTA (ethylene-diamine-tetra-carboxylic acid) and, especially, with cyclic anhydrides such as dodecenyl-succinic anhydride or phthalic anhydride, or polymers containing these structures such as the OMAs (section 4.3.3a). Such additives have also been described as wax crystal growth inhibitors [22] and anti-agglomerants [40] [98].

Cyclic anhydrides react readily with two mole equivalents of a 2HT-amine to give a half amide/half amine salt structure. For example, phthalic anhydride reacts with 2HT-amine to give dialkyl-ammonium-2-(N,N-dialkyl-amido)-benzoate (Figure 4.21), in which the alkyls (R_2) are the HT mixtures of octadecyl and hexadecyl [98].

Another WASA is claimed as the product prepared by heating EDTA to 190°C with four molar equivalents of 2HT-amine [72]. The product of the reaction is, presumably, a mixture of several possible amides (mono-, di-, tri-, maybe not tetra because of steric hindrance) and amine salts. This was then described as being used as a WASA, in combination with EVAC or ethylene/vinyl propionate copolymer.

Another WASA, based on the 2HT- (or 2 C-) amine reaction with cyclic anhydrides, is derived from the 1-tetradecene/1-hexadecene (mixed) /maleic anhydride copolymer (C_{14}/C_{16} OMA). The OMA is reacted with a small proportion (5 mole% relative to succinic anhydride groups in the OMA) of polyether-primary-amine (to give cyclic N-polyether-imide groups), and a major amount (0.9 to 1.9 mole per mole succinic anhydride groups) of 2HT- or 2 C-amine [99]. The company describing this WASA has made other references to this structure type: the reaction product of C_{14}/C_{16} OMA with two molar equivalents of 2HT-amine per succinic anhydride group in the OMA (no polyether-amine) [53], and the 2HT-amine reacted with styrene /maleic anhydride copolymer [73]. There have also been research studies on the structurally similar $C_{18}/C_{20}/C_{22}$-mixed-alkyl-methacrylate copolymer with maleic anhydride, post reacted with 2HT-amine, as a WASA – said to combine the WASA-active structure with a wax crystal growth inhibiting, comb-polymer structure [100].

4.3.5 Nucleators

The nucleation event determines the number, hence size, of crystals in any crystallisation process. In wax bearing fuels, the empirical result is that the smaller the shape-modified wax crystals are, the less flow problems they cause.

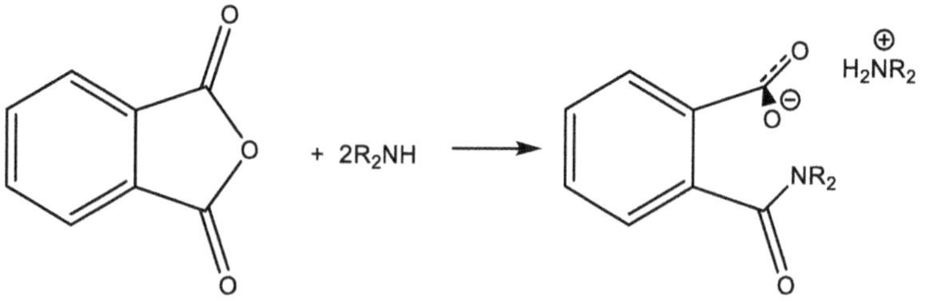

Figure 4.21 WASA Formed from the Reaction of Phthalic Anhydride with 2HT-Amine.

To pre-empt the next section (section 4.4.4a), the degree of super-saturation (chemical potential) needed to start and support the nucleation of crystals is much more than that needed for their continued growth. Thus, in a static tank of fuel, cooling slowly (real world situation), once the primary nucleation process gives way to crystal growth, super-saturation falls and there is no further nucleation (in the absence of settling or agitation). This underlines the importance of the nucleation stage in the modification of wax crystals by cold flow improvers. However, in the CFPP test, the main industry test for middle distillate fuel operability, below its cloud point, fuel is agitated by being repeatedly drawn through a fine mesh – such agitation is expected to cause secondary nucleation [101]. It must be remembered that the CFPP test does not attempt to imitate the real world but is designed to provide a reliable correlation in a practical, rapid test.

In a broad or normal fuel (i.e. one responding to the application of EVAC as a cold flow improver), nucleation depends upon the proportion of the longest n-alkanes present, often n-$C_{28}H_{58}$ to n-$C_{32}H_{66}$ (section 4.4.3). Fuel variations have already been discussed (section 4.2.5) in terms of their distillation characteristics; another way of quantifying this is to compare the amount of higher C-number n-alkanes in a fuel with the amount of the lower C-number n-alkanes that form the bulk of the wax. If the ratio of the total amount of n-alkanes above C_{22} to those C_{18} to C_{21} is less than 0.25, and the gradient of the % n-alkane against C-number from C_{18} to C_{26} is less than −0.3, the fuel is difficult to treat with current commercial MDFIs [28].

The terms 'nucleator' and 'wax nucleating agent' were first applied to an EVAC with a lower vinyl acetate content (17 wt.%), and higher molecular weight (Mn 3,000), than has the pour depressing, wax crystal 'growth inhibiting' EVAC ([21] and polymer A in [60]) (section 4.3.1). As already discussed, the combination of this EVAC with the growth-inhibitor EVAC was discovered, in the early 1970s, to be a highly effective MDFI, in EVAC-responsive fuels [21]. However, there remain a range of EVAC-unresponsive fuels, particularly, the 'difficult-to-treat', extreme narrow boiling distillates that lack the higher n-alkanes – in these, the EVAC 'nucleator' is ineffective. The sizes of wax crystals from such fuels indicate that nucleation is the main problem in such fuels [102] (section 4.4.3). A different kind of nucleator, one that provides wax crystal nuclei, was needed for such fuels. Adding the higher n-alkanes, as paraffinic waxes, was shown to be effective [103], but this presented application problems resulting from the low solubilities of such n-alkanes: they raise the cloud point and are difficult to handle as a concentrate (i.e. in an additive). Combining a PPD with petrolatum, a non-crystalline wax, reduced the amount of PPD needed in one fuel from 0.225% to 0.0125%, had the advantage over paraffinic wax of good solubility, but was needed at 0.6% [81].

It had been found that a number of compounds, in particular poly-ethoxylated long chain esters and alcohols (non-ionic surfactants), could improve the response of such fuels to EVAC-PPDs in the Enjay Programmed Fluidity Test (PFT [43]). Such non-ionic surfactants, bearing a high proportion of long alkyl chains (such as octadecyl) form micellar phases in hydrocarbon solvents, so have better solubilities than long chain n-alkanes. Initially, these materials were off-the-shelf surfactants: matching their descriptions in the PFT patent, they were the long standing products such as Span™ 65 and Tween™ 65 (trade names now owned and described by CRODA [104, 105], which were not optimised for this application.

Developed from this initial clue, a wax crystal nucleator suitable for such narrow boiling fuels was found that lead to a different class of MDFIs for application in such fuels: MDFIs that contained no EVAC. Of the non-ionic surfactants, behenic (docosanoic) acid di-esters of polyethylene glycol (PEG) with molecular weights of 200–1000, were found to be the most effective nucleators in NBD fuels [22]. While such compounds were found to be effective wax crystal nucleators, the small wax crystals formed in their presence were unmodified plates, which gelled the fuel. It was necessary to combine them with a PPD, ethylene/vinyl ester copolymer or a comb polymer, to provide the crystal shape modification needed for the pour point and CFPP depression needed [37]. In a slow cool test (the Distillate Operability Test), it was found that extreme narrow fuels required 1,500–3,000 ppm of the two part EVAC MDFI (i.e. combination of EVAC growth inhibitor and 'nucleator') to produce a passes at −10°C (narrow fuel CPs −5 to −1.5°C); for comparison, only 300–700 ppm of the PEG ester/C12-14 FVAC combination gave passes in the same fuels [37]. Such PEG esters have been included in the boilerplates[30] of many patents, such as [106].

Another variation on this approach was the use of behenate esters of poly-ethoxylated ammonia, reportedly combined with a variety of co-additive PPDs, including poly-alkylmethacrylate or a low molecular weight ethylene/propylene copolymer [107]. For the higher cloud point (−5 to 0 °C) Japanese NBDs, the nucleator/growth inhibitor dual component approach was proposed but with non-EVAC components [22, 37, 108]. The 2018 WDFQS shows that, even now, some of Japanese Grade 2 (the major grade) fuels are very narrow cut and, from the cloud point/CFPP difference, are unresponsive to MDFIs. However, it appears that such additives are not widely used, possibly because of a performance debit or cost, when compared with the inexpensive EVAC.

Quite a different structure has also been found to be an effective co-additive to wax crystal growth inhibitors and is described as a wax crystal nucleator alongside the established EVAC nucleator [109]. This is a hydrogenated di-block polymer made up of a linear poly-butadiene block and, either a poly-isoprene or a branched poly-butadiene block [109, 110]. Such structures appear to have a lot of patent and scientific literature coverage. These hydrogenated block copolymers have useful and scientifically interesting interactions with n-alkanes [111]. The block of hydrogenated, anionically polymerised 1,4-butadiene is crystalline, just as is low molecular weight, linear poly-ethylene ('PE') – it is a long chain n-alkane; the molecular weight of this block is about 1,500, so approximately $C_{100}H_{202}$. The second block with molecular weight of about 5,000 is non-crystalline, as a result of extensive branching resulting from being either hydrogenated poly-isoprene, whose structure is like a poly-ethylene/propylene ('PEP') copolymer, or from hydrogenated poly-butadiene made under conditions promoting 1,2- monomer additions, which has a structure like a poly-ethylene/1-butylene copolymer ('PEB') [109, 110]. Hence the acronyms PEPEP and PEPEB are used for these polymers; as with the PEG esters, these too have entered the boilerplate of coadditives for use in MDFI blends.

There are also claims for block co-polymers of hydrogenated poly-butadiene block with polystyrene [53, 112]. Here, the poly-styrene blocks are insoluble, forming micelles with the poly-butadiene blocks (after hydrogenation) projecting into fuel. The molecular weights of these polymers are claimed to be in the region of 1,000 to 100,000 but those in the examples are much higher (24,000–90,000) than those just discussed, so would be expected to give problems of solubility, viscosity and compatibility with other components in additive blends. While interest has been shown in such polymers over the years, it is sporadic and not followed up by groups of patents and widespread references as blend components in other patents. So it is probable that such polymers are not widely used commercially.

These di-block polymers are made by anionic polymerisation (Figure 4.22), which is a living process. Polymerisation is usually initiated by butyl lithium, which produces an anion with the first monomer, here butadiene, which adds another monomer, which then bears the anion, and this continues until all the monomer is used up. At this point, a second monomer, for example isoprene, can be added and the anionic end group, still active ('living'), grows the second polymer (or block) on the end of the first. Hydrogenation, to obtain the final product, is a standard process using hydrogen under pressure, at an elevated temperature, over a hydrogenation catalyst (Raney nickel, Raney cobalt, or palladium or platinum on a support). Styrene is much more difficult to hydrogenate so would be unchanged by the conditions used.

In hydrocarbon solvents, the poly-butadienyl and poly-isoprenyl lithium growing chains are mainly dimeric and the polymerisations proceed 90%–95% 1,4-addition and 5%–10% 1,2-addition of monomer [113]. When the lithium ion is solvated by adding an amine or an ether, such as

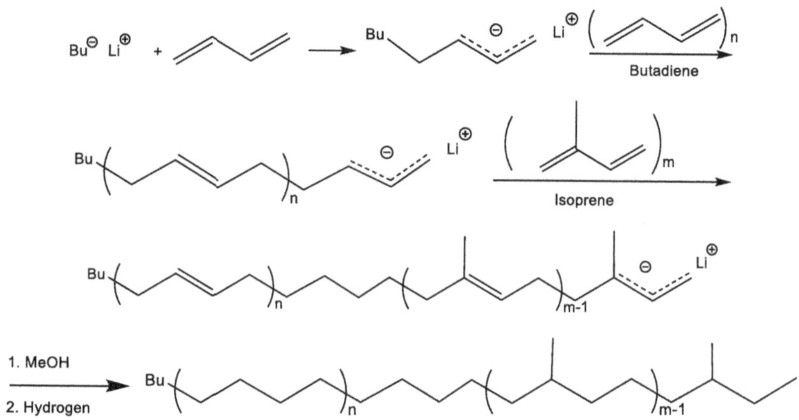

Figure 4.22 Anionic Polymerisation of Butadiene-block-Isoprene and Hydrogenation to PEPEP.

tetra-hydro-furan (THF), and the temperature is lower, the proportion of 1,2-addition increases. For butadiene, it can increase to 73% with amine or 37% with an ether, in hydrocarbon solvent (at 30°C). So, the branched, non-crystalline co-block may be made by adding a tertiary amine, such as triethylamine, to the polymerisation after the mainly 1,4-butadiene polymerisation stage. More butadiene is then added for the branched structure resulting from about 50% 1,2-additions. Polymerisation is followed by hydrogenation to give PEPEB [110].

Interestingly, PEPEP, PEPEB and PEG esters are also described as wax crystal nucleators and components for use in cold flow blends for jet fuel [33].

4.3.6 Alkylphenol-Formaldehyde Condensates (Apfcs)

Alkylphenols react readily with formaldehyde (acid or base catalysis) to give, initially, low molecular weight polymers in which the phenols are linked through methylene groups[31]. APFCs with long n-alkyl groups, for example octadecyl, are claimed to be effective pour depressants for lubricating oil [114]. However, such polymers do not seem to have gained commercial use as MDFIs. It is, then, surprising to find the, apparently, widespread use of APFCs as standard, commercially used cold flow improvers, as indicated by their inclusion in cold flow patent boilerplates[30] [79]. These APFCs are nonyl-phenol-formaldehyde condensates in which the nonyl is the branched tri-propylene (a commodity refinery product) which would be expected to have no WCM activity. The process is an acid or base catalysed condensation

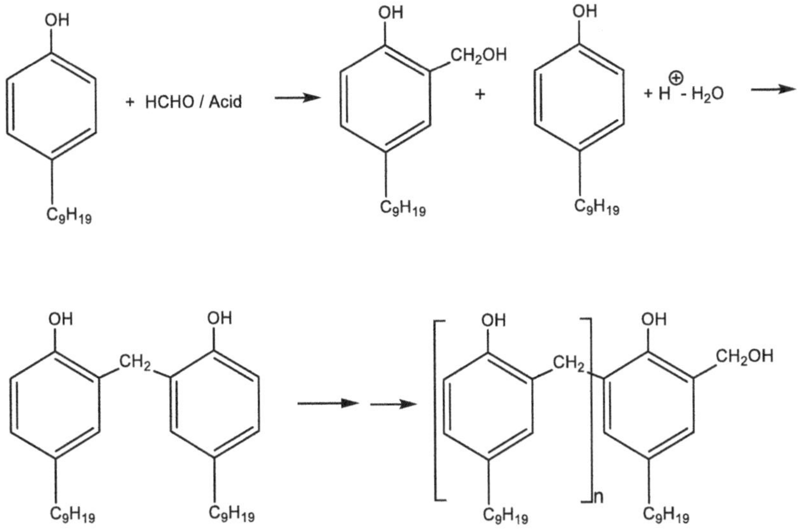

Figure 4.23 Alkylphenol/Formaldehyde Condensation Polymer.

reaction [114]; when acid catalysed, one mole of formaldehyde, protonated, undergoes nucleophilic attack by the reactive 2-position of the alkyl phenol to give the phenol substituted methanol (Figure 4.23). After protonation, water is displaced from the hydroxy-methylene group by another alkyl-phenol to give a methylene bridged di-phenol. This process repeats several times to give an oligomeric structure (n in the region of 7 to 12, giving a molecular weight of preferably about 2,000, as used in examples [45]) known as an alkyl-phenol-formaldehyde condensate.

In an extensive patent on APFCs [45], with many tests in eight fuels, the combination of EVAC (standard MDFI) and APFC gave no significant improvement in CFPP over the EVAC alone. However, the APFC/EVAC combinations did appear to improve WAS over EVAC alone in some fuels, but the major advantage was gained when the APFC was combined with both EVAC and WASA ('polar nitrogen compound'), improving both CFPP depression and WAS in comparison with EVAC or EVAC/WASA without the APFC. It was claimed that the effect of the APFC was to solubilise the polar nitrogen compounds in paraffinic and aromatic solvents. APFCs are most often referred to in combinations with EVAC and polar nitrogen compounds in WAFI blends, which may also be used to enable north American fuels to past the stringent low temperature filterability test (LTFT) [52].

APFC derivatives are also used as demulsifiers (Chapter 7), to prevent the take up of water into fuels as an emulsion [115].

4.4 Mechanism of Wax Crystallization and Modification

4.4.1 Wax Crystal Compositions and Structures

4.4.1a Compositions

N-alkanes are, individually, the most plentiful components in middle distillates. This is clear from middle distillate Gas Liquid Chromatograms (GLC, for example Figure 4.24), from which their proportions can be determined. GLCs of waxes separated from diesel fuels (or heating oils), 10°C below the fuel CPs, show that they contain up to 95% n-alkanes that come, preferentially, from the high carbon number end of their distribution in the fuel [116, 117] (for example Figure 4.24).

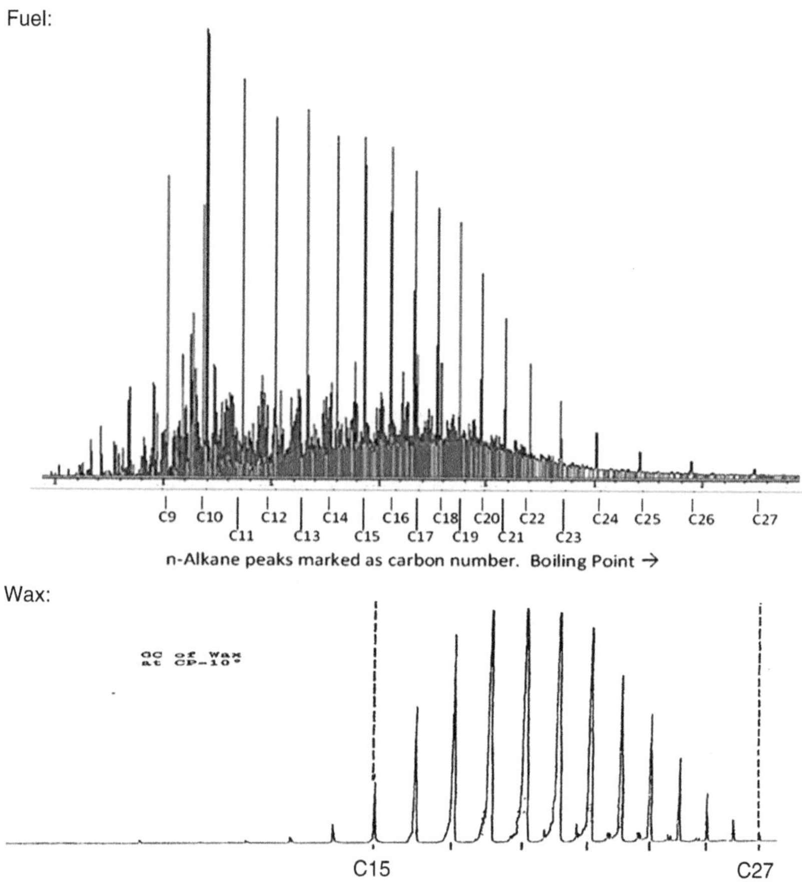

Figure 4.24 Diesel Fuel and Wax Separated, Gas Liquid Chromatograms.[32]

In the earlier discussions of middle distillate fuel characteristics, the distillation range was used to describe fuels by their low and high FBPs, broad and narrow distillation ranges and distillation cut-offs (Figure 4.5). When considering the details of cold flow and wax, it is useful to consider the range of n-alkanes in the fuel. The approximate relationship between the n-alkanes and fuel distillation type is indicated mainly by the highest n-alkane in the fuel, which will also separate in the wax (Table 4.4) [9].

The lowest n-alkanes are usually about C_9, as found in kerosene, which is a usual diesel blend component. Exceptions are certain high FBP fuels, with high cloud points, that may be found in hot countries. Variations in distillations and n-alkane distributions of blended diesel fuels result from differences in the distillations and proportions of the blend components. This may be illustrated by an example of a simple refinery that would make diesel fuel blends out of its components: heavy naphtha, kerosene, Light Gas Oil and Heavy Gas Oil (n-alkane ranges Table 4.5 [118]); these components could all contribute to a low CP winter blend, based on the three lowest boiling components, with sufficient of the HGO to provide MDFI response; in summer, the heavy naphtha may all end up in the gasoline, while the LGO and HGO might have raised FBPs (to increase volume) and, together, form the basis of the diesel blend, with the kerosene controlling CP. In another refinery, in a different geography, this HGO might be called a middle gas oil, and its HGO would have a higher boiling (and n-alkane) range.

In detail, blends of these components (Table 4.5) showed that more of both LGO and HGO increased the cloud point and the CFPP[33] (with or without MDFI) [118]. Also, it was shown that the best CFPP response to MDFI is for the fuel blends with about 10% of HGO, which gives the fuel an n-alkane distribution that has a long, low tail at the high carbon number end. CFPP response to the fuel blend with no HGO is not given but, from previous experience, it would be an unresponsive extreme NBD (FBP $< 308°C$); indeed, a blend with only 5% HGO shows no CFPP response to 500 ppm of Keroflux M™ (an EVAC).

Table 4.4 Highest n-Alkane in Different Middle Distillate Fuel Types.

Fuel Distillation Description	Highest n-alkane (C number)
Extreme Narrow:	≤ C26
Narrow/Normal:	C27-30
Broad Boiling/Normal:	C30-32
High FBP:	C32-36
Extreme FBP:	≥ C36

Table 4.5 n-Alkane and Distillation Ranges of Simple Refinery Middle Distillate Fuel Components.

Component	n-Alkane Range	IBP, °C	T90%, °C	FBP, °C
Heavy naphtha	C8 to C14	132	170	182
Kerosene	C9 to C16	165	219	229
Light gas oil	C10 to C20	237	293	308
Heavy gas oil	C13 to C28	282	370	(T95%, 379)

Overall, this reference illustrates the ability to match cloud point and CFPP (with MDFI) to specifications while optimising the proportions of HGO and LGO (hence diesel fuel yield) in the blend.

Waxes at 10°C below the fuel cloud point contain a range of the highest n-alkanes in the fuel, from the highest carbon number present down to around hexadecane (Figure 4.25). As a result, the waxes from the higher boiling fuels usually have a broader range of n-alkanes and, because of the much lower solubilities of long chain n-alkanes than of the shorter chain n-alkanes, such wax is a lower percent of the fuel (at 10°C below the fuel cloud point [9]). Wax n-alkane distributions in the various diesel and heating oils have been described [9] and can generally be illustrated as normal distributions (Figure 4.25).

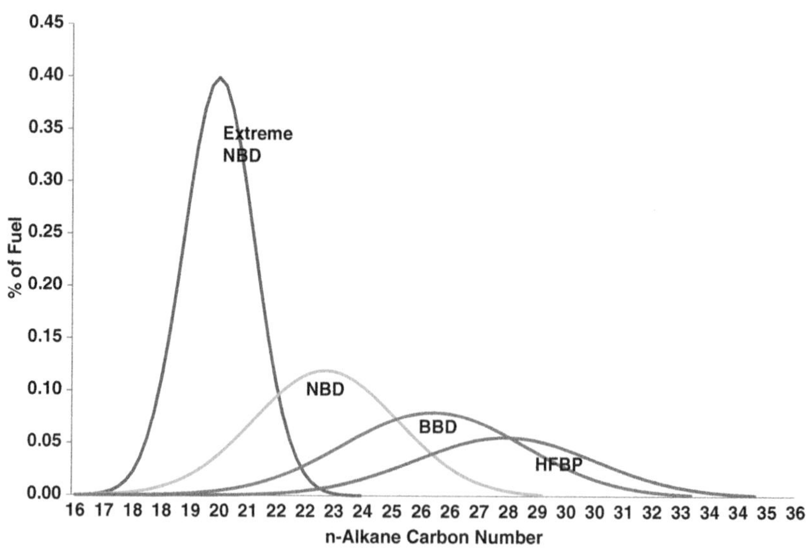

Figure 4.25 N-Alkanes in Waxes from Different Middle Distillate Types.

These are reasonable representations of middle distillate fuel n-alkane distributions – which vary according to the distillation cut points (boiling point ranges) of each of the components, their proportions in the fuel blend and distillation efficiency (overlap of adjacent component distillations). Often, such distributions are skewed, showing a sharper cut off at the low C-number end and an extended tail (kurtosis) at the high C-number end, depending upon the efficiency of the refinery's pipestill and the blend makeup. Jet fuel n-alkane distributions have a similar pattern to those in an extreme NBD but with C_{10} to C_{15} n-alkanes [119].

4.4.1b Structures

The structures of crystals of n-alkanes and their mixtures, including paraffin waxes, have been much studied (see reviews [120–122]). In summary, individually, n-alkanes with even carbon numbers (C_{16}–C_{30}) take up monoclinic or triclinic structures, while those with odd carbon numbers, higher carbon numbers and mixtures of close homologues ('waxes') take up orthorhombic structures. The mixtures of homologous n-alkanes in a wax are randomly dispersed in a single orthorhombic structure, termed a solid solution [117, 121, 123]. A solid solution is a single crystal structure in which different, but similar molecules (as middle distillate wax n-alkanes are) randomly occupy the crystal sites. Wax does not contain separate crystals of nonadecane ($C_{19}H_{40}$), eicosane ($C_{20}H_{42}$), heneicosane ($C_{21}H_{44}$) etc. but crystals of a single, average structure containing all (or most) of those separating as wax. Thus, the X-ray diffraction pattern of a wax, containing $C_{18}H_{38}$ to $C_{28}H_{56}$ n-alkanes, produces only one low angle diffraction band (and overtones) due to the crystal long axis [117].

The lower solubilities of the higher carbon number n-alkanes present result in them forming the distribution of n-alkanes in the solid solution wax that separates first, with falling temperature. As temperature falls further, this distribution shifts down in average carbon number, causing the n-alkane distributions of wax separating at lower temperatures to contain progressively lower carbon numbers (Figure 4.26, data [116]). Mathematical models of the compositions and quantities of waxes separating from diesel and jet fuels, based upon the solid solution assumption and using the n-alkane distributions in the fuels (obtained by GLC), give good correlations with experimental results [124, 125]. Discussions of wax crystals will assume this single structure.

The orthorhombic crystal structure is one in which the n-alkane molecules, whose structures resemble cylindrical rods, stack up next to each in an approximate hexagonal close packing, in layers. Layers form on top of each other by a BCF mechanism, which is described below under growth (section 4.4.5). The region between the layers is a region of low crystallinity as it contains the

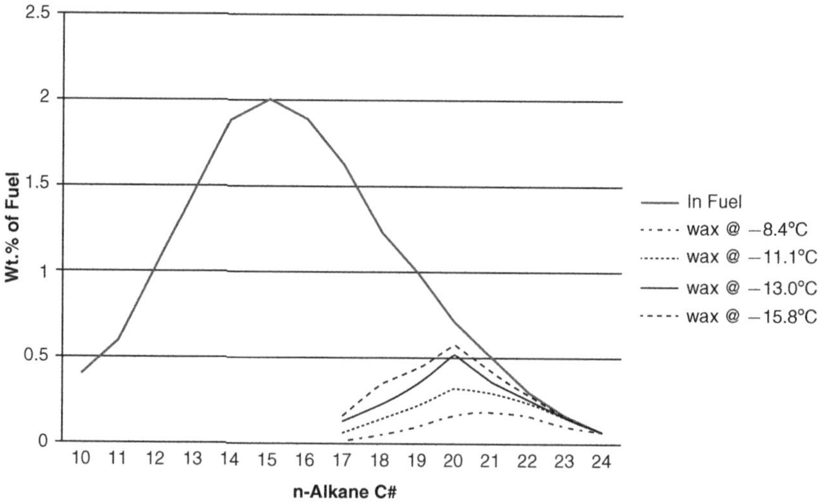

Figure 4.26 n-Alkanes in a Diesel Fuel and Waxes at Various Sub-Cloud Point Temperatures.

excess lengths of alkyl chains over the average for the solid solution; it also contains segments of molecules moving between lamella (interchain mixing), and the effects of chain twisting (away from trans–trans) and of chain folding [121]. While the layers form by hexagonal close packing, so would be expected to form hexagonal shapes, the untreated wax crystals usually appear as rhombi (parallelograms, schematic in Figure 4.30, and actual in Figure 4.29) because the [010] faces bounding the AB plane grow out very quickly, leaving the [110] and [111] faces (A and B directions) which grow quite quickly, and the [001] face (C direction), which grows slowly [117].

Attractive interactions between n-alkane molecules are weak van-der-Waals (dispersion) forces. However, because of the regularities of their structures and the multiplicity of methylene-methylene interactions, the attractions in the plane perpendicular to their molecular axes (A and B directions in Figure 4.30) are quite strong; for example, there are twenty methylenes for each docosane, $C_{22}H_{44}$, molecule with which it interacts with its neighbours. At the same time, the attractions between the layers (C direction) come only from single, end methyl-methyl interactions, so are weak. Growth rates depend upon the strength of attraction that an incoming molecule has to an existing crystal face ('attachment energies' [126, 127]), so n-alkane crystals grow rapidly in the plane perpendicular to their molecular axes (the AB or 001 plane) and slowly in the direction parallel to these axes: n-alkane crystals grow as thin plates, in layers (lamellae).

4.4.2 The Crystallisation Process

Bulk crystallisation is the result of two processes: nucleation and crystal growth. Nucleation determines the number of crystals hence, for a given quantity of wax, their sizes. Rates of growth in the different directions, on each crystal face, determine crystal shapes.

The driving force for crystallisation is supersaturation [101]: the actual concentration of n-alkanes (C_T) in the solution (fuel) at a temperature (T) is compared with the thermodynamic saturation concentration (C_{Tsat}) at that temperature:

- S, super-saturation ratio, = C_T/C_{Tsat}
- б, relative or absolute super-saturation, = $(C_T\text{-}C_{Tsat})/C_{Tsat}$. (x 100 for %).

Before any wax crystallises, as the temperature of the fuel falls, so too does C_{Tsat} while C_T remains constant – given by the character of the fuel. Below the saturation temperature T_{sat} at which C_{Tsat} is less than C_T, S becomes greater than 1.00 and the fuel is super-saturated in wax. This provides the thermodynamic driving force for crystallisation. For crystallisation driven by falling temperature, the degree of undercooling, below the saturation temperature, or 'super-cooling', are also used to describe the level of super-saturation:

Super-cooling (°C) = Solution saturation temperature (T_{sat}) – observed temperature (T)

Once crystallization is underway, it continues while cooling continues. When the temperature stops falling, super-saturation moves to zero as crystal growth takes up any solute concentration in excess of the saturation concentration, at that temperature, then growth stops.

4.4.3 n-Alkane-Wax Nucleation

As the temperature of static fuel falls slowly – for example, overnight cooling in the fuel tank of a static vehicle is about 0.5 to 2°C/h – and it moves below the wax saturation temperature, wax molecules come together, reversibly, in clusters [117]. During further cooling (to about 2 to 5°C below Tsat.), super-saturation increases to a relatively high level, providing the driving force for these clusters grow as wax crystal nuclei. Their growth is limited by the increase in surface energy that occurs as the surface area of the cluster increases, which depends upon the second power of the cluster's linear dimension (diameter, d). However, the energy of attraction between the molecules in the cluster depends upon its three-dimensional bulk, i.e. to the third power of the cluster's linear dimension. As a result, while surface energy dominates at the very smallest sizes, the bulk energy of attraction increases more rapidly with size until a critical size, after which the clusters continue to grow, taking up the wax in solution that is in excess of its solubility, its supersaturation.

Once these wax nuclei are larger than a critical size (critical nuclei sizes are estimated to be 100–300 molecules [128]), the clusters become ordered and they start to grow as crystals. Crystal growth requires a much lower level of super-saturation than does nucleation, so growth is, initially, rapid until the excess super-saturation is much reduced. On observing this process through a microscope, one sees small wax crystals suddenly appear everywhere in a fuel sample, growing rapidly for a few minutes [117]. Growth then slows down to a rate that closely follows the falling solubility of wax, in the fuel, with falling temperature. Once crystal growth is established, super-saturation is too low to support further nucleation, so crystal sizes are determined by the number of viable crystal nuclei that first form, the proportion of wax-forming n-alkanes in the fuel, and how far cooling continues below Tsat (close to the cloud point).

The nucleation process is the key to understanding the difference between a MDFI-responsive BBD and an unresponsive, extreme NBD. While the wax content (ten degrees below the cloud point) of an extreme NBD is only twice that of a BBD, the wax crystals from such a fuel are about 1,000 times larger, in volume (about 10x in each dimension) [102]. Arithmetically, this translates to about 500 times more crystals in a gram of BBD fuel than in a gram of NBD fuel[34] (at 10°C below cloud point). N-alkane distributions of these fuels differ, mainly, in the highest n-alkane carbon numbers: the extreme NBD wax has C_{16} to C_{25} and the BBD has C_{16} to C_{32} n-alkanes (Figure 4.25). This is related to the difference between their rates of nucleation.

An equation has been developed that relates the rate of cooling to the degree of super-cooling below T_{sat} before crystallization is observed [101]:

Log (rate of cooling) = constant + m.Log (super-cooling)

Such data was obtained for fuels by Differential Scanning Calorimetry (DSC) [102]. Wax separation from an unresponsive NBD, in the DSC experiment, shows a relatively high level of super-cooling (plot is to the right of the graph, Figure 4.27, using reported data [102]) which changes only a small amount for a large change in the rate of cooling (m, gradient = 40). A responsive BBD, however, shows lower levels of super-cooling (plot to the left of that for the NBD) which change much more with change in rate of cooling (m = 7). Such a difference is similar to that discussed by Mullin for unseeded (high m) and seeded (low m) nucleation of ammonium sulphate from aqueous solution [101], from which it is proposed that the BBD is showing seeded nucleation while the NBD is not.

The obvious candidates for such seeding are the higher n-alkanes present in the tail of the BBD n-alkane distribution, but not present in the extreme NBD

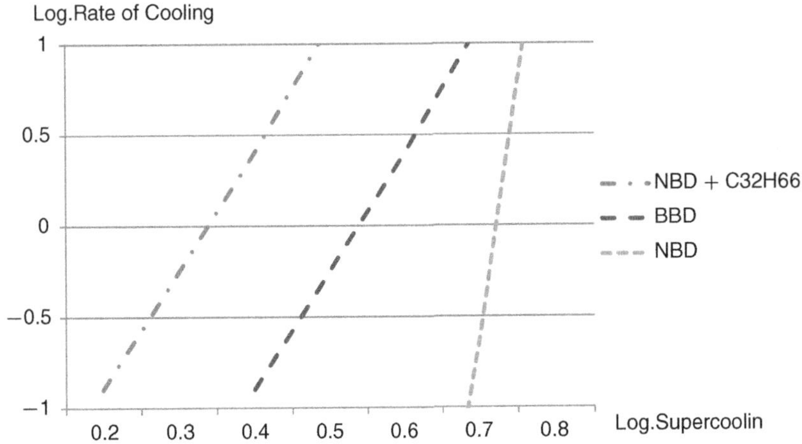

Figure 4.27 Nucleation Kinetics for Wax Separations from Extreme NBD and BBD Diesel Fuels. (diagrammatic representation of reported plot [102]).

(Figure 4.25). In support of this theory, when 200 ppm[35] of n-do-triacontane (n-$C_{32}H_{66}$) is added to the extreme NBD fuel, the rate of cooling/super-cooling plot shows much lower levels of super-cooling, with a gradient similar to that of the BBD (m = 7) (Figure 4.27). Relating this to cold flow improvement, n-$C_{32}H_{66}$ itself has been shown to provide cold flow improvement to EVAC MDFIs in diesel fuel [103]. It has been reported that in a 'normal' (BBD) middle distillate fuel (jet or diesel) the higher n-alkanes are mostly present in the earliest separating wax (as the temperature falls). This what may form seed nuclei, followed by a gradual shift down in the n-alkane range during crystal growth. However, in the NBDs, the n-alkane distribution varies little with falling temperature, showing no initial higher n-alkane separation that might seed nuclei formation [124].

4.4.4 Effects of Additives on Nucleation

4.4.4a EVAC Nucleator

The term 'wax nucleator' was first applied to an EVAC 'nucleator' in 1976 [21] (section 4.3.5). In the EVAC-nucleator patent, the advantage claimed for an EVAC, having 16 wt.% vinyl acetate and 3,000 Mn, was its ability, *in combination with the existing growth inhibiting EVAC*, to improve the treatment of a test fuel (no distillation data given[36]), as measured by the CFPP test. Also, it was claimed to advance the wax appearance temperature, as measured by DSC (at 10 °C /min cooling), over that of the untreated fuel and of fuel with growth inhibitor alone. As already discussed, (section 4.4.1a),

EVAC-MDFI response depends upon the n-alkane distribution of the fuel. Responsive fuels have relatively broad n-alkane distributions containing higher n-alkanes (C28 and above) while unresponsive fuels lack these higher n-alkanes. The mechanistic difference between responsive and unresponsive fuels is that of a difference in nucleation, which depends upon the presence these higher n-alkanes.

MDFI nucleators could have one of two effects: they could interact with the higher n-alkanes in the BBD fuel, to produce more nuclei from them, or they could provide a further source of seeding hetero-nuclei (as was shown for n-dotriacontane). It is possible that the nucleation claimed for EVAC copolymers is of the former type, since EVAC is generally a wax crystal growth inhibitor so should inhibit the growth of nuclei and extend the supersaturation period, which could enable the birth of further nuclei (a mechanism previously described [129]). One author says that EVAC delays nucleation kinetics [130], others state that EVAC can either provide nuclei for n-alkane crystals or inhibit their growth[37] [131]. Often, the interpretation of the effect of EVAC depends upon advancements or depressions of the onset of crystallisation by cooling in a DSC, at rates much greater than usually experienced by fuels: one uses 30°C /h [131] and another uses 120°C /h [132] (2°C/min). Under these conditions, advancement or depression by an EVAC depends upon fuel type [132]. Unfortunately, the DSC can only observe the heat flow during crystal growth so extrapolation to the nucleation stage is only tentative.

Comparing the nucleator EVAC polymer structure with that of the growth inhibitor EVAC: while the nucleator EVAC contains less acetate groups per 100 backbone methylenes (three) than growth inhibitor EVAC (eight) it is, however, made at a higher temperature (140 to 150°C [30]) so it has a higher degree of alkyl side branching of eight [59], which gives a total of eleven interruptions per 100 backbone methylene groups. A 'nucleator' EVAC is, therefore, expected to have a similar degree of crystallinity as the growth inhibitor EVAC. However, all EVACs used in MDFIs have low levels of crystallinity because of their high degree of branching: 10–12 interruptions per 100 backbone methylenes, alkyl plus ester groups, which suggests a crystallinity of less than 10%[38] [133]. In ethylene/n-alkene (such as n-butylene) copolymers, with more than 4.5 mole% co-monomer branches, no crystal lamellae can be observed in the polymer by transmission electron microscopy – as they can be at 2.2mole% branches [133]. In other words, such polymers are unlikely to form crystalline particles that can act as seed nuclei.

The conclusion reached in a study of n-alkane/growth-inhibitor-EVAC interactions is of an interaction between EVAC and n-alkanes at the nucleation stage and 'Co-crystallisation provides a strong enough EVAC-paraffin interaction that could be the microscopic origin of the EVAC activity'[39] [132]. The higher molecular weight of the 'nucleator' EVAC makes it less soluble – phase separation of polymers from a solvent is a consequence of their high molecular weights, regardless of crystallinity [77]. Altogether, this suggests that the nucleator EVAC is not a source of primary wax crystal nuclei but, by inhibiting the growth of first-formed crystals of the less soluble, higher C-number n-alkanes, it increases the number of nuclei, hence more, smaller wax crystals.

Turning back to NBD fuels, that lack the high C-number n-alkanes, such fuels have a problem with nucleation when compared with BBD fuels. However, the EVAC 'nucleator' is ineffective in such fuels, as was demonstrated in slow cool testing, in comparison with a PEG ester additive [37], and by its inability to provide CFPP depression (at 500 ppm) in extreme narrow fuels [22].

4.4.4b Nucleator Additives with Crystallinity, PEG Esters and PEPEP

An EVAC nucleator, combined with an EVAC growth inhibitor, is effective in fuels with higher FBPs that contain higher n-alkanes, but is ineffective in NBD fuels that lack higher n-alkanes. Fuels that are designed to be responsive to EVAC type MDFIs contain higher n-alkanes the distillation tail. In the refinery, this may be the result of experienced blending of components, or the efficiency of fractionation in the pipestill[40], especially of the highest boiling component, or both. In the absence of higher n-alkanes, extreme cases, such as that of a Chinese waxy extreme NBD (distillation FBP 358°C and highest n-alkane C_{24}), show no response even to very high concentrations of standard EVAC-MDFI; such fuels as this can, however, be made responsive by adding n-alkanes directly [134]:

- CFPP both untreated and with 1500 ppm EVAC: −1 °C
- CFPP with 1500 ppm EVAC plus 0.5% of a paraffinic wax: −12 °C

(the wax had C_{20} to C_{38} n-alkanes, distribution peaking at C_{26})

This is consistent with the picture of EVAC interacting with the higher n-alkanes to provide more nuclei. Such a picture has led to the proposal of the balance of higher n-alkanes to the lower, bulk-wax-forming n-alkanes in a fuel as a measure of fuel responsiveness [28].

Extending the n-alkane distribution at the high carbon number end in an unresponsive NBD with the missing higher n-alkanes provides fuel responsiveness to MDFIs. So, a synthetic material that provides seed nuclei for wax crystals should also aid response to MDFIs, where response is lacking. Such a material would need n-alkane character but should also have reasonable solubility for use in additive concentrates, which individual n-alkanes as do-triacontane or higher

C-number paraffinic waxes do not have. As described above, the non-ionic surfactant system of PEG esters or poly-ethoxylated long chain alcohols provides such properties. At ambient temperatures, in solution, such non-ionic surfactants are soluble as micellar structures, while at lower temperatures they form structures like n-alkane crystals [135].

PEG esters and poly-ethoxylated long-chain alcohols, in solution, take the form of cylindrical micelles that have a structure much like that of an n-alkane crystal or nucleus (Figure 4.28). As such, they form a good template for the adsorption of sufficient wax n-alkanes to form wax crystal nuclei (heneicosane, n-$C_{21}H_{44}$, is chosen in the figure as this is about the average chain length of many middle distillate waxes). The nucleation kinetics experiment of Figure 4.27 shows the PEG ester to have the same effect on the NBD as does do-triacontane.

The PEPEP or PEPEB additives are also described in cold flow practice as nucleators which are particularly effective at improving the CFPP depressing effectiveness of growth inhibitor EVAC [109, 110]. Such polymers are not non-ionic surfactants but are all-hydrocarbon. Much detailed research has been carried out on this system using small angle neutron scattering (SANS) [111, 136]. Such molecules self-assemble in solution to form 'hairy platelets' (the polyethylene chains form the platelets and the PEP or PEB chains form the hairs) which act as seed nuclei for the crystallisation of n-alkanes in the fuel. The structural association of PEPEP and n-alkanes, in solution, has been demonstrated with the aid of neutron scattering [111]. In summary of the average structural details of a PEPEP, based on a 1% solution in decane below 35°C, were found to be:

- Platelets are 300 nm across, with the crystalline polyethylene core 5 nm thick and the non-crystalline PEP chains forming a 10–25 nm layer on either side.
- The platelets stack to form needles, whose lengths are 'several microns' and have a repeat distance of 40–100 nm.

A conference presentation explains the structure and effect of these polymers on paraffin crystallization, as determined by neutron scattering [137]. In this presentation, the polymer is referred to as a PARAFLOW™, a commercial additive nomenclature.

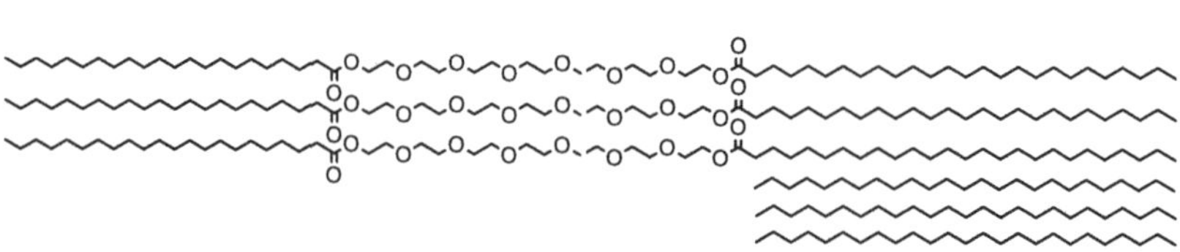

Figure 4.28 PEG Di-behenate Micelle with Adsorbed n-Heneicosane Molecules.

4.4.5 n-Alkane-Wax Crystal Growth

4.4.5a Comparison of Untreated and WCM Treated Wax Crystals

Most wax crystal modifiers (WCMs) are wax crystal growth inhibitors. Apart from their effects on nucleation, just discussed, the main effect of EVAC and other ethylene copolymers or comb polymer WCMs is to control the shapes of the growing wax crystals. Without WCMs, n-alkane waxes that separate from middle distillate fuels form large, thin plates – about 0.3 to 0.5 mm across in a normal diesel fuel. With WCMs present (shown to adsorb onto the wax crystals [131]), the wax crystals are smaller, compact prisms, usually less than 50 microns long and 5–10 microns thick (Figure 4.29) or even cubes (see WASA effect, Figure 4.34). While wax crystals are usually smaller in the presence of WCMs because of more nucleation, the change in shape from thin plates to

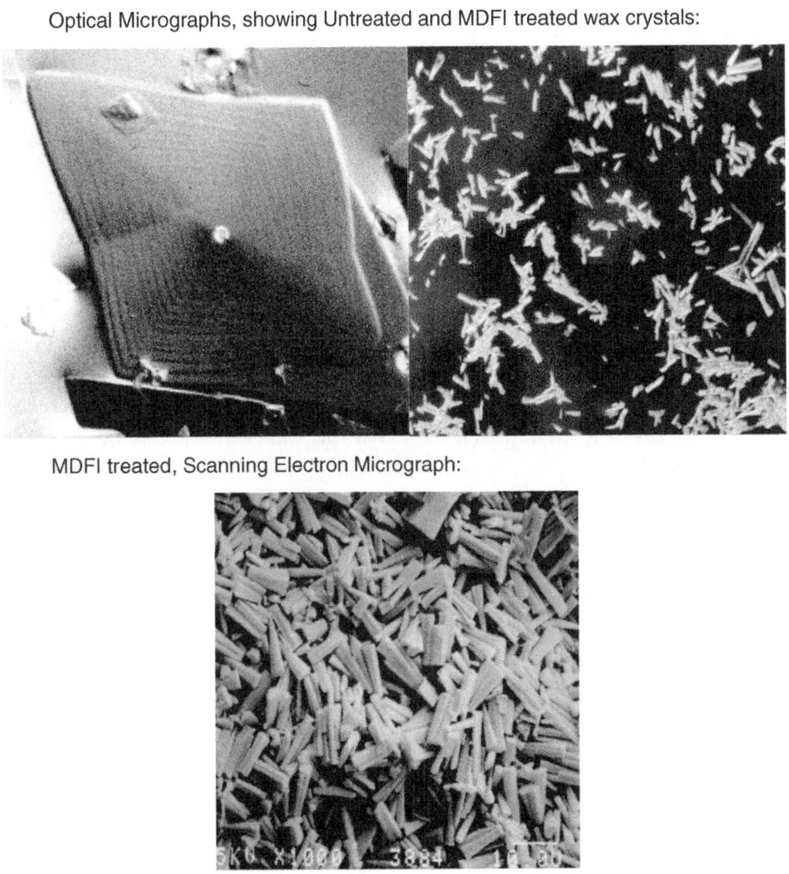

Figure 4.29 Untreated and MDFI Treated Wax Crystals in Cold Diesel Fuel.

compact needles itself is enough to prevent gelling and thus reduce the pour point. A mass of thin crystal plates growing in a liquid, at random orientations, very soon grow into each other, immobilising the liquid in boxes composed of these crystal plates. As a result, pour points of untreated middle distillate fuels are about 5°C below the cloud point (i.e. less than 1% wax). Such a gel structure has difficulty in forming when the crystals are compact prisms, as they are in a treated middle distillate, then the pour point may be 30°C, or more, below the cloud point.

The change in shape confers another advantage to the wax crystals separating from a middle distillate fuel. Above the pour point, the wax crystals are carried along with the fuel, in suspension, to the fuel filter. Vehicle fuel systems are designed to remove particles that are 2 microns across and larger – such particles of grit would cause significant wear in the high-pressure fuel injection pump. So, all wax crystals are held up at the fuel filter. With crystals in the form of thin plates, wax rapidly coats the surface of a filter with a continuous, impermeable glaze, which prevents the flow of fuel. In the form of small prisms, however, wax forms a porous filter cake (SEM in Figure 4.29) that allows fuel to pass through easily[41]. As a result, a diesel test vehicle, using untreated fuel, failed at only 1°C below the cloud point [31], while a diesel vehicle using fuel adequately treated with MDFI may operate at 10–15°C below the fuel cloud point [9]. Filters in heating oil systems are less stringent than diesel vehicle filters, though they would also be blocked by wax plates in untreated fuel.

The space in a fuel filter is limited. If the MDFI-modified crystals are large (as EVAC treated NBD wax, Figure 4.36), they form a porous filter cake, but they also take up a lot of volume and may fill the vacant space between the filter and its housing. When this happens, the wax will back some distance down the narrow fuel pipe, causing a blockage. Smaller prisms take up less space while still forming a porous filter cake (Figure 4.29), which increases the tolerance of the filter for a larger amount of wax.

4.4.5b Mechanism of Crystal Growth

Much work has gone into the understanding of the mechanisms by which crystals grow – especially at low super-saturations, in static systems. Middle distillates usually experience slow rates of cooling in car, truck and storage tanks in the open, as overnight temperatures drop by 5–20°C over 12 hours. Under such slow cooling conditions, super-saturations (after nucleation) are typically low. Earlier theories were that crystal faces grow by a process of 2-D nucleation on their surfaces, followed by growth of these outcrops. However, it was shown that growth by such a process can only take place at much higher levels of super-saturation than those usually experienced during controlled crystal growth [4,129].

Understanding of crystal growth at low supersaturations came from the introduction of the premise that crystals are not, in fact, perfect – they contain various imperfections called dislocations. Dislocations, at which adjacent layers of molecules are not crystallographically aligned, give rise to permanent steps on a crystal face that extend as molecules of solute are added (Figure 4.30). The formation of such dislocations on n-alkane crystals, grown from solution, and their effects on growth in monomolecular layers, has been investigated by electron microscopy [1–3]. This investigator observed the appearance steps of monomolecular-height (i.e. of the n-alkane extended chain length), in a helical arrangement, on crystal plates a few microns across. From his observations, a dislocation is formed in the early stages of growth of the n-alkane crystal, caused by a slip of the n-alkanes along one crystal plane, in part of the crystal, giving rise to a screw dislocation [1,138] (the screw dislocation is at the centre of the growing rhombus-shaped plate in Figure 4.30). All the time, all exposed edges (the base plate and all steps) are growing outwards, limited by their crystallographic shapes. At each level, a plate starts to grown out from the region of the dislocation, but at each point, growth is occurring later than on an earlier step; this gives rise to helical, layer growth – towards the shape of a very shallow cone. Such a process, unlimited, would lead to circular spirals but, while rounded corners to the layers are often observed, they are mostly limited by the preferred, crystallographic edges[42].

So, as new layers are formed, the edge of each layer (a step) sweeps across the layer below. In the presence of a screw dislocation, the steps progress in a helical fashion, which may be observed as growth spirals on the crystal surfaces of higher n- alkanes [2, 3], giving crystal growth perpendicular to the observed face (Figure 4.30). Such growth spirals are observable on wax crystals forming in responsive fuels [117] (see BBD wax crystal, Figure 4.35), and were

Figure 4.30 n-Alkane Crystal: Spiral Growth around a Screw Dislocation.

first observed for paraffin wax crystals, grown in fuel, in 1937 (referred to by Franck [6]). This mechanism of crystal face growth, at low super-saturations, is known as the Burton-Cabrera-Franck (BCF) process, after the scientists who developed this theory [4, 5].

To understand how such low concentrations of a WCM additive can affect crystal growth, we need to understand how the individual molecules attach themselves to a growing plate edge (step). These steps are not perfectly linear, except at absolute zero, because of a process known as thermal or kinetic roughening [7], which causes some molecules to sit proud on the step edge, and others to leave gaps in the step edge (Figure 4.31). These positions on the step, called kink sites, have four close neighbours, so are energetically favoured to be filled by incoming n-alkane molecules. As these sites are filled, the empty position moves along the step, forming a new layer of molecules on the step. These are reversible processes until the n-alkane molecule is fully incorporated as other molecules are incorporated next to it. This process of crystal growth from solution [139, 140] is depicted in Figure 4.31:

a) Diffusion of solute molecules from solution to the crystal surfaces, adsorption onto the crystal steps, on the ledge.
b) Diffusion along the ledge to a position on the edge of a growing layer of n-alkane molecules, a kink site.
c) Incorporation into this kink site (now having four near neighbours in the AB plane) and setting up another kink site.
d) This process continues adding further n-alkane molecules.
e) Until the kink site is blocked by the strongly adsorbed crystal modifying molecule, such as the WASA of Figure 4.32 (here presented diagrammatically) which, as an overall consequence slow down growth in the AB plane.

With this crystal growth process, an impurity that interferes with any of these stages – in particular, incorporation at a kink site – can slow down the growth of a crystal face [129]. And, if it is specific for a specific crystal face, it changes the shape of the crystal formed.

4.4.5c Effects of Additives on Crystal Growth

It follows from the mechanism of crystal growth, just described, that the adsorption of impurity molecules on a wax crystal face, blocks the incorporation of further n-alkane molecules, with the result that the advance of that crystal face will slow down. Adsorption on the step, at kink sites (Figure 4.31d), is particularly effective at slowing down growth of that face [139, 140]. An n-alkane molecule must cause the impurity to desorb from the kink site before it can occupy that site, for that crystal face to continue growing. For impurity molecules to adsorb competitively, they should have similarities

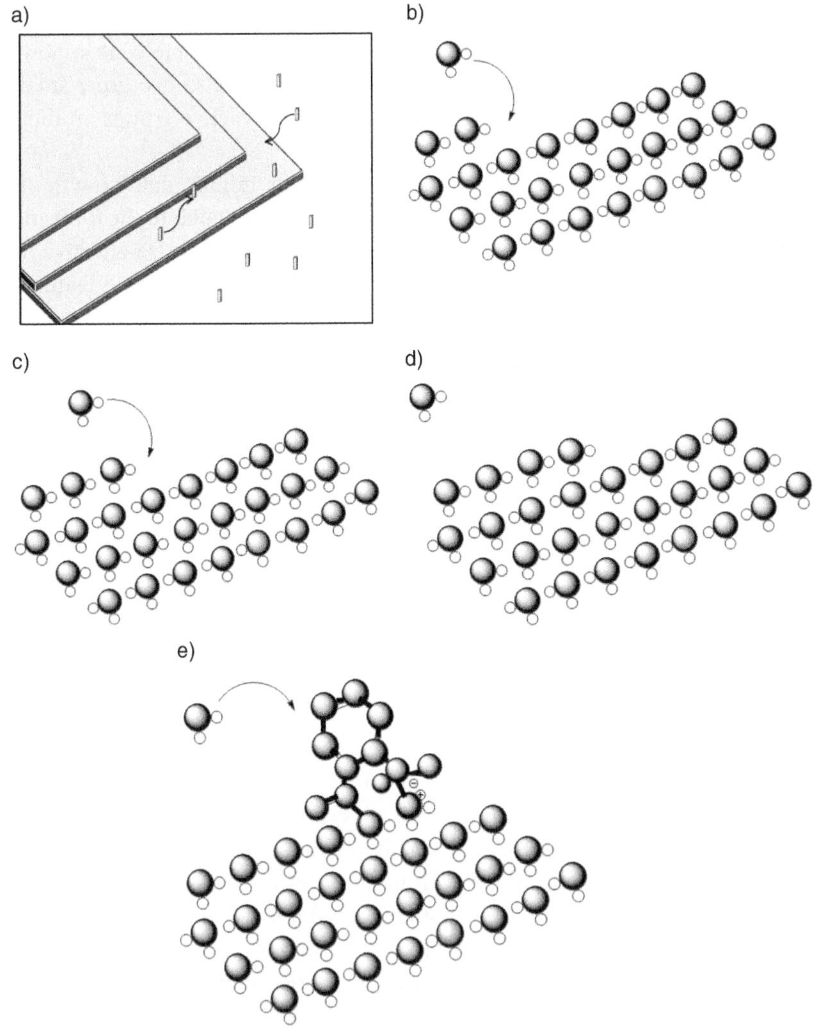

Figure 4.31 Growth of n-Alkane Crystal in Layers by Adsorption and Incorporation at Kink Sites. (a) Diffusion of n-alkane molecules from fuel to steps on the wax crystal surface. (b) A slice through the methylenes of n-alkane molecules in the AB plane, arranged in the crystal at the edge of a layer, is viewed down the C-axis. A newly arrived n-alkane molecule is diffusing along the step to find a kink site. (c) The molecule in (b) has docked into the kink site, setting up another one. Another molecule is now diffusing to that new kink site. (d) So this part of the crystal edge grows as further molecules diffuse to and dock into the kink sites as they are formed. (e) Until a wax crystal growth inhibitor (here, a WASA) docks into the growth site first and cannot be dislodged, then the process stops as approach of the n-alkane molecule to the kink site is blocked.

with the wax molecules, so should carry n-alkyl groups. To be effective, the impurity should be more strongly adsorbed than n-alkane molecules; however, it is possible for the incoming n-alkane molecules to overgrow small impurity molecules, so the impurity should have obstructive groups on their non-adsorbed sides.

As described above, wax crystal growth inhibitors inhibit plate growth, in the AB plane, by slowing growth in the direction perpendicular to the n-alkane molecule long axis, (Figure 4.30). Looking at molecules that are effective as wax crystal shape modifiers (section 4.3), it is now clear why certain features are needed:

1) n-Alkyl groups that can fit into the growth/kink sites on those alkyl polymethylene faces that are relevant to plate growth. Such groups are polymethylene segments of ethylene copolymers, the long alkyl side chains in comb polymers, and the alkyls attached to the amines in WASAs.
2) Multiple n-alkyls, present in all of these, allow strong multipoint adsorption in more than one neighbouring site on the crystal face, compared with an n-alkane molecule that can only adsorb into one crystal site.
3) Large non-n-alkyl groups attached to the n-alkyls to prevent overgrowth by the n-alkane crystal, such as loops bearing the acetate and butyl groups in EVAC (Figure 4.12), the poly-ester backbones of comb polymers (Figure 4.19) and the poly-carboxylic acid parts of WASA molecules (Figure 4.21).

The value of structures that carefully fit into the wax crystal structure is best illustrated by the WASA molecules. While these have a small number of n-alkyl groups (compared with the ethylene copolymers and the comb polymers), they are very effective because their alkyl groups are arranged in such a fashion that they fit exactly into two adjacent n-alkane crystal sites, on the fast growing faces [141]. As depicted in Figure 4.32, the alkyls on the nitrogen atoms are so spaced that they can line up with n-alkane molecules on the crystal surface (a, viewed along the AB plane); while the view down the C axis (b) is of a slice through the AB plane, showing the arrangement of alkane molecules in the crystal face (solid black circles) and the two dialkylamine chains (shaded circles). The connected phthalic group, in the middle of the alkylamine chains, is held out into solution, inhibiting overgrowth by approaching n-alkane molecules.

Selective inhibition of wax crystal growth in the AB direction presupposes that the wax crystal needles (prisms) in MDFI treated middle distillate fuels (Figure 4.29) are the result of the major direction of growth being in the C direction, parallel to the n-alkane molecular axes, rather than in the AB

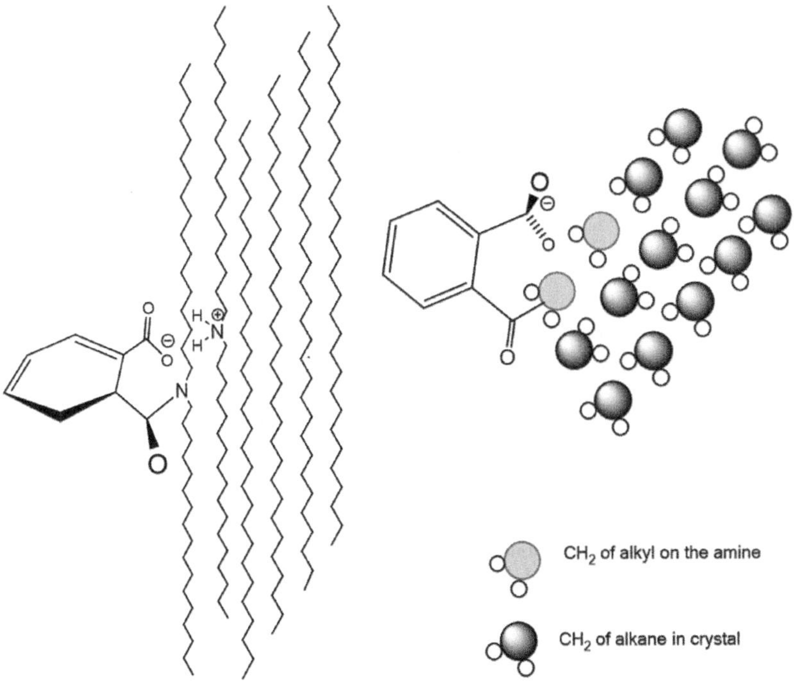

Figure 4.32 Adsorption of a WASA Molecule onto an n-Alkane Crystal Surface.

plane as in the untreated wax crystal [141]. This is supported by the microscopy of wax crystals:

- Optical microscopy, illuminating crystals by transmission of plane-polarised light, with a crossed polariser before the objective lens, shows untreated wax crystal plates with little contrast from their surroundings (Figure 4.29); this is because the n-alkane molecule arrangements, viewed down the C direction (Figure 4.30), are symmetrical, so have little selective interaction with polarised light. Under the same viewing techniques, the MDFI treated prisms, now lying with their C direction perpendicular to the direction of light in the microscope, show up brightly (Figure 4.29) – with contrast depending upon their orientation relative to the plane of light polarisation. In MDFI treated wax needles, viewed in this orientation, the rod-like n-alkane molecules are far from symmetrical and act as a polarised light analyser.
- In scanning electron micrographs of the wax crystal prisms, their ends can be seen to be the rhombi of the plates from which they grew (Figure 4.33). In some samples, such prisms have seen some dissolution during their preparation, and their scanning electron micrographs show the prisms to be stacks of plates (the dissolution is selective for the low crystallinity phase between the lamellae).

Figure 4.33 MDFI Treated WAX Prisms (Aka Needles) at Higher Magnification.

4.4.5d Very Small Wax Crystals and Wax Anti-Settling

It is possible, in low wax fuels with broad n-alkane distributions, to produce wax crystals that are small enough to pass through older-style diesel fuel filters and heating oil filters, for several degrees below the fuel cloud point. Cold flow additives that produce such an effect are usually combinations of multiple components: EVAC or EV2-EH terpolymers (or both), comb polymers and WASAs. It has been demonstrated that such an additive can give such small wax crystals, less than 2 μ, that they pass through the fuel filter at as much as 10°C below the fuel cloud point [60]. Such wax crystals are found to be approximately cubic – an effect of the same modified growth as for the longer prisms normally seen, but at an earlier stage of C-direction growth, so much shorter prisms (Figure 4.34). Clearly, for each crystal to be so small, there need to be many more of them to accommodate the amount of wax separated: there need to be ten times as many prism crystals for them to be one tenth as large, with the same amount of wax. So, more nucleation must be occurring, as well as growth inhibition.

Wax crystals of such a size do not settle even over days. However, it is not necessary for wax crystals to be quite so small to show a practical level of wax anti settling over several hours. Wax crystals several microns long will settle slowly, especially in a fuel that is made more viscous by the fact that it contains a high number concentration of wax crystals. Settling of a spherical particle in a viscous medium is described by Stokes Law:

Rate of settling = (2xRadius^2) (particle density– fluid density) g/(9xviscosity of fluid)

Figure 4.34 WASA Treated Wax Crystals, Here, Cubes about 5 Microns Across.

g = gravitational acceleration

Stokes law needs to be further adjusted to allow for particle concentration (increases apparent viscosity) and particle shape (plates settle much more slowly than spheres of the same mass). In practice, these sizes are of crystals 10°C below their first appearance (at fuel cloud point) and, during the hours their cooling, they grow from zero and gain about one tenth of their final size for every 1°C of cooling. As a result, during the early part of the sub-CP cooling, they are too small to settle at all, and may even be invisible for some °C below the CP. One more factor to aid wax anti-settling is the mixing effect of convection currents in the fuel: cooling occurs at the tank walls so, during cooling, a layer of fuel next to the tank walls is constantly falling as a result of its greater density; then warmer fuel, in the middle of the tank, rises as it is displaced at the bottom by cold fuel.

A detailed study indicates that this may not explain all variations in wax crystal settling and proposes that the mechanism of anti-agglomeration also plays a part [98] – a mechanism proposed early in the introduction of WASA technology [40]. Anti-agglomeration is necessary as several small particles may stick together and form larger particles – this has been described as a problem in the wax-crystal-rich settled layers [40]. The rate of agglomeration of two particles rises with the square of the particle number concentration so, clearly, this rises markedly as the wax crystal sizes fall to ten times below that with MDFI treatment alone. As a result, agglomeration also provides an upper limit to the number of nuclei that survive to grow into individual crystals.

Stabilisation against agglomeration is a colloid stability problem. In non-aqueous systems, colloids are sterically stabilised – the adsorbed stabiliser (usually a polymer) forms a physical barrier between two particles [142][43]. In aqueous systems, the adsorption of charged surfactants produce ionic bi-layers that prevent two particles merging [140], but hydrocarbon solvents cannot readily support charged systems (low dielectric and no H-bonding nor dipoles). However, most WASAs do contain positively charged, oil soluble dialkylammonium cations with oil soluble 2-(N,N-dialkylamido)-benzoate anions, both of which can form micelles with charge-stabilising water-rich cores: fuels contain enough molecules of water, relative to molecules of WASA, to form localised, stabilising micro-environments. Such systems have been shown to provide additional, charge stabilisation to particle dispersions in hydrocarbons [143]. Charge stabilisation at the nucleation stage, preventing the agglomerations at the high number concentrations needed for small wax crystals, is probably a contributing factor.

4.4.5e Cloud Point Depression

Wax anti-settling and treatment for the LTFT require wax crystals to be as small as 2–7 microns across, 10°C below the fuel cloud point. It follows then, that these crystals will be less than 1 μ at 1–3°C below the cloud point. The wavelengths of visible light are of a similar size at 0.4–0.7 μ, so WASA treated wax crystals may be invisible, or difficult to detect, for a few degrees below the cloud point, in other words WASA treatment shows CPD (1 to 3°C only).

Compounds designed to be CPDs, however, are comb polymers. These, apparently, prevent wax crystal separation altogether. DSC shows that CPDs depress wax separation temperatures, and engine tests show that no wax is collected on the fuel filter at temperatures between the CPs of untreated fuels and the CPs of the same fuels treated with cloud point depressants [31]. The data also shows that for a fuel treated with more than about 0.1% of a cloud point depressant, the cloud point depression increases linearly with the amount of the cloud point depressant. So, there appears to be a quantitative relationship between the amount of wax held up and the amount of cloud point depressant used.

It was proposed that comb polymers, bearing a regular array of long alkyl groups, adsorb strongly onto wax crystals. It is, therefore, reasonable to suggest that n-alkanes may also adsorb onto the comb polymers and form a separate, disperse phase, effectively reducing their concentrations in solution. A detailed study of n-dotriacontane (n-$C_{32}H_{66}$) crystallisation from solution (in xylene) showed that the presence of a CPD comb polymer increased the n-alkane solubility while reducing its diffusivity by an order of magnitude [144]. Diffusivity is inversely proportional to molecular size, so these findings strongly suggest that n-alkanes, in the presence of CPD molecules, become part of much larger particles – though these are particles that are smaller than the wavelengths of visible light, less than 1 μ. These facts support the proposal that the n-alkane molecules adsorb onto the CPD polymer molecules, forming a stable, dispersed phase.

Above, comb polymers were discussed as aids to cold flow improvement (section 4.3.3a) while CPDs were reported as being antagonistic to cold flow improvement, such as by CFPP (section 4.2.6). Comb polymer activity is overly sensitive to the chain length of the substituent side groups. For example, it was shown that the C_{12}/C_{14} FVAC is a helpful pour and CFPP depressant, in combination with a nucleator, while the C_{14} FVAC did not help, and the $C_{14}/_{16}$ FVAC is a CPD which is antagonistic to CFPP and PPD [37]. One may speculate that at a matching side-chain length (these are double combs with the alkyl side chains alternating on either side of the backbone, Figure 4.19), comb polymers can adsorb onto a growing n-alkane crystal and control its growth. A longer side-chain comb polymer can form a stable combined molecule, with adsorbed higher n-alkanes that completely inhibits their crystal growth. When these higher n-alkanes are thus tied up, they are not available to assist the controlled nucleation and growth that MDFIs need for their effectiveness.

4.4.5f Rapid Growth of Wax Crystals in Narrow Boiling Distillates

Wax crystals in untreated extreme NBD fuels can be up to a thousand times larger (approximately 10X in each dimension) than those in a responsive BBD fuel, even though the rate of wax separation in the NBD is only two or three times that in the BBD [102]. These wax crystals are malformed – showing rounded, non-crystallographic edges and much twinning (plates growing out of the first formed plate, at an angle). Compare this with the regular rhombi formed in a BBD, following a BCF growth mechanism (Figure 4.35).

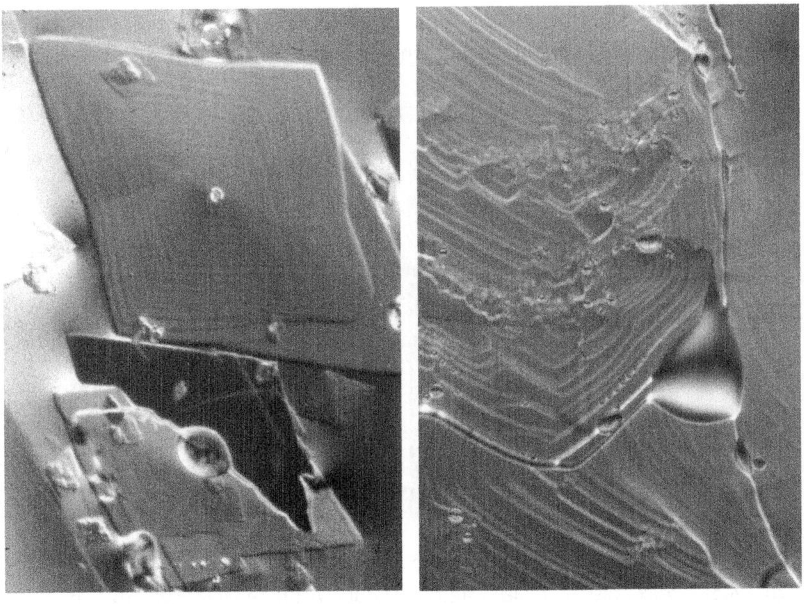

Figure 4.35 Wax Crystals from an Untreated BBD (Left) and a NBD(Right). (only a small part of the 5 mm NBD Crystal is shown). Area shown, 0.5 mm across.

The size difference highlights a nucleation problem (section 4.4.3): fewer crystals, same amount of wax, or more, must mean larger wax crystals. A further consequence of fewer crystals (for a given total amount of wax) is that the total crystal surface area available for adsorption of molecules, coming out of solution, and subsequent growth, is much reduced[44]. As a result, the influx of n-alkane molecules per unit crystal surface area is much increased – an effect equivalent to an increase in supersaturation. Combined with the two to three times faster total wax separation, the growth is no longer by the BCF mechanism, but the birth-and-spread mechanism of continuous 2-dimensional nucleation on the growing faces, caused by the rapid influx of solute (wax) molecules onto the growing surfaces. Orderly fitting into a limited number of kink sites is overcome by a continuous nucleation of new growth sites – such a situation is termed 'kinetic roughening' [8]. Such a change in mechanism, with increased supersaturation, has been demonstrated by growth kinetics for crystals of hexa-triacontane ($C_{36}H_{74}$), growing from solution [145]. When crystals are following this growth mechanism, a few molecules of a growth inhibitor, blocking key kink sites, are readily overgrown. Growth is then controlled by the rate of diffusion of n-alkane molecules to the crystal surface.

When extreme NBDs are treated with high treats of EVAC-MDFIs, the wax crystals separating during slow cooling may form dendrites, in contrast to the small prisms formed in the MDFI treated BBD fuel [54] (Figure 4.36). Studies have been made of dendritic crystal formation, which is a quite widespread phenomenon[146]. A simplified explanation is that the MDFI polymer layer, adsorbed on the wax crystal surface, impedes the approach of solute n-alkane

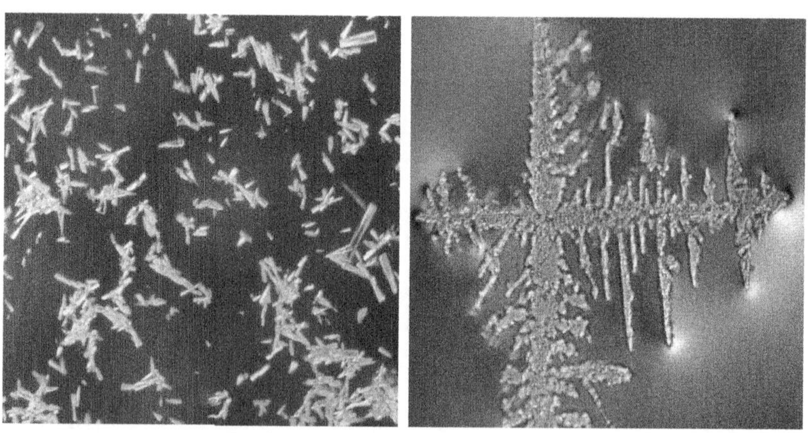

Figure 4.36 Wax Crystals in MDFI Treated Middle Distillate Fuels: Left, BBD and Right, NBD. (Micrographs about 1 mm across).

molecules, which leads to a build-up of a layer of higher super-saturation. Random thermal variations enable fingers of crystal to grow out rapidly into the super-saturated layer, sporadically, until the local super-saturation falls, and the polymer layer recovers.

4.6 Cold Flow Tests

Cold flow tests have been described in detail in many places, such as the Automotive Fuels Reference Book [9,147] and the ASTM standards themselves [24]. Here are summaries of the test methods for those tests most referred to in this chapter.

Cloud Point (ASTM D2500): The older method uses a 50 ml fuel sample in a 100 ml cylindrical, clear glass, flat-bottomed vessel (closed ended tube) with a thermometer, whose bulb is suspended near the bottom. It is placed in a fixed temperature bath (−34°C and moved to a −54°C bath when the sample temperature reaches −24°C) and visually observed every 1°C to determine when the fuel (usually at the bottom of the sample) starts to become cloudy. The temperature is noted as the cloud point. Modern, automated equipment uses a 1 ml sample in a capillary tube, in a temperature controlled environment, and an adapted photocell arrangement to observe the onset of wax separation.

Pour Point (ASTM D97, D5950): The older method uses the same arrangement for the sample in the 100 ml tube as for CP, except that the thermometer bulb is near the top of the sample. The sample is examined for gelling every 3°C by tilting the sample. The pour point is the last temperature at which the fuel flows. Automatic equipment is now used that has a paddle arrangement to judge when the fuel has gelled.

Cold Filter Plugging Point, (CFPP), (IP 309, EN116, ASTM D6371): The CFPP test is widely used and run on automated equipment, based upon the original manual method [9]. Again, the sample and container are as for the CP test, so too are the cooling bath temperatures, though the switch to the lower temperature bath takes place when the sample temperature reaches −20°C. During cooling, starting above the cloud point (usually at + 10°C), every 1°C a low vacuum is applied to a pipette, on whose lower end is a cylindrical brass fitting, which sits on the bottom of the tube[45] and which contains a 1 cm circle of a fine, 44-micron wire mesh. The mesh is held, by the brass fitting, about 1 cm above the bottom of the sample. An arrangement of photocells detects the filling (and emptying) of the pipette. If the pipette fills in less than 1 minute, then a pass is noted at that temperature, the vacuum is released and the sample returns to its vessel. This process is repeated every 1°C during cooling, until the pipette does not fill within 1 minute (due to wax blocking the mesh) when the process stops, and that temperature is recorded as the CFPP.

Low Temperature Flow Test, LTFT (ASTM D4539): A similar concept to the CFPP, but on a larger sample and much slower cooling: sample size is 200 ml, cooled at 1 or 2°C per hour; the test is for 180 ml to pass through a 1 cm circle of 17-micron mesh within 1 minute, under a higher vacuum (6 inches of mercury) than used in the CFPP. There is no repeat testing on a sample, so several samples need to be cooled together, taking one sample for testing at each temperature.

Engine testing on the Cold Climate Chassis Dynamometer (CCCD, Europe) or All-Weather Chassis Dynamometer (AWCD, North America) is not part of diesel fuel specifications, being far too slow and expensive for routine application. It has been applied in the validation of those tests (CFPP and LTFT) that are used to set specification requirements. CCCD testing is also used to check out the effects of new fuel variations (for example, gas-to-liquid fuel or biodiesel), engine and fuel-line arrangements (such as exposed fuel-lines, fuel-line heaters) and the effectiveness of new cold flow additives. Details of CCCD/AWCD procedure have varied but the basic processes are all somewhat similar. Many different vehicles have been used over time to ensure relevance – even diesel vehicle simulation rigs (built mainly from real diesel vehicle fuel systems) have been used to obtain more data in a consistent procedure (referred to in a presentation [41]). Reported comparisons between fuels and additives are usually made from data obtained in a self-consistent set of CCCD or AWCD tests [41]. The general procedure is to fill, or half fill, the fuel tank of a suitable diesel vehicle with the test fuel. The vehicle is on the static rollers of a chassis dynamometer, inside a temperature variable climate chamber. A range of sensors are placed in critical parts of the fuel system to monitor temperatures (fuel tank, fuel lines, before and after filter) and pressures (before and after fuel filter). The static vehicle is cooled slowly, usually linearly (1 or 2°C/hour), from several degrees above the cloud point, overnight, followed by a 'soak' period of two or three hours at the target test temperature. Then attempts are made to start the vehicle; if successful, the vehicle is accelerated and run under various conditions of speed and load. Assessments are made as to the degree of trouble-free engine performance during this process: it may not be possible even to start the engine due to wax completely blocking the fuel filter (clear fail); some uneven running may occur during the first 15 min which clears later to even running (pass or fail depending upon severity of problem); fully trouble free starting and motoring (clear pass). An extensive summary of low temperature diesel vehicle testing is presented in chapter 17 of the Automotive Fuels Reference Book, third edition [147].

Notes

1 Some 'waxy' crudes may contain up to 50% n-alkanes.
2 The cloud point is the temperature at which wax first visibly appears while cooling the fuel under conditions specified as ASTM D 2500.
3 This is the common arrangement in many places, such as Europe, though in N. American diesel engine designs this can lead to fuel overheating, so the return fuel enters the fuel tank at a distance from the take-off.
4 Cloud Point Depressants are dealt with in more detail below 4.2.6 and 4.4.5e.
5 Freeze Point in jet fuel is measured as the temperature at which the wax redissolves on reheating after cooling to a temperature below that needed for wax separation.
6 While the data comes from two patents, it is clear from the fuel properties given that the same fuel is used in both patents (and the same inventor).
7 The small elevation of CFPP relative to untreated fuel is thought to result from the wax crystals that first appear being thicker and sturdier when EVA-treated than the fragile thin plates from untreated fuel. The latter break into small pieces in contact with the CFPP mesh while the former resist fracturing.
8 Standard distillation measures the amount of fuel distilling as distillation temperature (T) rises (Figure 4.4). The results are expressed as % collected against T, T20% is temperature at which 20% was collected and so-on.
9 After discussions with author Coley, the value of 13°C for FBP-90% of HIGH FBP (Figure 4.16 of this book [9]) is a misprint for the 30°C used here in Figure 4.5.
10 A plot of the amount in each distillation fraction against BP or % n-alkane against C-number gives a skewed normal distribution; the spread of the distribution at the high BP side is called the distillation tail.
11 When referring to the chemical components, the more general term of cold flow improver is preferred as the components are usually claimed for all possible cold flow applications, while an MDFI is a specific, cold flow improver blend for middle distillates.
12 T95% is more reproducible than the FBP, which is about 10 °C higher.
13 CFPP and PP of Swedish fuels are only 1 to 3 °C below the CP, indicating absence of MDFI, while they are more than 10 °C below the CP in the Finish fuels, indicating MDFI use.
14 Light and heavy are widely used to mean low and heavy density, which correlates with low and high boiling. In the table, the distillations are shown for the average fuels in 2012 (left of the /) and in 2018 (right of the/) in each country.
15 Fuel properties and numbers were taken from the average fuel data tables in the 2018 WDFQS [27].

16 N-Alkane distributions are discussed in more detail under Mechanisms and Nucleators, below.

17 Oil companies use exchange agreements to provide fuel to service stations that are remote from one oil company's refinery or terminal but are close to one from another oil company. The second oil company may take supply from the first, where the logistics are reversed.

18 N-alkanes are the most plentiful wax forming hydrocarbons in middle distillate fuels, though in higher boiling fractions, especially heavy fuels, there are also long chain iso-alkanes, but these too depend upon the long, n-alkyl segments for their crystallinity.

19 Low solubility additives may not redissolve when the wax redissolves, just above the CP, so are not effective on a second cool down.

20 The patent [58] has many examples such as, at 250 ppm, EVAC III, 33% VA, made at 105°C, branching 3.5, gave CFPP of −18°C in a fuel, while EVAC, 32% VA, made at 150°C, branching 12, gave CFPP of −12°C, in the same fuel (base CFPP -10°C).

21 'Nucleator' EVAC had 17 % wt. Vinyl Acetate, Mn 3,000 [59] or 16% wt. VA and Mn 3,000 [21].

22 See Table 2.1 which identifies those positions from which hydrogen atoms are most readily lost.

23 The reactivity ratio is the rate of reaction of a polymer radical end of monomer Ma reacting with another Ma, is compared with its rate of reaction with another monomer, Mb; a low ratio shows a preference of monomer radical Ma to react with the other monomer, Mb. Similarly, for the second monomer, Mb. If both radical ends prefer to react with the other monomer present, then an alternating copolymer is the likely result. If both ratios are around 1.0, the copolymer will be made up of a random sequence of the two monomers.

24 Acryloid 150 was a commercial PPD for lubricating oils or dewaxing aid from Rohm & Haas (first 1946), a poly alkylmethacrylate with C10 to C22 alkyls [149].

25 Microcrystalline wax is a high molecular weight wax containing branched and cyclic alkanes, a by-product from the production of lubricating oils.

26 Evonik is a German company that has taken over Rohm & Haas/Rhomax specialty chemicals business.

27 Atom Transfer Radical Polymerisation, Reversible Addition Fragmentation Chain Transfer and Nitroxide Mediated Polymerisation.

28 The hexadecyl and octadecyl esters are prepared from the alcohol mixture derived from 'tallow', animal fat, by saponification and hydrogenation. Similarly, the behenyl alkyl mix originates from a variety of rapeseed/ canola vegetable oil.

29 Alumoxane is an oligomeric structure of Al-O-Al bonds containing methyl-Al groups, made by controlled hydrolysis of tri-methyl aluminium [150].

30 Boilerplate is a jargon expression for a generalised and oft repeated part of similar patents [151]. Here, it refers to the lists of readily available components with which a newly proposed component may be used.

31 May be described as an electrophilic attack on the phenol by protonated formaldehyde (acid catalysed) or a nucleophilic attack on the formaldehyde by the phenate anion (base catalysed).

32 Positions of n-alkanes in the chromatogram do not exactly match as fuels and waxes run on different instrument settings.

33 CPs and CFPPs ranged from −14 to + 3°C while, in a wide range of blends, LGO varied from 20 to 60% and the HGO from 10 to 60% of the total, in a wide range of blends (the 5% HGO blend was a single blend) [118].

34 Volume & weight of each NBD wax crystal is about 1,000x that of a BBD wax crystal so a gram of BBD wax has 1,000x more crystals than a gram of NBD wax. NBD fuel has about twice as much wax as BBD fuel, so a gram of BBD fuel would have about 500x as many crystals as a gram of NBD fuel.

35 An amount that might be found in a 1980s' BBD.

36 From the CFPP depression of 14°C by 200 ppm of a 3/1 mix of growth-inhibitor/nucleator EVACs [21], this test fuel was a responsive BBD. Unfortunately, no CFPP depression is given with either of these polymers, at 200ppm, separately.

37 This reference reports much detailed work, using DSC, turbidity, and X-ray diffraction. A proposed model for the n-alkane EVA interaction is given (Figure 4.11), though this assumes an average alkyl chain length in the EVA of C_{22} which only takes into account the vinyl acetate content and not the butyl short chain branching.

38 Crystallinities for 90 mole% ethylene/10 mole% hexene copolymers (about 9 methyls per 100 CH_2), which also have butyl branches (as do most of MDFI EVAs), are reported to be only 10%. The identity of short side branches has little effect on ethylene copolymer crystallinity, which depends upon the backbone [133].

39 The EVA used in this study had 28 wt.% vinyl acetate and Mn of 2400.

40 The less efficient the fractionation in the pipestill, the longer the distillation tail, but less efficient fractionation reduces yield at a given CP.

41 This is obvious to anyone measuring a fuel's wax content by the method of filtering off the wax from an MDFI treated fuel compared with an untreated fuel, in a cold room or cold box at 10°C below fuel CP.

42 Non-crystallographic surfaces are rough & irregular, so provide many kink sites and faster growth, until they reach the order of a more stable crystallographic face.

43 Steric stabilisation is discussed in the chapter on fuel detergents.

44 For spheres, volume = $a.r^3$, number of crystals = b/volume = $b/a.r^3$; total surface area = number x $c.r^2$ = b.c/a.r. Where a, b and c are constants, r is radius. So total surface area of particles, making up a fixed amount of material, is proportional to 1/r, i.e. is halved for every doubling of crystal linear size (radius of spheres); or reduced by 0.1x when r is increased by 10X.

45 The lower edges of the 1.5 cm diameter brass cylinder is gap-toothed to allow fuel to enter.

References

1. Anderson, N.G. and Dawson, I.M. (1953). The study of crystal growth with the electron microscope. III growth-step patterns and the relationship of growth-step height to molecular structure in nonatriacontane and in stearic acid. June Proceedings of the Royal Society of London. Series A, *Mathematical and Physical Sciences* 218 (1133): 255–268.
2. Dawson, I.M. (1952). *The Study of Crystal Growth with the Electron Microscope. II the Observation of Growth Steps in the Paraffin n-Hectane*. 72–79. London: The Royal Society. Proceedings of the Royal Society of London. Series A. Mathematical and Physical Sciences, Vols. 214, No 1116.
3. Dawson, I.M. and Vand, V. (1951). *The Observation of Spiral Growth-Steps in n-Paraffin Single Crystals in the Electron Microscope*. 555–562. London: Proceedings of the Royal Society of London. Series A. Mathematical and Physical Sciences. Vols. 206, No.1087.
4. Burton, W.K., Cabrera, N., and Frank, F.C. (1951). *The Growth of Crystals and the Equilibrium Structure of Their Surfaces*. 299–358. London: The Royal Society. June Philosophical Transactions, Series A, Mathematical and Physical Sciences, Vol. 243 (866).
5. Frank, F.C. (1949). Influence of dislocations on crystal growth. *Discussions of the Faraday Society* 5: 48–54.
6. Frank, F.C. (1958). Introductory lecture-growth and perfection of crystals. In: *Proceedings of International Conference on Crystal Growth* (eds. B.W. Roberts, D. Turnbull, and D.H. Doremus) pp. 1–10, J. Wiley & Sons N.Y.
7. Xiang, L. and Bennema, P. (1994). On the morphology of crystals of triclinic even normal alkanes: Theory and observation. Elsevier Science, *Journal of Crystal Growth* 135: 209–223.
8. Xiang, L. and Bennema, P. (1993). Kinetic roughening in relation to the roughening transition in odd-numbered alkane crystals. Elsevier Science, *Journal of Crystal Growth* 128: 69–73.

9 Owen, K., Coley, T.R., and Weaver, S.R. (1995). *Automotive Fuels Reference Book*, 2nd edn. Society of Automotive Engineers, Warrendale, P.A. ISBN 1560915897.

10 Davis, G.H.B. (1943). Diesel fuels. US Patent 2,403,267. Filed USA, 24 August Standard Oil Development Company.

11 Davis, G.H.B. (1930). Hydrocarbon oil and process for manufacturing the same. US Patent 1,815,022. Filed 3 May.

12 Rupar, C.B. and Ilnyckyj, S. (1962). Pour depressant for middle distillates. CA Patent 646,267. Granted Canada, 8 July.

13 Ilnyckyj, S. and Tiedje, J. (1962). Middle distillate fuels. CA Patent 646,456. Granted Canada, 8 July.

14 Ilnyckyj, S. and Rupar, C.B. (1959). Ethylene-vinyl ester pour depressant for middle distillates. US Patent 3,048,479. Filed 3 August .

15 Infineum Worldwide Winter Diesel Fuel Survey 2012. (2012). Infineum.com. https://www.infineuminsight.com/media/1095/infineum-wwdfqs-2012-intro-and-trends.pdf (Cited: 9 May 2020).

16 Coley, T.R. (1989). Diesel fuel additives influencing flow and storage properties. In: *Gasoline and Diesel Fuel Additives* Vol. 4. (ed. K. Owen), 105–132. John Wiley & Sons.

17 IGF3, CEC Investigation Group (1982). Low temperature operability of diesels. CEC, CEC Report P-171–182.

18 Cold Flow Operability testing- biodiesel in common rail trucks reveals unexpected operability issues (2009). Insight. December 44: 1–3. Infineum International Ltd.

19 ATC Europe (2004). Fuel additives and the environment. atc-europe.org. 1 January http://www.atc-europe.org/public/doc52.pdf (Cited: 1 October 2013).

20 Danilov, A.M. and Additives, F. (2007). Development and use in 2001–2005. Springer Science + Business Media Inc., *Chemistry and Technology of Fuels & Oils* 43 (2): 155–171.

21 Ilnyckyj, S. and Cole, C.O. (1974). Middle distillate compositions with improved filterability and improved process therefor. US Patent 3,961,916. Filed 28 February.

22 Tack, R.D., Brazier, J.R.T., and Lewtas, K. (1982). Flow improver additive for distillate fuels and concentrate thereof. EP 0,061,895. Filed Europe, 24 March.

23 Platts Methodology and Specifications Guide: European Products (2013). Platts.com. January http://www.platts.com/IM.Platts.Content/methodologyreferences/methodologyspecs/europeanoilproductspecs.pdf (Cited: 21 February 2013).

24 ASTM Committee D02 (2002). standard specification for fuel oils. D396-02. Annual Book of ASTM Standards. ASTM International, Vol. 5.

25 Petroleum Association of Japan (2015). Petroleum industry in Japan. Sept http://www.paj.gr.jp/english (Cited: 13 May 2020).

26 Yermakov, V., Henderson, J., and Fattouh, B. (2019). Russia's heavy fuel oil exports: Challenges and changing rules abroad and at home. OIES paper WPM 80. https://www.oxfordenergy.org/publications/russias-heavy-fuel-oil-exports-challenges-changing-rules-home-abroad (Cited: 13 May 2020).

27 Infineum Worldwide Winter Diesel Fuel Survey 2018 (2018). Infineum.com. https://www.infineuminsight.com/en-gb/resources/winter-diesel-fuel-quality-survey/2018-winter-diesel-fuel-survey (Cited: 9 May 2020).

28 Morton, C., Tack, R.D., and Armitage, P.D. (2006). Fuel oil compositions. US Patent 9,051,527. Filed 6 February.

29 Tack, R.D., Lewtas, K., More, I. et al. (1987). Liquid fuel compositions. EP 239,320. Filed Europe, 18 March.

30 Ilnyckyj, S., Cole, J.C.O., Feldman, N., et al. (1968). Enhancing low-temperature flow properties of fuel oil. US Patent 3,640,691. Filed 16 September.

31 Metzelaar, R.J. and Tack, R.D. (1985). *Distillate Fuel Cloud Point Depressants: Technology and Economics*. Houston: National Petroleum Refiners Association. Fuels and Lubricants Meeting. Vols. FL-85-83.

32 Deng, F., Carey, W.S., Eldin, S. et al. (2003). Aviation fuel cold flow additives and compositions. US Patent 2004/0,250,465. Filed 12 June.

33 Tack, R.D., Jackson, G., and Armitage, P. (2003). Jet fuel compositions. EP 1,357,169. Filed Europe, 17 March.

34 Tack, R. and Spencer, D. (2004). *Jet Fuel Compositions Comprising a Comb Polymer*. EP 1,524,311. Filed Europe, 3 September, Infineum UK Ltd.

35 Arnold, J.C. (1944). Improvements in or relating to fuels for high compression ignition engines. GB Patent 573,364. Filed UK, 30 June.

36 Letoffe, J.M., Claudy, P., Vassilakis, D. et al. (1995). Antagonism between cloud point and cold filter point depressants in a diesel fuel. 12. Butterworth-Heinemann, *Fuel* 74: 1830–1833.

37 Tack, R.D., Pearce, S.L., and Rossi, A. (1985). Middle distillate compositions with improved low temperature properties. US Patent 4,863,486. Filed 20 February.

38 Lewtas, K. (1985). Middle distillate compositions with improved cold flow properties. EP 0,156,577. Filed Europe, 11 March.

39 Innospec Fuel Specialties (2020). Fuel additives product summary. https://www.innospecinc.com/our-markets/fuel-specialties/refinery-specialties/cold-flow-improvers (Cited: 14 May 2020).

40 Feldman, N. (1978). Combination of ethylene polymer, polymer having alkyl side chains, and nitrogen containing compound to improve cold flow properties of distillate fuel oils. US Patent 4,211,534. Filed 30 August.

41 Brown, G.I., Tack, R.D., and Chandler, J.E. (1988). *An Additive Solution to the Problem of Wax Settling in Diesel Fuels*. Portland: SAE. International Fuels and Lubricants Meeting and Exposition. Society of Automotive Engineers, Warrendale, P.A. SAE Technical Paper Series 881652.

42 Tack, R.D. (2007). Fuel oil compositions. CA Patent 2,613,953. Filed Canada, 12 December.

43 Kober, A.E. and Rossi, A. (1970). Fuel oil composition containing oil-soluble pour depressant polymer and auxiliary flow-improving compound. US Patent 3,762,888. Filed 16 November.

44 Innospec Fuel Specialties Fuel additives product summary. Innospec Inc. http://login.innospecinc.com/assets/-files/documents/dec-07/cm-1197641587-FS-Product-Summary-US.pdf (Cited: 4 January 2013).

45 Krull, M., Feustel, M., Reimann, W. et al. (1998). Flowability of mineral oils and mineral oil distillates using alkylphenol-aldehyde resins. US 5,998,530. Filed 5 January.

46 Jackson, G., Ledeore, C., and Tack, R.D. (1999). Additives and oil compositions. WO Patent 99/61,562. Filed 13 May.

47 Krull, M., Volkmer, T., and Reimann, W. (2001). Copolymer blends and their use as additives for improving the cold flow properties of middle distillates. US Patent 6,565,616. Filed 13 March.

48 Clariant Cold Flow Improvers (2020) https://www.clariant.com/en/Business-Units/Oil-and-Mining-Services/Refinery-Services/Cold-Flow-Improvers (Cited: 16 May 2020).

49 Afton Chemical Cold flow improvers. AftonChemical.com. http://www.aftonchemical.com/Solutions/FuelAdd/Specification_Distribution/Diesel/Pages/ColdFlowImprovers.aspx (Cited: 22 August 2014).

50 Brewer, M.L., DeBalthasar, F., and DeWilbrand, K. (2009). Fuel composition. US Patent 8,273,137. Filed Canada, 8 January.

51 Maehling, F.O., Troetsch-Schaller, I., Lubojanski, H. et al. (2009). Use of solubilizers for homogenizing additive concentrates. US Patent 2009/0,307,964. Filed 10 June.

52 Jackson, G. and Tack, R.D. (2002). Fuel additive. EP 1,314,771. Filed Europe, 5 November.

53 Krull, M., Hoffmann, F., and Redlich, K. (2004). Fuel oil additives and additized fuel oils having improved cold properties. US Patent 2005/0,016,060. Filed 21 July.

54 Holder, G.A. and Winkler, J. (1965). Crystal-growth poisoning of n-paraffin wax by polymeric additives and its relevance to polymer crystallization mechanisms. No.4998: Nature Publishing Group, August Nature, 719–721.

55 Ilnyckyj, S. (1968). Copolymers of ethylene with unsaturated esters and oil compositions containing said copolymers. US Patent 3,642,459. Filed 1 April.

56 Ilnyckyj, S. (1969). Increasing low temperature flowability of middle distillate fuel. US Patent 3,660,057. Filed 17 March.
57 Hollyday, W.C. and Feldman, N. (1972). Fuel oil with improved flow properties. CA Patent 916,445. Published Canada, 12 December.
58 Wisotsky, M.J. and Tunkel, N. (1975). Process for preparing copolymers of ethylene and vinyl esters or mixtures with other ethylenically unsaturated monomers. US Patent 3,981,850. Filed 19 August.
59 Tack, R.D., Lewtas, K., and Davies, B.W. (1982). Two component flow improver additive for middle distillate fuel oils. US Patent 4,481,013. Filed 23 March.
60 Lewtas, K., Lehmann, E.W., Tack, R.D. et al. (1987). Chemical compositions and use as fuel additives. EP 261,957. Filed Europe, 23 September.
61 Danilov, A.M. (2001). Fuel additives: Evolution and use in 1996-2000. Plenum Publishing Corp., *Chemistry and Technology of Fuels and Oils* 37 (6): 444–455.
62 Yu-hui, G. and Ben-xian, S. (2006). QSAR research of the activity of span surfactants as wax antisettling additives for diesel. s.l.: American Chemical Society, *Energy & Fuels* 20: 1579–1583.
63 Shimada, T., Isaka, T., Okada, N. et al. (1986). Fuel oil additive and fuel oil having improved flowability. US Patent 4,802,892. Filed 18 September.
64 Botros, M.G. (1999). Wax anti-settling agents for distillate fuels. US Patent 6,206,939. Filed 13 May.
65 Brod, R.J., Davies, B.W., and Ibrahim, T. (1993). Oil additives and compositions. EP 739,971. Filed Europe, 29 June.
66 Reimann, W., Feustel, M., and Hobes, J. (1991). Terpolymers of ethylene, their preparation, and their use as additives for mineral oil distillates. US Patent 5,254,652. Filed 23 December.
67 Krull, M., Siggelkow, B., and Hess, M. (2004). Cold flow improvers for fuel oils of vegetable or animal origin. US Patent 7,500,996. Filed 25 October.
68 Brown, G.I. and Goberdhan, D.G. (1995). Fuel oil compositions. US Patent 5,906,663. Filed 5 December.
69 More, I. (1998). Additives for oil compositions. US Patent 6,106,584. Filed 4 August.
70 Bock, J., Brod, R.J., Davies, B.W. et al. (1995). Oil additives, compositions and polymers for use therein. US Patent 6,306,186. Filed 1 September.
71 Jackson, G. (2004). Fuel oil composition. US Patent 7,497,880. Filed 28 May.
72 Oppenlaender, K., Wegner, B., Barthold, K. et al. (1990). Reaction products of aminoalkylene polycarboxylic acids with secondary amines and middle distillate compositions containing the aforesaid. US Patent 5,071,445. Filed 18 May.
73 Oppenlaender, K., Hartmann, H., Wegner, B. et al. (1989). Low-temperature-stable petroleum middle distillates containing polymers as paraffin dispersants. CA Patent 2,032,246. Filed Canada, 16 December.

74 Hovemann, F. and Nottes, G. (1975). Combustibles et carburants a base de distillats de petrole contenant des copolymeres oleosolubles de l'Ethylene. FR Patent 2,261,333. Filed France, 12 February.

75 Gropper, H., Brandstetter, F., Schwartz, E. et al. (1982). Petroleum middle distillate with modified flow properties. EP 084,148. Filed Europe, 20 December.

76 Hideo, N., Nobuhiro, A.O., and Minoru, N. (2002). EP 1,391,498. Filed Europe, 8 May.

77 Billmeyer, F.W. (1984). *Textbook of Polymer Science*. New York: John Wiley. ISBN 0471031968.

78 Deckers, A., Wenderoth, B., Ruhl, T. et al. (1997). Ethylene-vinyl formate copolymers, process for their preparation, their use as flow improvers, and fuel propellant compositions comprising them. US Patent 6,099,601. Filed 20 February.

79 Siggelkow, B., Reimann, W., and Krull, M. (2007)_. Additives for improving the cold properties of fuel oils. US Patent 8,961,622. Filed 17 July.

80 Morton, C. and Tack, R. (2005). Additives for oil compositions. EP 1,690,896. Filed Europe, 22 December. Infineum International Ltd.

81 Martin, E.J.(1948). Diesel fuel. US patent 2,615,799. Filed 3 April.

82 Choo, Y.M., Ma, A.N., Basiron, Y. et al. (2006). Palm-based biodiesel foundation. US Patent 2006/0,288,637. Filed 9 June .

83 Hess, B., Cybert, R., Malitsky, M.A. et al. (2010). A composition having improved filterability. EP 2,531,579. Filed Europe, 15 December.

84 Scherer, M. and Souchik, J. (1999). Copolymers obtainable by the ATRP method and a method for their preparation and their use. US Patent 6,391,996. Filed 30 November.

85 Janowicz, A.H. (1986). Pentacyanocobaltate(II) catalytic chain transfer agents for molecular weight control in free radical polymerisation. US Patent 4,722,984. Filed 14 October.

86 Huybrechts, J. (1999). Low molecular weight (meth)acrylate copolymer emulsions. WO Patent 9,942,505. Filed 17 February.

87 Tutwiler, G.S. and Hollyday, W.C. (1956). Copolymers of vinyl acetate and fumarate. GB Patent 804,029. Filed UK, 3 October.

88 Hollyday, W.C. (1956). Copolymers of vinyl acetate and fumarate. GB Patent 806,906. Filed UK, 4 December. ER&E.

89 Young, D.W. and Clough, T.J. (1963). Fuel oil composition of reduced pour point. US Patent 3,413,103. Filed 29 July.

90 Tack, R.D. (1985). Polyesters as flow improvers for hydrocarbons. EP 0,183,447. Filed Europe, 14 November.

91 Panzer, J. (1973). Combination of asphaltenes with flow improver polymers to improve the flow properties of high boiling fuel oils. GB Patent 1,480,909. Filed UK, 12 October.

92 Krull, M., Siggelkow, B., Hess, M. et al. (2003). Cold flow improvers for fuel oils of vegetable or animal origin. US Patent 7,041,738. Filed 11 June.

93 Krull, M. and Nagel, W. (2000). Copolymers and their use as additives for improving the cold-flow properties of middle distillates. US Patent 6,458,174. Filed 19 January.

94 Roesch, J., Mueller, H.-J., Schweier, G. et al. (1995). Process for preparing olefin oligomers. WO Patent 9,623,751. Filed Worldwide, 1 February.

95 Rossi, A., Chandler, J.E., and Barbour, R. (1997). Polymers and additive compositions. US Patent 5,807,949. Filed 24 October.

96 Deng, F., Carey, W.S., Eldin, S. et al. (2003). Aviation fuel cold flow additives and compositions. US Patent 2004,250,467. Filed USA, 12 June.

97 Simon, B., Grassi, A., and Boistelle, R. (1974). Kinetics of growth of the (110) face of the paraffin C36H74 in solution, II Growth in the presence of an inhibitor, di-octadecyl-amine. s.l.: North-Holland Publishing Co., *Journal of Crystal Growth* 26: 90–96.

98 Marie, E., Chevalier, Y., Brunel, S. et al. (2004). Settling of paraffin crystals in cooled middle distillate fuels. s.l.: Elsevier, *Journal of Colloid and Interface Science* 269: 117–125.

99 Krull, M., Feustel, M., and Erdmann, M. (1995). Polyetheramines with polymers of alpha-beta-unsaturated dicarboxylic acids. US Patent 5,705,603. Filed 22 June.

100 El-Gamal, I.M., Khidr, T.T., and Ghuiba, F.M. (1998). Nitrogen-based copolymers as wax dispersants for paraffinic gas oils. s.l.: Elsevier, *Fuel* 77 (5): 375–385.

101 Mullin, J.W. (2001). Crystallization. 4th. Edition. s.l.: Butterworth-Heinemann, ISBN0750648333.

102 Tack, R.D. and Lewtas, K. (1990). *Wax Crystallisation in Diesel Fuel: Additive Control of Size and Shape of n-Alkane Crystals*. Exeter: IOPP Ltd., British Crystallization Association Spring Meeting, Structural Aspects of Crystal Growth.

103 Company British Petroleum (1960). Improvements relating to hydrocarbon oils. GB Patent 893,175. Filed UK, 30 March.

104 CRODA (2020). Product finder select span™ 65. CRODA. https://www.crodahomecare.com/en-gb/products-and-applications/product-finder# (Cited: 21 May 2020).

105 CRODA (2020). Product finder select tween™ 65. CRODA. https://www.crodahomecare.com/en-gb/products-and-applications/product-finder# (Cited: 21 May 2020).

106 Deneker, V.J.P.R. and Harffey, P. (2002). Additive compositions. US Patent 6,790,913. Filed 7 October.

107 Ishizaki, T., Kimura, T., and Yamazaki, S. (1984). Method for improving cold flow of fuel oils. EP 117,108. Filed Europe, 13 February.

108 Ishizaki, T., Nagai, T., and Yamazaki, S. (1982). Method for improving cold flow of fuel oils. US Patent 4,491,455. Filed 21 September.

109 Panagiotis, D. (1998). Fuel oil additives and compositions. US Patent 6,254,650. Filed 2 December.

110 Goberdahn, D.G., Tack, R.D., Lewtas, K. et al. (1996). Fuel oil additives and compositions. US Patent 6,767,374. Filed 13 March.

111 Radulescu, A., Schwahn, D., Richter, D. et al. (2003). Co-crystallization of poly(ethylene-butene) copolymers and paraffin molecules in decane studied with small angle neutron scattering. International Union of Crystallography,, *Journal of Applied Crystallography* 36: 995–999. ISSN00218898.

112 Makoto, O., Koji, S., and Masahiro, F. (1999). Fluidity improver for middle distillate. JP 2000,256,684. Filed Japan, 4 March.

113 Young, R.N., Quirk, R.P., and Fetters, L.J. Anionic polymerization of non-polar monomers involving lithium. In: *STETAnionic Polymerization,* 56, 2005, pp. 1–90. Springer-Berlin Heidelberg. ISBN 9783540387367

114 Martella, D.J. and Jaruzelski, J.J. (1988). Alkyl phenol-formaldehyde condensates as fuel and lubricating oil additives. EP 311,452. Filed Europe, 7 October.

115 Frost, K.A. (1973). Low-temperature flow improvers in fuels. US Patent 3,850,587. Filed 29 November.

116 Van-Winkle, T.L., Affens, W.A., Beal, E.J. et al. (1987). Distribution of N-alkanes in partially frozen middle distillate fuels. Butterworth Co. Ltd., July *Fuel* 66: 947–953.

117 Lewtas, K., Tack, R.D., Beiny, D.H.M. et al. (1991). Wax crystallization in diesel fuel: habit modification and the growth of n-alkane crystals. In: *Advances in Industrial Crystallization* (eds. R.J. Davey, A.G. Jones, and J. Garside), 166–179. Oxford: Butterworth-Heinemann Ltd.

118 Rossemyr, L.I. (1979). Cold flow properties and response to cold flow improver of some typical fuel oils. Bergen: American Chemical Society, *Industrial and Engineering Chemistry. Product Research and Development* 18 (3): 227–230.

119 Van, W.T.L., Affens, W.A., Beal, E.J. et al. (1987). Determination of liquid and solid phase composition in partially frozen middle distillate fuels. Butterworth & Co Ltd., *Fuel* 66: 890–896.

120 Turner, W.R. (1971). Technical review – Normal alkanes. American Chemical Society, *Industrial and Engineering Chemistry. Product Research and Development* 10: 238–260.

121 Craig, S.R., Hastie, G.P., Roberts, K.J. et al. (1998). Investigation into the Structures of Binary-, Tertiary- and Quinternary- Mixtures of n-Alkanes and Real Diesel Waxes Using High-Resolution Synchrotron X-Ray Powder

Diffraction. Royal Society of Chemistry, *Journal of Materials Chemistry* 8 (4): 859–869.

122 Craig, S.R., Hastie, G.P., Roberts, K.J. et al. (1999). In-situ study of the solid-solid phase transitions occurring in real diesel wax crystalline systems using differential scanning calorimetry and high-resolution synchrotron x-ray powder diffraction. Royal Society of Chemistry, *Journal of Materials Chemistry* 9: 2385–2392.

123 Coutinho, J.A.P. and Ruffier-Meray, V. (1997). Experimental measurements and thermodynamic modelling of paraffinic wax formation in undercooled solutions. American Chemical Society, *Industrial and Engineering Chemistry Research* 36: 4977–4983.

124 Coutinho, J.A.P. and Thermodynamic, A. (2000). Model for predicting wax formation in jet and diesel fuels. American Chemical Society, *Energy and Fuels* 14: 625–631.

125 Reddy, S.R. (1986r). A thermodynamic model for predicting n-paraffin crystallization in diesel fuel. *Fuel* 65: 1647–1652.

126 Bennema, P., Liu, X.Y., Lewtas, K. et al. (1991). Morphology of orthorhombic long chain normal alkanes: Theory and observations. Elsevier Science Publishers B.V., *Journal of Crystal Growth* 121: 679–696.

127 Clydesdale, G., Roberts, K.J., and Docherty, K. (1994). Modelling the morphology of molecular crystals in the presence of disruptive tailor-made additives. Elsevier Science B.V., *Journal of Crystal Growth* 135: 331–340.

128 Lovette, M.A., Browning, A.R., Griffin, D.W. et al. (2008). Crystal shape engineering. American Chemical Society, *Industrial and Engineering Chemistry Research* 47 (24): 9812–9833.

129 Boistelle, R. and Astier, J.P. (1988). Crystallization mechanisms in solution. Elsevier Science Publishers B.V., *Journal of Crystal Growth* 90: 14–30.

130 Petinelli, J.C. (1979). Influence of additives on the crystallization of n-paraffins in a hydrocarbon medium. *Oil & Gas Science and Technology* 34 (5): 771–790.

131 Marie, E., Chevalier, Y., Eydoux, F. et al. (2005). Control of n-alkane crystallization by ethylene-vinyl acetate copolymers. Elsevier Inc., *Journal Colloid Interface Science* 290: 406–418.

132 Zhang, J., Wu, C., Li, W. et al. (2003). Study on performance mechanism of pour point depressants with differential scanning calorimeter and X-ray diffraction methods. Elsevier Science Ltd., *Fuel* 82: 1419–1426.

133 Isasi, J.R., Haigh, J.A., Graham, J.T. et al. (2000). Some aspects of the crystallization of ethylene copolymers. Elsevier Science Ltd., *Polymer* 41: 8813–8823.

134 Costello, J.K. (1987). Liquid fuel compositions. EP 0,255,345. Filed Europe, 28 July.

135 Thetford, D. (2003). Applications of oligomeric surfactants in polymer systems.Ch.5. In: *Surfactants in Polymers, Coatings, Inks & Adhesives*, Vol. 1. pp 120-151 (ed. D.R. Karsa), Applied Surfactant Series, Blackwell/CRC, Boca Raton. ISBN 9780367812416

136 Richter, D., Schneiders, D., Monkenbusch, D. et al. (1997). Polymer aggregates with crystalline cores: The system polyethylene-poly(ethylenepropylene). American Chemical Society, *Macromolecules* 30: 1053–1068.

137 Richter, D., Radulescu, A., Schwahn, D. et al. (2014). *Wax Control by Self-Assembling Polymers – A Scientific Approach*. Washington: Oak Ridge National Laboratory, Neutrons for Energy for the Future.

138 The Study of Crystal Growth with the Electron Microscope IV (1955). The nucleation and growth of n-propyl-pentacontanoate. In: *Proceedings of the Royal Society of London* Vol. 228.(eds. N.G. Anderson and I.M. Dawson), Series A, 539–548. Mathematical and Physical Sciences.

139 Davey, R.J. (1976). The effects of impurity adsorption on the kinetics of crystal growth from solution. North Holland Publishing Co., *Journal of Crystal Growth* 34: 109–119.

140 Hohoh, H., Ueno, S., and Sato, K. (2018). Fundamental aspects of crystallization of Lipids. In: *Crystallization of Lipids* (eds. K. Sato), Wiley-Blackwell, Chichester, West Sussex. Ch. 4 pp. 105–141.

141 Lewtas, K. (1987). Improved fuel additives. GB Patent 2,231,584. Filed UK, 24 September.

142 Tadros, T.F. (2007). General principles of colloid stability and the role of surface forces. *Colloid Stability and Application. Wiley VCH* 1 (Chapter): 1.

143 Poovarodom, S. and Berg, J.C. (2010). Effect of Particles and Surfactant Acid-Base Properties on Charging of Colloids in Apolar Media. Elsevier Science B.V., *Journal of Colloid and Interface Science* 346: 370–377.

144 Beiny, D.H.M., Mullin, J.W., and Lewtas, K. (1990). Crystallization of n-Dotriacontane from hydrocarbon solution with polymeric additives. Elsevier Science Publishers B.V., *Journal of Crystal Growth* 102: 801–806.

145 Madsen, H.E.L. and Boistelle, R. (1979). Growth kinetics of the (001) faces of hexatriacontane (C36H74) in solution. North Holland Publishing, *Journal of Crystal Growth* 46: 681–690.

146 Hollander, F.F.A., Stassse, O., Van-suchtelen, J. et al. (2001). Recrystallization phenomena of solution grown paraffin dendrites. Elsevier Science B.V., *Journal of Crystal Growth* 233: 868–880.

147 Richards, P. (2014). *Automotive Fuels Reference Book*, Third Edition. Society of Automotive Engineers, Warrendale, P.A.

148 West, T.H. and Martella, D.J. (1986). Solvent dewaxing using combination poly (N-C24) alkylmethacrylate-poly (C8-20 alkyl) methacrylate dewaxing aid. US Patent 4,728,414. Filed 21 November.

149 Kaminsky, M. and Buschermohle, M. (1987). Process for the preparation of a 1-olefin stereo-block polymer. US Patent 4,841,004. Filed 25 November.

150 Fish and Tsang (2014). Intellectual property law, boiler plate. fishiplaw.com. http://www.fishiplaw.com/~fishipla/chapter7a.html#boilerplate (Cited: 20 April 2014).

5

Protection of Metal Surfaces in Fuel Systems

Lubricity Improvers and Corrosion Inhibitors

The purpose of lubricity improvers is to protect the surfaces of metals in fuel pumps that move over one another, against physical erosion (wear). Corrosion inhibitors protects metal surfaces against chemical erosion, whether they are moving or not. Steel is the most widespread metal of concern in fuel systems, including fuel pumps; others are copper and its alloys (such as brass), silver (in sensors) and aluminium alloys (in aeroplanes). Both of these additive types act by adsorbing onto the metal surfaces and forming a protective layer; to provide this effect, molecules of these additives carry polar, metal-loving groups to ensure their adsorption and hydrocarbon groups to provide fuel solubility and to create a protective film. Many of the long-standing corrosion inhibitors are also effective lubricity improvers, while the most common lubricity improvers are not the best corrosion inhibitors – especially for non-ferrous metals. In addition, the scope of corrosion inhibitors must be widened to include those needed in systems that burn heavy fuels.

Details of the problems of lubricity and corrosion are quite different, so these will be discussed separately. Common additive chemistries will act as a bridge between the two before the divergence into the more effective or specialised corrosion inhibitors.

5.1 Lubricity: Introduction

Fuel pumps, in passenger cars and small commercial vehicles, rely upon the fuel to provide them with their necessary lubrication. In the absence of adequate lubrication, metal surfaces moving over each other, in close contact,

Fuel Additives: Chemistry and Technology, First Edition. Robert D. Tack.

suffer surface roughening and wear. This leads to early failure of fuel pumps resulting from increased clearances, limiting their ability to provide adequate pressure, or even seizure. Wear processes also expose clean metal surfaces which are then more susceptible to corrosion. The ability of a fuel to provide adequate lubrication is known as lubricity.

Lubricity problems in fuel systems were first noticed in the 1960s and 70s as 'severe aviation fuel pump problems' caused using highly hydrofined [1] aviation kerosene, used as jet fuel [1] [2]. These problems of excessive wear in various parts of the fuel pumps were, then, addressed by changes to fuel specifications and by redesigning jet fuel pumps. After the rapid rise in OPEC crude oil prices in the 1970s, there was much interest in increasing fuel economy of automotive engines, for which reduction of friction in engine moving parts was one approach. A contribution to this approach was the use of ashless friction modifiers in gasoline engines. Friction modifiers, such as fatty acid esters, were added to crankcase lubricants and to gasoline [2]. Glycerol mono-oleate was one such friction modifier [3], an additive that had earlier been proposed for use as a corrosion inhibitor for diesel fuel and jet fuel [4].

Before 1990, fuel pump lubrication by the fuel itself was sufficient for automotive engines that relied upon fuel injection. Up to that time, diesel fuel and gasoline were allowed to contain quite high levels of sulphur, such as 0.1%–0.2% as in 'conventional winter-grade diesel fuel' [5], from which harmful sulphur acid aerosols were produced in the exhaust fumes. As part of a long-term European programme to reduce such harmful emissions, permitted sulphur levels in these fuels were reduced, initially to 50 ppm and then to 10 ppm; now, the removal of sulphur from automotive fuels occurs worldwide. Unfortunately, in Sweden, the first country to apply such low sulphur levels to diesel fuel (1991), it was found that these desulphurised fuels led to rapid failures of fuel injector pumps in light vehicle diesel engines [6]. In the 1990s and 2000s, diesel engines, with their superior fuel economies, became the engines of choice in Europe, displacing gasoline engines in most light commercial and half of all new passenger cars, raising the need for lubricity additives (section 1.2.2).

Heating oil usually contains the less severely hydrotreated middle distillates that cannot be used in ultra-low sulphur diesel (ULSD) blends. In addition, fuel pumps in heating systems do not have to supply the same pressures as those needed for diesel fuel injection. So, lubricity is not usually a problem in heating oils. However, sometimes heating oils are blended from the same, low-sulphur, low aromatic refinery streams as those used in diesel fuels. In addition, sulphur levels in heating oil specifications are being reduced towards those in diesel fuel in richer countries. As with most fuel additives, when needed, heating oil uses the same lubricity additives that are already in use in diesel fuel, for the same purposes [2]. Because the needs of automotive fuels determine what materials are used as lubricity improvers, there will be little separate discussion of lubricity in heating oils.

5.2 Detailed Lubricity Problems

The ability of a fuel to provide lubrication to moving parts, in fuel pumps and injectors, is termed lubricity. Lubricity depends upon a fuel being able to provide both hydrodynamic and boundary lubrication (section 5.4). Hydrodynamic lubrication requires the fuel to be viscous enough to maintain a liquid layer between moving metal surfaces. To provide boundary lubrication, the fuel must contain compounds that can adsorb on metal surfaces, and remain there when metal surfaces are pressed together, expelling most of the liquid fuel [7, 8]. Poor lubricity can arise out of the fuel having too low a viscosity to provide hydrodynamic lubrication – a problem found with the low boiling grades of arctic diesel fuel, and with fuels such as methanol and ethanol when substituted for gasoline [9] or diesel fuel [10]. The severe hydrotreating necessary to produce ultra-low sulphur fuels removes compounds that contain oxygen and nitrogen and poly-aromatics, which are naturally occurring providers of boundary lubrication in straight-run fuels.

5.2.1 Jet Fuel

Jet fuel is subject to special scrutiny because of the catastrophic nature of any engine failure during flight. In the 1960s and 70s, it was found that while straight-run jet fuels usually gave adequate lubrication of fuel pumps, there were problems with hydrotreated, low sulphur fuels: excessive bore wear, piston seizure, failures in control valves and wear in fuel injector plungers [1, 11]. To overcome the wear problems, pumps were redesigned and lubricity specifications were put in place to allow the use of kerosene blends containing hydrotreated[1] fuels. Lubricity specifications were developed for jet fuel using the Ball-On-Cylinder Lubricity Evaluator (BOCLE) test (ASTM D6078) [12]. BOCLE testing produces a small wear scar on a steel ball when it is held against a rotating steel ring, which is partly immersed in fuel[3]. The jet fuel specification states that this wear scar should have a maximum diameter of 850 microns [13].

If the jet fuel contains a high proportion of hydrotreated kerosene, it may be treated with a lubricity improver to provide the required lubricity [7, 13] – fortunately, it was discovered that materials already approved for use in jet fuel as corrosion inhibitors also provided lubricity [14]. However, worries about the possible effects of these additives on jet engine parts have led manufacturers of civilian aircraft to rely upon engineering solutions and fuel specifications [15]. Jet A or A-1, the main grades for commercial use, do not usually contain lubricity/corrosion additives, but rely upon a careful selection of suitable refinery blend components to provide lubricity, as measured by the BOCLE test. Because military jet fuels (JP grades) are more highly stressed, they are required to contain lubricity additives which are, in fact, the approved corrosion inhibitors [7, 13] (section 5.7.1).

5.2.2 Gasoline

The story for gasoline is in two parts, sandwiching the rising need for lubricity improvers for diesel fuel. Investigations into alternative fuels came to the fore after the oil shock of the early 1970s when methanol and ethanol were considered as alternatives for gasoline. These alcohol-fuels pose lubricity problems because of their low viscosities: the poor lubricity of unadditised methanol as a fuel, and its need for lubricity improvers, was demonstrated by the wear of the piston rings and cylinder bore in a gasoline test engine [12]. At that time, several patents were filed that described lubricity additives for use with methanol and/or ethanol [16]. However, for various reasons, the alkanols made little impact on gasoline, at that time.

More importantly, in the 1970s, the USA government passed the Corporate Average Fuel Economy (CAFE) standards act, requiring automobile manufacturers to double passenger vehicle efficiencies (miles per gallon) over the next ten years [17]. As a result, any efficiency improvements were welcomed, including those resulting from improvements in engine lubrication. One such improvement was the addition of friction modifiers to the engine lubricating oil (usually referred to as the crankcase lubricant) – advertised as extra mileage ingredients. However, the friction modifiers most favoured were carboxylic acids and esters which have short life spans in the tough environment of a crankcase lubricant. To overcome this problem, the friction modifiers, such glycerol mono-oleate [3] (Figure 5.1), became gasoline fuel additives which would continuously top up lubricity in the crankcase lubricant [18].

Wear caused using ethanol, in place of regular gasoline, appears to be less than by the use of methanol [12] and any problems of lubricity, caused by the

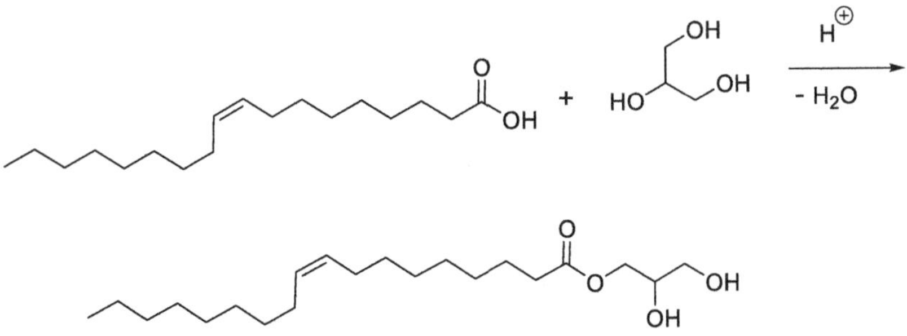

Figure 5.1 Glycerol Monooleate from Oleic Acid and Glycerol.

use of ethanol, responded to lubricity improvers. This is fortunate as ethanol has become a major component (25% in E25) of gasoline in Brazil[4] [19] and, since 2000, has grown to be a substantial part (10%–15%) of the gasoline mix in the USA, termed gasohol [20].

Pre-1990, gasoline lubricity in automotive engines [21] was not a problem because fuel vapour and air were mixed in a carburettor before entering the combustion chamber, drawn in by the negative pressure of the intake-stroke (the piston moving down in the cylinder). In turbocharged engines, air is pressurised by a small compressor so that more air and gasoline vapour can be forced into the combustion chamber. In such engines, fuel pumps did not need to provide significant pressures and sulphur levels were not then constrained[5], so they did not suffer from lubricity problems.

To provide more precise control of gasoline entering the engine cylinders, electronically controlled fuel injectors were introduced in the last decades of the twentieth century and are now the norm for gasoline engines. Most of the first wave of gasoline injection engines used indirect injection, known as port fuel injection (PFI), whereby gasoline is injected into the air stream, at the inlet port, just before the inlet valve. Compared with carburetted engines, PFI engines produce lower emissions, better power output, better response, and better fuel economy. While PFI needs fuel pumps to provide higher pressures (0.25–0.45 MPa) than for carburettors (30–40 kPa), these are much lower pressures than those required in diesel engines in which fuel is injected almost at the top of the compression stroke (up to 200 MPa or more [22]).

Since the late 1990s, in response to the need for continued improvements in performance, fuel economy and lower emissions, Gasoline Direct Injection (GDI) has become the norm. With GDI, the gasoline is injected into the cylinder at various times through the stroke. At low and medium loads, injection is just before spark ignition (top of the compression stroke) needing higher pump pressures (4–14 MPa) than for PFI so, for GDI, injector pump wear problems were anticipated [8]. While GDI technology is more expensive than PFI, Bosch, a major manufacturer of fuel injectors, predicted in 2013 that 25% of new vehicles in 2020 would have GDI[6] [23]. However, partly for reasons of fuel economy, 51% of new gasoline vehicles had GDI in 2018 [24].

The main changes for gasoline from the late 1990s onwards have been the stepwise reductions in permitted sulphur levels to 10 ppm today, in Europe and the USA, being followed by the rest of the world. Gasoline with such low levels of sulphur could produce significant wear in GDI fuel pumps and injectors, without the use of lubricity improvers. However, gasoline fuels may already contain friction modifiers, lubricity improvers by another name, and corrosion inhibitors [2].

5.2.3 Diesel

Fuel injector pumps in Heavy Goods Vehicles (HGVs) are lubricated by the engine oil, so do not rely upon the lubricity of their diesel fuel. However, smaller diesel engines in lighter vehicles – vans and passenger cars – use rotary injector pumps that rely upon the fuel for their lubrication. Unfortunately, naturally occurring, lubricity-providing compounds are mostly removed from fuel by the hydrogenation processes used to remove sulphur – as discovered in 1991, in Sweden, when such fuels were quickly found to lead to premature failure of rotary injection pumps. The occurrence of this problem in a rapidly growing sector of passenger vehicles in Europe, plus the plans for reductions of permitted sulphur levels to a maximum of 10 ppm Europe-wide (now part of the EN590 Ultra Low Sulphur Diesel, ULSD, specification), led to rapid growth in the demand for and use of diesel fuel lubricity improvers.

Following the European process, reductions in diesel fuel sulphur levels have been taking place around the world, leading to continued uptake of lubricity additive technology. While some countries were moving slowly, in the 2014 WDFQS [25][7], only in eight countries (out of fifty-one surveyed) were the average diesel fuel sulphur levels above 100 ppm (some samples as high as 0.5% sulphur). However, in the 2018 WDFQS [26], the sulphur levels in these eight countries had dropped significantly, though for six of them the average was still between 100 and 500 ppm sulphur, except Kuwait with 0.1%; in the other two countries, China and Argentina, sulphur levels had dropped to less than 50 ppm At the same time, in both WDFQSs, all 219 and 211 (2014 and 2018 respectively) samples collected in Europe plus North America had sulphur levels less than 10 ppm.

Following on from the jet fuel and gasoline friction modifier experiences, additives known as corrosion inhibitors, lubricity improvers or friction modifiers, were introduced for use as lubricity additives in low sulphur diesel fuel. Early in this process, a simple, quick fuel lubricity test was needed for routine, everyday use and an associated specification of what constituted adequate lubricity [27, 28]. As with the use of most fuel additives, the running of automotive vehicles to failure is far too slow and expensive a method to provide an ongoing measure of lubricity of refinery produced fuel, on a continuous basis. There were several possible wear tests, such as the Ball-On-Cylinder Lubricity Evaluator (BOCLE), already in use for jet fuel, and the High-Frequency Reciprocating Rig (HFRR). After much evaluation and test development, it was found that the HFRR – with a clearly specified operating procedure – gives the best prediction of adequate vs inadequate lubricity for the protection of rotary fuel injection pumps [15, 27, 28].

In the HFRR test as applied to diesel fuel (ISO 12156-1, CEC F-06-A-96, ASTM D6079), a steel ball is held fixed on the moving arm of a vibrator and is pressed by an applied load (200 g) against a static steel plate that is covered by the test fuel (2 ml) at a controlled temperature of 60°C (Figure 5.2). The arm is vibrated over a stroke length of 1 cm at a frequency of 50 Hz for 75 minutes. At the end of the test, the diameter of the wear scar on the steel ball is measured, under a microscope, in the same linear direction as that of its movement on the plate and again perpendicular to this direction; the average of these two diameters is taken to be the Wear Scar Diameter (WSD).

The Coordinating European Council accepted the HFRR as the standard test for evaluating lubricity and anti-wear characteristics of a diesel fuel – this test is now used in many countries. In Europe, the maximum WSD for acceptable lubricity of diesel fuels is specified as 460 microns (EN590), in the USA

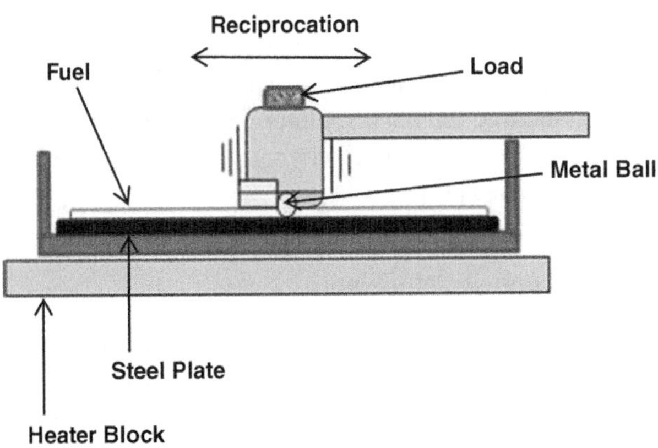

Figure 5.2 The HFRR Equipment (Diagrammatic).

it is 520 microns (ASTM D975), while the Worldwide Fuel Charter[8] recommends 400 microns [25]. In the 2018 WDFQS [26], in 64% of diesel fuel samples the WSD was below 400 microns and 89% below 460 microns; almost all samples from the USA had WSDs below the specification of 520 microns; only in the middle east were there samples showing WSDs greater than 550 microns.

5.3 Chemistries of Lubricity Improvers

For the severe environment of the crankcase, surface active agents are routinely added to the lubricating oil to adsorb onto metal surfaces and produce a barrier that provides boundary lubrication between those surfaces in contact – a process that was identified as far back as 1922 [29]. Compounds such as zinc dialkyl-dithio-phosphates (ZDDP) have been widely used as anti-wear additives in crankcase lubricants since the 1940s [3]. Such lubricating oil additives were recognised as too chemically reactive for use in fuels [6]; in addition, they add sulphur to low sulphur fuels and the zinc and phosphorus in ZDDP produce detrimental deposits in fuel systems. Consequently, fuel lubricity additives have been developed that are ashless organic molecules that rely upon oxygen and nitrogen containing groups to provide the polarity needed to adsorb onto metal surfaces. Most simply and practically, this means alcohols, ethers, carboxylic acids and esters, amines and amides.

Metal parts of a fuel system that need the presence of lubricity improvers are usually made of steel alloys. Consequently, their surfaces are, mostly, iron and iron oxides. A lubricity improver is a molecule that combines two, necessary chemistries. One is a polar group that will adsorb strongly onto metal and metal oxides, anchoring the molecule on such surfaces. And second, an oil-soluble, hydrocarbon group large enough to keep the two contiguous metal surfaces apart and to lubricate their movement over each other. These requirements are most readily fulfilled by fatty acids, dimer acids and their amine salts, amides, and partial esters of polyols. Lubricity improvers, proposed and commercial, have been summarised [30].

5.3.1 Carboxylic Acids as Lubricity Improvers

These two structural requirements are like those for inhibitors of steel corrosion (section 5.5). For corrosion inhibition, the hydrocarbon group is needed to keep water away from the metal surface, while for lubricity it is to keep the metal surfaces apart. The similarities of the molecular requirements have often led to additives with the same chemistries for these two applications. Indeed, as already pointed out, it was found that the corrosion inhibitors already approved for use in jet fuel also provided lubricity [14]. These are dimer/trimer acids and the di-acids dodecenyl succinic acid (DDSA) and tetra-propenyl succinic acid (TPSA) (Figures 5.6 and Figure 5.7). The chemistries of corrosion additives that are also used as lubricity improvers, and their preparations, are described below (section 5.5 and on).

Since its introduction, ULSD has been the focus of attention for lubricity improvers, following the clear need for such additives in the diesel engines of light vehicles. As already stated, gasoline has less need for lubricity improvers and often already contains some of these materials as friction modifiers, such as glycerol mono-oleate (Figure 5.1), and corrosion inhibitors, such as DDSA [2]. While further discussion of lubricity improvers will centre on the ULSD application, any such additives may also be used in gasoline or heating oil.

Vegetable oils themselves act as lubricity improvers but the proportions needed are high, for example 5% has been recommended[9] [31]. Simple vegetable oils have low volatilities, high molecular weights and high viscosities so are not recommended for use as diesel fuel components because they tend to produce nozzle coking, engine deposits and ring sticking [32]. However, the preferred bio-diesel components, Fatty Acid Methyl Esters (FAME), provide good lubricity at 1%–2 % in ULSD [33–35] and FAMEs are being added at up to 20% of diesel fuel as a sustainable option, giving only minor additional problems (the 2018 WDFQS found that almost half of the samples contained 1%–21% FAME [26]). Such proportions are blend components – FAMEs are not effective at additive concentrations (see section 1.3.5). The relevance of the lubricity of FAME, though, is that there is usually no need for lubricity improvers to be added to biodiesel blends containing 2% or more of FAME.

Fatty acids are the most widely used fuel lubricity improvers [2]. Commercial examples are a 'mono-acid' diesel fuel lubricity improver by Afton [36], and 'fatty acid based ... lubricity enhancing additives' from Lubrizol and Infineum [37]. For good solubility in middle distillates, at low temperatures, these acids need to be low melting[10], hence unsaturated – for example, the melting points of stearic (octadecanoic) and oleic (octadec-9-enoic) acids are 69°C and 13°C, respectively. For availability and low cost, naturally occurring fatty acid mixtures, containing high proportions of unsaturated fatty acids, are used such as olive and rapeseed (canola) oils, and tall oil fatty acids (TOFA[11]). A variation on the use of simple fatty acids as lubricity improvers is the use of dimer-acids (Figure 5.6), trimer-acids, DDSA or TPSA (Figure 5.7), as multiple adsorbing carboxylic acid groups improve the strength of adsorption of the additive. However, as these diacids require more processing than the simple mono-carboxylic acids they are more expensive so, as lubricity improvers, they are not as widely used as TOFA and similar low melting fatty acids.

5.3.2 Carboxylic Esters and Amides as Lubricity Improvers

Fatty acids are not without problems, they have been found to be responsible for fuel filter deposits, injector deposits and top ring corrosion [38]. Zinc-based deposits are thought to be caused by acid lubricity improvers leaching zinc from brass fuel-line fittings (section 3.5.1b). Additive companies provide ester-based lubricity improvers to avoid such problems, for example Innospec [39]. In particular, the patent literature points to polyol esters – not least glycerol mono-oleate (Figure 5.1), described as a friction modifier for gasoline [2]. Glycerol monooleate was identified as effective for the specific purpose of diesel fuel lubricity in the early 1990s [40] and is identified as a commercial Infineum product [37]. Such chemistry is still the subject of patents, such as a preparation of a soya mono-glyceride [41]. The general principle is to replace the deposit-forming, reactive carboxylic acid as the polar group, necessary for adsorption on the metal surface, by a less reactive ester group along with two hydroxyl groups for strong, multipoint adsorption. Pentaerythritol mixed esters of oleic or tall oil acids, which also contain ester groups and free hydroxyls, have also been proposed as lubricity improvers, having improved low temperature properties [42].

The esterification reaction used to prepare lubricity improving esters is relatively simple organic chemistry – it is a reversible, acid catalysed condensation reaction (elimination of water) between a carboxylic acid and an alcohol, driven to completion by removal of water (Figure 5.1). When the alcohol is a polyol, such as glycerol or penta-erythritol, while some control may be had from the molar proportions of the two reactants, it is not easily possible to produce just a mono-ester or di-ester – the product maybe a mixture of mono-, di- and tri-esters [18]. The mixture is richer in the monoester if a 1:1 molar mixture of polyol and acid are reacted and richer in di-ester if 1:2.

A diesel fuel additive blend patent describes, and uses in tests, six commercially available lubricity improvers: three are TOFA; one is another mixture of fatty acids; one is a mixture of glycerol mono- and di- (minor proportion of tri-) mixed oleate/linoleate esters; one is an alkyl salicylic acid; and one is an amide of TOFA with diethanolamine [43]. Another patent claims a lubricity potency advantage for mixtures of an alkylsalicylic acid and hydroxyethyl-alkylsalicylate (Figure 5.3) [44], providing a quantitative HFRR comparison of various preparations[12] with a commercial glycerol mono-oleate/linoleate (mixed esters) (Table 5.1). HFRR WSDs were obtained in a middle distillate test fuel treated with various concentrations of the preparations to determine the minimum needed for WSDs of 350–355 microns[13]: the results demonstrate a competitive lubricity improving potency in this fuel.

Several patents have been taken out by Ethyl that demonstrated the anti-wear or lubricity improving properties of diethanolamine amides of carboxylic acids and diacids. The reaction product of diethanolamine with oleic acid was described (Figure 5.4) and its anti-wear activity demonstrated in gasoline, using

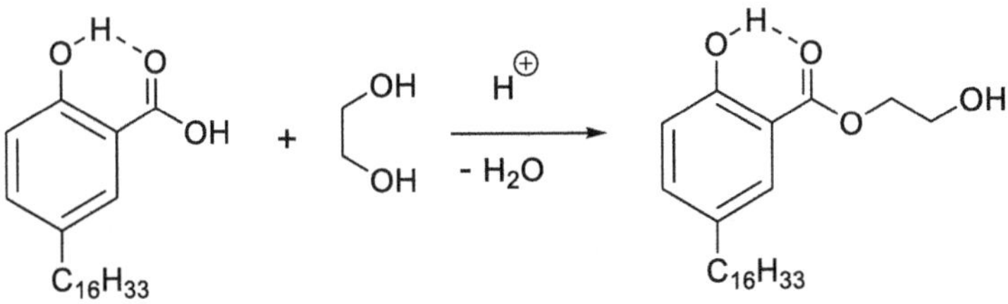

Figure 5.3 Alkyl Salicylic Acid and Its 2-Hydroxyethyl Ester.

Table 5.1 Comparison of the HFRR Potencies of Alkyl-Hydroxyethyl-Salicylate with Commercial Glycerol Mono-oleate.

Lubricity improver	Ppm needed for HFRR wear scar of 350–355 microns
None	Wear scar 548 microns
Alkyl salicylic acid	172
Glycerol monoester (commercial)	60
Alkyl salicylic acid, 35% hydroxyl-ethyl ester	15–25
Alkyl salicylic acid, 80% hydroxyl-ethyl ester	= < 15

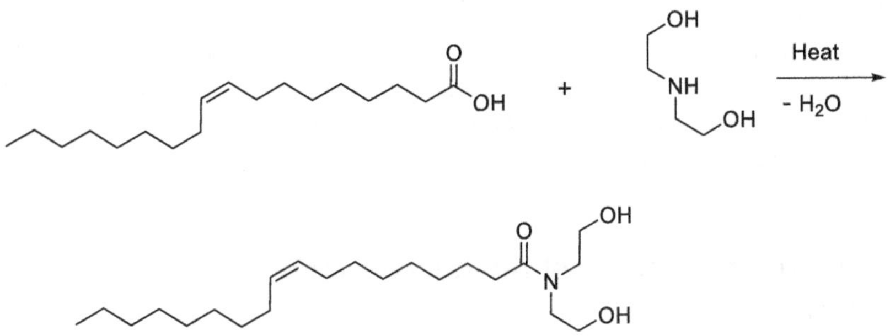

Figure 5.4 N,N-bis(2-hydroxyethyl)-Oleamide.

the four ball wear test, and in a low sulphur biodiesel fuel, using both the BOCLE and HFRR tests [45, 46]. Replacing the oleic acid by the less expensive TOFA is the source of the TOFA/diethanolamine product [43]. Such a material is believed [37] to be available as a commercial lubricity improver Hitec™ 4848A, from Afton (who retained the related patents on the reorganisation of Ethyl).

The principle of esterifying a deposit forming acid with a polyol has been carried through to diacids such as dodecylsuccinic and the dimer acids. Since attempts to esterify mixtures of polyols (even diols) and such di-carboxylic acids would produce polymers with high viscosities, their preparation should include monoacids such as oleic acid and/or mono-ols as dodecanol[14] [47]. Dimer acid esterified with ethylene glycol (1,2-dihydroxy-ethane) was found particularly effective in providing lubricity to 100% methanol, when used as an automotive fuel in a gasoline engine [9]; such compounds had previously been claimed to improve the lubricity of gasolines, jet fuel, kerosene and diesel fuel – the lubricity was demonstrated in jet fuel using the Ryder gear scuff test [48]. A recent patent claims an improved process for the manufacture of this lubricity improver: reacting ethylene carbonate with dimer acid, using a tertiary amine catalyst, which avoids the established sulphonic acid (sulphur-containing!) catalysed esterification of dimer acid and an alkylene diol[15]; lubricity testing of the ethylene glycol ester at 100 ppm in an American ULSD, presumably by the HFRR, brought the off-specification WSD of almost 600 microns into specification (below 520 microns in the USA) [1].

Assessing the relative degrees of effectiveness of the various lubricity improver chemistries is fraught with difficulties. Published data is almost entirely within patents, some of which make comparisons only with a fuel having no lubricity improver while others may only make comparisons with untreated fuel and with fuel containing a precursor to the claimed improvement. In addition, since the histories of many of the materials in use stretch back over decades, original tests used may bear little resemblance to current day tests; indeed, a whole range of test data has been obtained using a wide range of anti-wear tests. Finally, the fuels in which the lubricity improvers of interest were tested were all unique and a lubricity improver's effectiveness varies from fuel to fuel.

5.4 Understanding of Boundary Friction and Lubricity

When two pieces of metal in close contact are moving past each other, there is friction, causing resistance to motion and a gradual wearing away of the metal. As the pressure is increased, by an increase in load, friction and wear increase causing surface roughening. Surface roughening leads to further increases in friction and, eventually, seizure and pieces of metal and metal oxide being dislodged at the point of contact, exposing the clean metal surfaces to corrosion. The dislodged particles can cause further wear, such as scoring, in the underlying metal surfaces. This general process is often referred to as 'scuffing' [49].

On the microscopic scale, metal surfaces are irregular, not smooth (Figure 5.5a [50]). When the pressure between the moving metal parts is low ('low load') the presence of the lubricating fluid, which is initially held in place by surface tension, provides hydrodynamic lubrication between the metal surfaces [49]. As one surface moves over another, the fluid between the surfaces gathers up beneath the moving surface allowing the moving surface to skim over a film of the fluid and 'the solid surfaces are completely floated apart by the lubricant' [29]. When water is the fluid, the process is known as aquaplaning – a problem when car wheels lock trying to brake on wet roads but applied for pleasure when water skiing.

Hydrodynamic lubrication, by fuel alone, may be sufficient in non-severe conditions. However, in many parts of an engine, such as a fuel injection pump, the environment is severe as there is significant pressure between moving metal parts. Asperities on the metal surfaces then come into contact and the fluid is squeezed out (Figure 5.5b). Contact area between the asperities

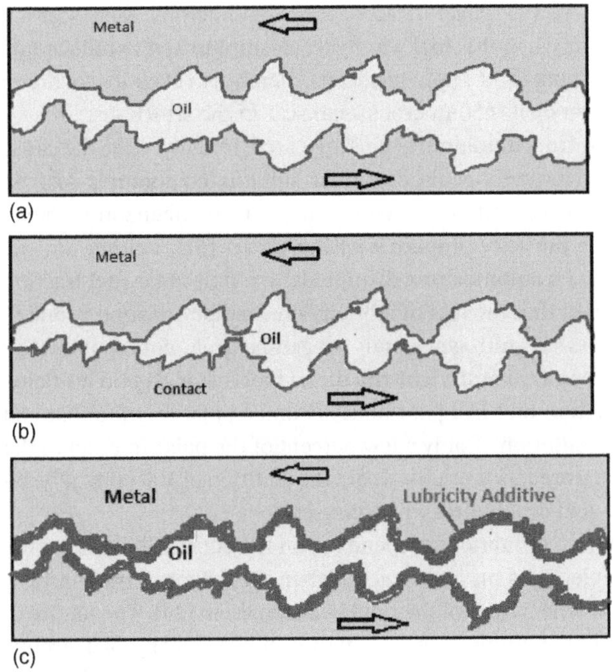

Figure 5.5 Lubrication of Moving Metal Surfaces in Contact (a) Moving Metal Surfaces, Hydrodynamic Lubrication. (b) Metal Surfaces under Pressure, Asperities in Contact, No Continuous Oil Layer. (c) Metal Surfaces with Adsorbed Lubricity Additive, Boundary Lubrication.

is much less than may appear from the macroscopic picture – it may be as little as 0.01% of the apparent area of contact [49] – so the pressure at the points of contacts can be quite large. Plastic deformation of the contacting asperities occurs and attraction between the atoms and molecules of the two surfaces can cause adhesion; as the surfaces continue moving apart, parts of the adhering surfaces are torn away, leading to surface roughening, which is observable by scanning electron microscopy of such surfaces [11]. Hydrodynamic lubrication is insufficient to prevent seizure and/or extreme wear under this condition, which is known as boundary friction. In addition to this, there is a component of wear known as 'fretting', which is caused by small amplitude vibrations [27].

Under these more severe conditions, the fluid needs to contain surface active agents, Lubricity Improvers, that chemically adsorb onto the metal surfaces and produce lubrication by means of a boundary film of the LI, with its solvating fuel, between them (Figure 5.5c). Prior to severe hydrodesulphurisation (HDS), there were no problems of rotary injection pump wear. Naturally occurring compounds, bearing polar groups, were providing the necessary lubricity. For example, a straight-run light gas oil, with 1203 ppm of sulphur (and 15 ppm of nitrogen[16]), gave an acceptable 460-microns wear scar on HFRR testing [51]. But when this fuel was hydro-desulphurised to reduce fuel sulphur content to 10 ppm (and 8 ppm nitrogen), there was a clear loss of diesel fuel lubricity, as shown by its 650-microns wear scar in the HFRR test.

Removal of sulphur from automotive fuels appeared, initially, to be the cause of loss of lubricity. However, replacement of sulphur compounds lost on hydrodesulphurisation of diesel fuel, by benzothiophene or dibenzothiophene, did not replenish the lubricity properties of the ULSD [51]; neither did the addition of up to 1% of a sulphide or a disulphide to a non-polar fuel fraction [52]. It is now accepted that the loss of lubricity on desulphurisation is due to the concomitant loss of nitrogen- and oxygen-bearing compounds [8]. Chromatographic separation of straight-run diesel fuel that had good lubricity, gave a non-polar fraction that had poor lubricity and a polar fraction that had very good lubricity – addition of only a few percent of the polar fraction to the non-polar fraction restored its lubricity, as did the addition of 100 ppm of hexadecanoic (palmitic) acid or 8-hydroxyquinoline [52].

Boundary lubrication, or lubricity, depends upon the chemical nature of the surfaces – for example, glass on glass has a higher coefficient of friction than steel on steel – and on the nature of the fluid between them [29]. The nature of the fluid has two contributions: one, the nature of the fluid's polar groups determine its strength of adsorption, so carboxylic acids more effectively reduce friction for steel on steel (or glass on glass) than alcohols which are, in turn, more effective than n-alkanes; second, the longer the alkyl group attached to the polar groups, the lower the friction [29]. Friction between surfaces under pressure is due to adhesive attractions between the atoms or molecules on the two interacting surfaces, so is different for different materials. For a given material, the greater the separation between the surfaces, the less their mutual attraction; so longer alkyl chains on the adsorbed species, for example on stearic acid ($C_{17}H_{35}COOH$), lead to lower attractions between the surfaces, hence lower the friction between them, than do shorter alkyl chains, for example on lauric acid ($C_{12}H_{23}COOH$) and on acetic acid (CH_3COOH) [29]. For steel or glass, the negative slopes of the coefficient of friction against the molecular weight of the fluid were found to be the same for homologous series of alcohols and carboxylic acids (though with different intercepts on the ordinate), demonstrating a constant relationship between the increased lengths of the alkyl groups of the adsorbing fluid ('lubricant'), the increased separation of the surfaces and the reduced friction between the surfaces.

Carrying these findings through to the action of lubricity improver molecules: a polar group, such as a carboxylic acid group, enables the molecule to adsorb onto the metal surface, and adsorption continues until a mono-layer of the carboxylic acid is formed on the metal surface, with the alkyl groups projecting normal to the surface (demonstrated by x-ray diffraction of surface layers [53]). These alkyl groups should be large enough to separate the two surfaces sufficiently that the attractive forces are minimised. The same alkyl groups must also confer good solubility in cold fuel, so long, straight chains, such as those on stearic acid, are not favoured while unsaturated versions, as oleic acid, are (section 5.3). Good solubility, as provided by unsaturation or short alkyl side branches, also suggests that the adsorbed molecules will carry solvating fuel molecules with them into that boundary layer, further improving lubricity. Choice is limited by economic availability and potency, which limits the returns for increased alkyl chain length: the polar group enables adsorption as a monolayer and doubling the length of the attached alkyl group doubles the amount that needs to be added to the fuel for that monolayer; so once there is an adequate chain length (about C_{18}), the potency falls off with increasing that chain length.

Turning to the adsorbing polar group, given that an outer steel surface is, in fact, hydrated ferric oxide (see section 5.8), the polar group needs to adsorb onto this. As in the discussion of corrosion inhibitors, carboxylic acids are well suited to adsorb onto such surfaces because they complex well with ferric ions and can hydrogen bond strongly with the oxide and hydroxide groups associated with the ferric ions – remembering that this is an ionic crystal structure, it is less easy to envisage the structure with adsorbed species[17]. The initial physical adsorption is expected to be reversible but adsorbed carboxylic acids go on to form a metal carboxylates on reacting with the ferric oxide surface — the formation of a 'metallic soap' is supported by the behaviour of the monolayers of carboxylic acids on a metal surface [53]. The resulting electrostatic attraction between the carboxylate ions and the ferric ions is stronger than physical adsorption by the dipole–dipole and hydrogen bond interactions. Further reactions to iron di- or tri-carboxylates are also possible, which could carry iron carboxylates into the fuel; since part of the fuel in a fuel injection system is returned to the fuel tank, and iron carboxylates can catalyse atmospheric oxidation, this would lead to unwanted destabilisation of the fuel (Figure 2.5, Fuel Stability), a potential negative factor to the use of the acid lubricity improvers.

Readily available alternatives to carboxylic acids are ester, amide and hydroxyl groups that adsorb by dipole–dipole and hydrogen bonding interactions with a steel surface. These are weaker interactions but are strengthened by multiplicity – as in glycerol mono-oleate, which has a polar end group containing two hydroxyls and one ester group, and the TOFA/diethanolamine, which has two hydroxyls and an amide group. Multiple adsorption sites increase the total strength of adsorption and reduce the chance of being completely desorbed, since the random desorption of one group leaves two groups adsorbed, so the molecule is held in place for re-adsorption of the desorbed group. An advantage of these non-acid lubricity additives is their low abilities to solubilise iron into the fuel.

While all of these considerations also apply to the actions of corrosion inhibitors in preventing the rusting of steels, there are subtle differences. A complete monolayer is not essential to providing lubricity as the purpose is to lubricate the passage of one surface over the other, which is also aided by the entrained fuel. The inhibition of rusting requires complete, cohesive surface coverage and the formation of a hydrophobic layer without holes, so reliably strong adsorption of surfactants bearing bulky alkyl groups are needed (section 5.8).

5.5 Introduction: Corrosion in Fuel Systems

Corrosion "In broad terms, is the interaction between a material and its environment that results in a degradation of the physical, mechanical, or even aesthetic properties of that material. More specifically, corrosion is usually associated with oxidation and/or dissolution of a metal, oxide or semi-conductor" [54]. For fuel systems, the concern is with the corrosion of metals that are used in the construction of engines, heaters, furnaces, exhaust systems, and all fuel movement and control systems, such as the pumps, filters and pipes found in automotive vehicles, aeroplanes, ships and heating installation. Corrosion problems are also found in long-range pipelines and storage tanks of all sizes.

Metals in contact with fuels are various steels, with smaller quantities of copper and zinc (pipes and brass connections) and silver (in fuel level sensors); jet fuel is also in contact with aluminium and its alloys – the major aircraft construction metals. Corrosion in automotive and jet fuel systems can lead to problems such as [39]:

- Storage vessel and pipeline leakage.
- Contamination of fuel with particulates of friable rust.
- Blockage of filters and screens from corrosion material.
- The requirement of expensive maintenance.

Corrosion is caused or aided by impurities in the fuels themselves. Fuels may contain acidic impurities, such as products of atmospheric oxidation; the introduction of biodiesel, esters of various fatty acids, can lead to problems from free fatty acids that may be present initially or are produced during the degradation of such fuels during storage. In exhaust systems, corrosive attack is from fuel oxidation products. Water and oxygen are always present and unhindered, cause rusting in storage tanks and fuel pipelines.

Sulphur and its compounds attack copper so when fuel sulphur levels are significant, corrosion inhibitors are needed to protect items made of copper-containing alloys, primarily brass. Silver connections in fuel level sensors, in the tanks of modern vehicles, are extremely sensitive even to the very low levels of 10–15 ppm sulphur in automotive fuels. As discussed under lubricity, automotive fuels have been cleaned up considerably over the last thirty years or so. In particular, sulphur levels have been reduced to around 10–15 ppm in automotive diesel and gasoline in Europe (specified as EN 590 and EN 228 [55]) and the USA [56] with countries in the far east [57], including Japan, following suit in most diesel specifications [26]. Good news for the reduction of copper and silver corrosion while bad news for lubricity.

While high sulphur levels are found in most marine fuels, some heating oils and some off-road applications, they are also being reduced in these applications [58, 59] and, as of 2020, these lower limits are now in place [60]. Combustion of marine diesel that contained significant levels of sulphur produce, now reduced to a maximum of 0.5%, produces the very acidic, hydrated sulphur oxides that are corrosive in marine engines – in addition to the organic acids from incomplete oxidation of fuel. Residual fuels that are used in marine engines and in power stations also contain vanadium compounds that produce vanadium oxides which are very corrosive to exhaust systems.

Other corrosion problems that have less impact than the above, but still need addressing, are zinc being leached from brass fittings and 'metal dusting' in oil fired heating burners, when using low or ultra-low sulphur fuels.

Corrosion shortens the lifetime of equipment in contact with fuels by increasing well engineered clearances, reducing the strength of stress bearing parts and creating holes through which fuels or exhaust gases leak. Furthermore, metals dissolved into fuels catalyse their oxidative degradation.

5.6 Corrosion Issues in Various Fuels

5.6.1 Automotive Gasoline and Diesel Fuels

A widespread corrosion problem, applicable to all petroleum fuels, is that of corrosive rusting in the distribution system that can create holes and cause premature failure in steel pipes, pumps and storage vessels. Such failures can cause major problems, such as refinery, power plant or pipeline shutdowns, and loss of products that generate hazards, such as diesel fuel spills, fire risk and environmental contamination [61]. Rust particles can block, or partially block, pipelines and filters [22].

Rusting of steel needs both oxygen and water, so water levels in fuels are specified at low levels: 0.02 % in euro-diesel (EN590) or 'non-visible' in gasoline (EN228). But fuels pick up water on transportation and storage due to random ingress of rain and tank 'breathing'[18], during which water from the air condenses into cold fuel. Water has a limited solubility in petroleum fuels (about 50 ppm) but biodiesel can absorb 1500 ppm [62] and methanol and ethanol are fully miscible with water. Such corrosion is catalysed by acids [63] which may arise from oxidative fuel degradation, especially when the fuel contains oxygenates, as in biofuels – alcohols in gasoline [64] or fatty acid methyl esters in diesel [62].

Engine parts that are in contact with the lubricant are protected from corrosion by several additives that are routinely used in lubricant formulations. Various steel alloys are widely used in engine construction and many engine parts are now made corrosion resistant. However, fuel storage and supply systems, and some parts of the engine, are susceptible to corrosive agents in the fuel so need protection.

Sulphur compounds are naturally present in all petroleum sources and, until the early 1990s, were present at significant levels in all automotive fuels. Copper corrosion, caused by these sulphur compounds, was a serious problem which was controlled by the introduction of corrosion inhibitors: copper and, to varying degrees, steel are significantly corroded by sulphur compounds in fuels at levels of 100 ppm and higher. Silver is used in the contacts of fuel tank level sensors because of its resistance to water and acids, but it is extremely sensitive to corrosion by sulphur compounds in fuel 'even after hydroprocessing' [65]. In most developed countries, after a series of downward steps, the permitted levels of sulphur are now extremely low in fuels for highway vehicles, though several samples in the 2018 WDFQS had sulphur levels from 100 to 1000 ppm [26]. Fuels for off-road vehicles may still contain higher levels of sulphur in some countries, while both automotive and off-road fuels with high sulphur levels are still in use in most African countries and some South American countries [66, 67]. One independent survey found that in many African countries, automotive fuels' sulphur contents are significantly more than the 10–15 ppm level – 17 out of 25 samples at the pump were found to contain more than 1,500 ppm (0.15%) of sulphur [67]. India, however, has been in the process of reducing sulphur in fuels since 1995 [68], specifications require automotive diesel and gasoline to have a maximum 50 ppm sulphur from 2017 (BS IV) and 10 ppm from 2020 (BS VI), nationwide; three Indian fuel samples in the 2018 WDFQS [26] contained an average of 27 ppm sulphur.

To prevent corrosion during fuel storage, transmission and use in automotive vehicles, corrosion inhibitors are routinely added to gasoline and diesel fuels at 6–100 ppm [69]. Biofuels, such as ethanol for gasoline and fatty acid methyl esters in biodiesel, increase the corrosive nature of these fuels, resulting from their tendency to attract and dissolve water and their easy degradation to organic acids. Such fuels need more careful consideration of corrosion and the use of corrosion inhibitors (DDSA or TPSA in particular) [70–72].

5.6.2 Jet Fuels

Sulphur levels in jet fuels are currently much higher than in automotive fuels: 0.3% maximum total sulphur (30 ppm max. mercaptan sulphur) in the main international grades Jet A and A-1 [13]. The potential for copper corrosion, in brass fittings, is controlled by the fuel having to meet a specification set by the copper strip corrosion test. Corrosion inhibitors, also thought of as lubricity improvers, are optional. In jet fuel, the reliability of which is critical, only certain corrosion inhibitors/lubricity improvers are permitted and must come from an approved list of those that have been found to be effective and have no adverse side effects [13].

Corrosion by water is, to some extent, obviated by the stringent requirements that 'Fuel shall be free from water' [13], since any water in jet fuel will certainly freeze and block fuel lines and filters at the very low temperatures experienced by wing tanks (outside temperatures can fall as low as -50°C during a flight (section 7.6, Anti-Icing). To minimise the risk of icing, jet fuel is passed through a dewatering system just before it is loaded into the wing tanks. Also, in commercial airlines, jet fuel is generally used within a short time of its production – from refinery to airport by pipeline and onto aeroplane – and is not generally stored for long periods in tanks. However, aluminium is susceptible to carboxylic acid corrosion – acids that are the products of oxidative degradation of hydrocarbons. Such acids are usually produced during prolonged storage, which can apply to fuel stored at small airports and, certainly, to some military jet fuels for which the use of corrosion inhibitors is mandatory (such as JP-4 and JP-5 [13]).

5.6.3 Heating Oils

Heating oils[19] are usually middle distillates with a similar or higher boiling range than diesel fuels or are varieties of kerosene. Their corrosion problems during storage and transportation – caused by air, water and adventitious acids – are the same as for diesels fuels. However, they may have higher sulphur levels, though specifications for sulphur levels in heating oils have fallen towards those in automotive fuels in richer countries. In the USA, heating oil recently had maximum of 0.5% sulphur [73] though some eastern states had started the process of reducing this, in stages, to be the same as ultra-low sulphur diesel (ULSD, 15 ppm S) by 2018 [74]. The UK specification for premium kerosene, used as domestic heating oil, now has 10 ppm sulphur, in line with that for ULSD (though standard C1 kerosene has a maximum of 0.04% sulphur) [75]. In Japan, the kerosene is specified as having 80 ppm maximum sulphur and fuel oil A, also used for heating, has a light grade with up to 0.1% sulphur, and a heavy grade[20] having up to 0.5% sulphur [76]. However, as discussed above, even automotive fuels still have high sulphur levels in some countries. It is, therefore, highly probable that less expensive, high sulphur gas oils are used for heating, off-road vehicles, and power generation in many developing countries, so copper and zinc corrosion in related equipment is always a concern.

As sulphur levels are reduced in some domestic heating oils, a further problem has arisen. The high temperature regions of some oil fired hot water heaters that were using ULSD, instead of heating oils, have been found to suffer from 'metal dusting' – a type of corrosion of steel alloys in hot, carbon-rich environments, more commonly known in refineries and synthesis gas production plants [77–79]. It is known that metal dusting can be controlled by adding sulphur to the metal surface environment in the form of hydrogen sulphide [78], so the appearance of metal dusting during the use of fuels from which the sulphur has been removed is not surprising. Application of a sulphiding agent would make the treated ULSD off-specification, so it appears that this problem will need to rely upon the development of a suitable alloy.

Some heating oils are heavy fuels, containing residual fuels, and they suffer from the same problems as those found in heavy marine fuels and heavy fuels used in electricity and industrial steam generation.

5.6.4 Distillate Marine Fuels and Off-Road Fuels

As a group, these are more commonly known as gas oils. Sulphur contents of off-road diesel fuels are now also limited to 10 or 15 ppm sulphur in many countries though there are countries in which these fuels are permitted to contain sulphur at higher levels. Fuel oil A, Japan, has already been mentioned as having up to 0.5% sulphur. Distillate marine fuel specifications have recently been changed and maximum permitted sulphur levels are, from 2015, 0.1 % in emission control areas (near coastlines, such as in the English Channel, previously 1%); elsewhere, in the open seas, the maximum sulphur levels, from January 2020, are 0.5% (previously 3.5% since 2012) [80]. However, even at the level of 0.5% sulphur, there is still a need for measures to control sulphur related corrosion.

5.6.5 Heavy (Residual) Fuels

Water in marine fuel oils/bunker fuels[21] is specified at a maximum of 0.5% [81] though such fuels at sea pick up larger amounts than this. Excess water may be separated by settling out in the bunkers (storage tanks), where rusting in the salty water layer is a likely problem; excess of water is removed from fuel by on-board centrifuges.

Heavy fuels are more acidic and often have higher levels of sulphur than distillate fuels, though from January 2020, the maximum of 0.5% applies to these fuels also. Power plants are recommended to treat heavy fuels that they burn with additive packages that contain corrosion inhibitors [82]; such power plants have been or are gradually being phased out, but are still in use in Russia and in some African countries, where sulphur levels may still be greater than 0.5%. At these high levels of sulphur, mercaptans and hydrogen sulphide may be present, which not only corrode copper but can also cause significant corrosion of steel; hydrogen sulphide, however, is now tightly specified in the revised ISO 8217:2012 to a maximum of 2 mg H_2S/Kg (2 ppm) of fuel [83]. The corrosive properties of mercaptans and hydrogen sulphide are of particularly relevance to, and well understood by, the oil and gas exploration and recovery industry [84] and are, therefore, part of a wider problem in refineries processing high sulphur content ('sour') crudes.

A major source of corrosion in ships, furnaces (some refineries) and power plants that use heavy fuels is the sulphuric acid that is produced when the sulphur in the fuel is burned. Even the lower levels of sulphur in these fuels (0.5%) will produce significant amounts of sulphur acids, given the high volumes of fuel burned in a ship's engine [85][22]. When burning high sulphur fuels, the sulphuric acid produced corrodes the piston, cylinder liner and exhaust valves. Acids are neutralised, primarily, by the high alkalinity (high total base number) of the lubricating oil. In the smaller, four-stroke trunk piston engines, running on distillate fuels, the oil is pumped around the piston and cylinder to maintain an acid-neutralising, lubricating layer. In the larger two-stroke crosshead engines, one lubricant is circulated to protect and lubricate the main bearings and camshaft, while a high base-number oil is injected inside the cylinder liner, which has holes to allow this oil into the cylinder to neutralise the acid (see engine description [86]).

Until recently, this problem was mitigated by running the engine above the condensation temperature of sulphuric acid. Now, however, as a result of improvements to engines and the way that they are operated when they reach the emission control areas (ECAs) near coastlines, the engines are run at lower temperatures, allowing the sulphuric acid to condense on the cylinder liner, causing 'cold corrosion' [85]. Cold corrosion is mostly countered by reformulated marine lubricating oils and operating procedures [85]. However, any help given by fuel-borne corrosion inhibitors to that provided by the lube oil should be beneficial and, as with all fuels, such corrosion inhibitors are used to protect storage tanks and fuel lines from corrosion by the fuel itself and any water that it may contain.

A quite different corrosion problem arises in the exhaust systems of diesel engines and burners using heavy fuels that contain petroleum distillation residues. This problem is caused by the relatively high levels of vanadium found in crude oils, which is concentrated in the distillation residues [87]. A heavy fuel can contain up to 0.05% vanadium, though 0.02% is more common[23] [88]. The vanadium is oxidised to vanadium pentoxide which has a relatively low melting point (675°C) and forms lower melting (as low as 300°C) complexes with sodium salts (also present in residual fuel) [88]. These partial melts are corrosive in themselves, forming fluxes with metal oxide layers which, when the flux falls off, leave the metal unprotected against further corrosion by exhaust gases. This process corrodes exhaust valves and turbocharger nozzles in the large, heavy-fuel-burning diesel engines used in ships or in power generation. In furnaces burning heavy fuels, to generate steam in some refineries or to produce electricity in steam turbines, the heat exchangers, turbine blades and parts of the exhaust system can also be corroded to premature failure.

5.7 Chemistries of Fuel Corrosion Inhibitors

From the above discussion, it may be seen that there are five main corrosion problems to be treated:

- Corrosion of steel by water and oxygen, rusting, aided by the presence of acids.
- Corrosion of metals by acids present in the fuel.

- Corrosion by sulphur in the fuel, in the form of mercaptans and sulphides.
- Corrosion by sulphur acids from the burning of high sulphur fuels.
- Exhaust system corrosion by vanadium oxides from residual fuels.

5.7.1 Corrosion by Water/Oxygen and by Carboxylic Acids

Additives to protect metals from corrosion by water and air, or adventitious acids, have surfactant structures: a polar group to anchor the molecule onto the metal surface and a bulky hydrocarbon group that presents and oleophilic/hydrophobic exterior in order to repel water [89]. The properties needed for effective corrosion inhibitors are [90]:

- An ability to strongly adsorb onto the metal surface – an ability invariably provided by polar anchoring groups containing nitrogen, oxygen, or both.
- Preferably, the molecule will contain multiple polar groups to increase the strength of adsorption.
- Bulky organic groups attached to the anchoring groups– the bulkier the better to provide good water-repelling surface coverage.

A number of commercial corrosion inhibitors that have been and often still are in use in gasoline, at 6–30 ppm, for protection from general corrosion of fuel distribution systems and engines (pipes, tanks, pumps, both brass fittings and steel parts)[24]:

- Fatty acids such as tall oil fatty acid (much of which is unsaturated).
- Dodecenyl succinic acid (DDSA) and a branched alkyl version tetrapropylene succinic acid (TPSA).
- Dimer and trimer acids, made by thermal oligomerisation of naturally occurring unsaturated fatty acids.
- The acids combined with an amine, which may be a polyamine (such as poly-ethylene-diamine) or a simpler monoamine or diamine.
- An alkenyl succinic acid or anhydride, which could be DDSA, 3 molar equivalents, reacted with triethanolamine, 1 molar equivalent.
- Polyamines and alkyl ethers of polyamines.
- Amines themselves such as hydrogenated tallow amine (consisting of about 70% octadecylamine, 25% hexadecylamine, 5% octadecenylamine).

Refined tall oil fatty acid (TOFA, for example product HiTEC 4140A™) has the advantage of its low cost (a by-product of the wood pulp industry) and good oil solubility, composed mostly of the unsaturated acids oleic and linoleic (48% and 36%, respectively, in one well-analysed sample [91]):

$$\text{Oleic acid: } HOOC\text{-}(CH_2)_7\text{-}CH=CH\text{-}(CH_2)_7\text{-}CH_3$$

Linoleic acid: HOOC-$(CH_2)_7$-CH = CH-CH_2-CH = CH-$(CH_2)_4$-CH_3

However, long chain ('fatty') carboxylic acids possess just one anchor group so have limited strength of adsorption so are not suitable for treating corrosion of any severity; di- and tri- carboxylic acids are preferred as they provide stronger anchoring from their multiple polar groups – an entropic advantage over two or three separate molecules of carboxylic acid. DDSA has two anchoring carboxylic acid groups and is the only corrosion inhibitor featured in the booklet 'Fuel Additives: Use and Benefits',[25] for use in gasoline, diesel fuel and heating oil [2]. DDSA and the more soluble TPSA (Figure 5.6) are made from 1-dodecene and tetrapropylene, respectively, in an ene reaction with maleic anhydride [92] (process Figure 3.7). Half esters of DDSA or TPSA with various alcohols have also been claimed – particularly those including alkene oxide derivatives (PEG, PPG or ethoxylated alcohols [93]); the extra advantages of multiple adsorbing groups such as carboxylic acids, carboxylic esters and multiple hydroxy- and polyether- groups, are apparent.

Dimer- and trimer-acids have been widely used in fuels as corrosion inhibitors and lubricity additives over many years. Dimer-acids were shown to inhibit rusting in tests in gasoline, kerosene, and diesel fuel at concentrations of around 10 ppm while the precursor mixture of linoleic and oleic acids failed in these tests at 74 ppm (1953 [94]). In general, dimer-acids are made by heating unsaturated fatty acids (or esters), preferably a mix of both mono- and di-unsaturated acids such as linoleic and oleic acids, as found in TOFA, to around 200–260°C, under pressure and over an acid clay catalyst [95]. There appear to be a number of reactions going on in the process; in particular, the linoleic acid isomerises under these reaction conditions to give conjugated double bonds, which are able to take part in a Diels–Alder reaction with any other double bond, either on the oleic acid or another linoleic acid molecule. This produces a number of isomers and further reaction to a trimer acid also

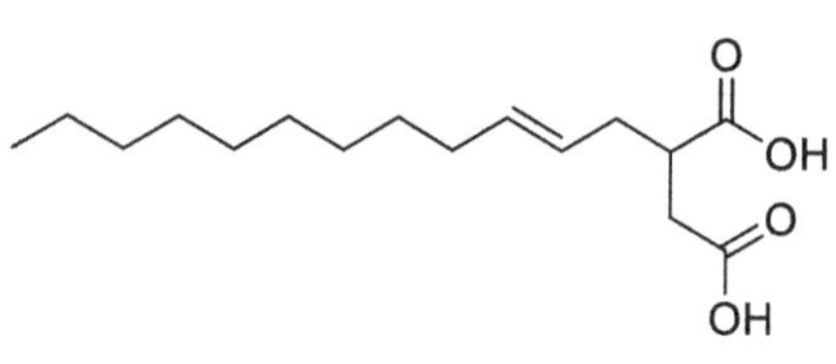

DDSA

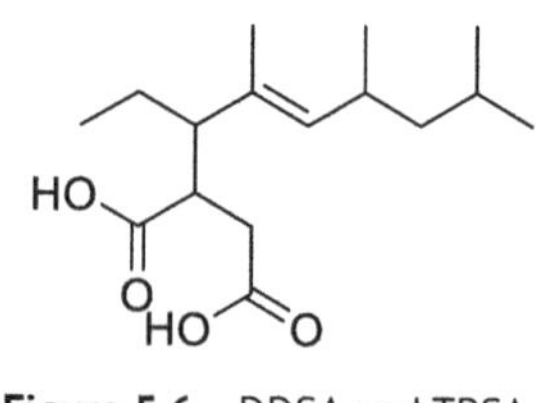

TPSA, mixed isomers

Figure 5.6 DDSA and TPSA.

occurs. The product is then distilled to give unreacted mono-carboxylic acids, a dimer acid concentrate and trimer acid, leaving a residue of oligomeric acids. Any description of the structure of dimer acid can only be a selection of the isomers present (Figure 5.7). In addition (and often ignored), it is likely that cationic alkene oligomerisation to dimer and trimer is also occurring over the acid catalyst. There is often not enough linoleic acid to react with all the oleic acid, in the natural mixture, but this is not a problem because oleic acid can be dimerised in the absence of linoleic acid under such conditions [96].

Concentrations used for corrosion inhibitors in gasoline, diesel or domestic heating oil are quoted as 5–100 ppm [2, 70] or less – suppliers recommend lower levels, in fact, such as 6–11 ppm in middle distillates [39]. However, lubricity improvers in ULSD or friction modifiers in gasoline are recommended for use at higher concentrations of 25–400 ppm, so if the low sulphur fuel is treated with a lubricity improver that may be a long chain mono-carboxylic

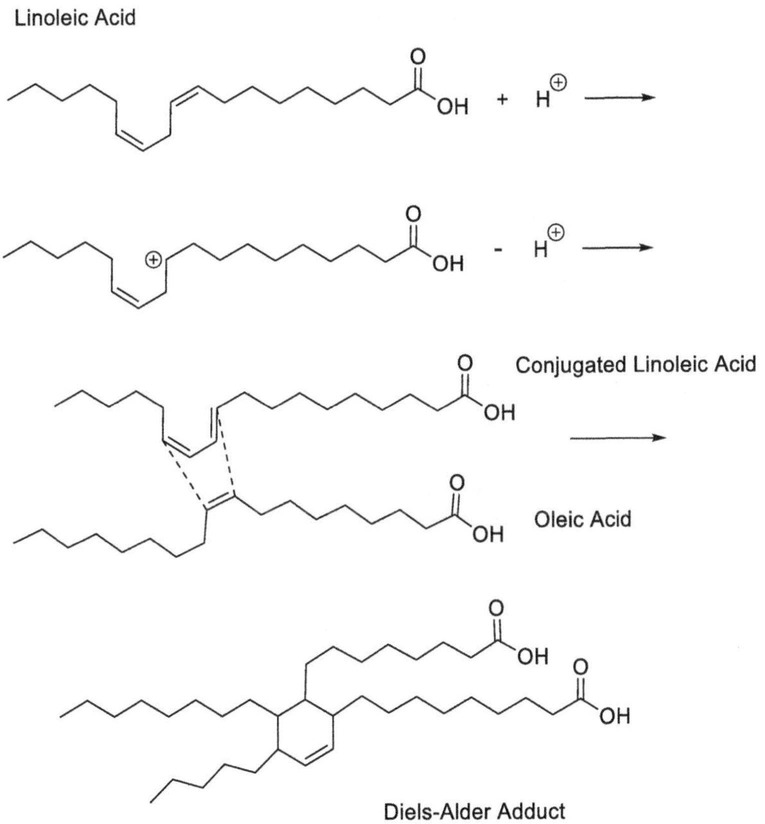

Figure 5.7 Rearrangement and Dimerisation of Linoleic Acid.

acid, it may not also need to have a corrosion inhibitor, depending upon the severity of the corrosion environment. Copper and silver, however, may need additional protection against sulphur corrosion.

DDSA and TPSA (Figure 5.6), PIBSA and dimer- and trimer-acids (Figure 5.7) often feature as comparators, as established corrosion inhibitors, in formulations described in patent applications [71, 93, 97]. They are often identified as commercial corrosion inhibitors for fuel (such as DCI-30 and DCI-4A [70]). Their rust preventing abilities were discovered much earlier than their use in fuels: DDSA and dimer acids were identified as corrosion inhibitors for use in lubricating oils in 1935 [98] and TPSA in 1956 [92].

Commercial corrosion inhibitors for jet fuel appear on the approved list as either the dimer/trimer acid chemistry – 'CI/LIs' such as DCI-6A and Tolad 245 that contain dimer acid [99] – or DDSA chemistry in products such as Specaid 8Q123ULS [70], both of which are commodity chemicals. They were originally added to military fuels to inhibit corrosion in the fuel distribution system and jet engines. Later, it was realised that lubricity improvers were also needed in military aircraft and that this need was also fulfilled by the materials classified as corrosion inhibitors [14]. Approved corrosion inhibitors are now required in military specifications, at concentrations of 9–23 ppm; in civilian jet fuel, the use of corrosion inhibitors is by agreement between supplier and user.

Amines, especially di- or tri-amines, form complexes with many metals – including iron, copper, zinc and silver. So, it is not surprising to find that amines are present in lists of commercial corrosion inhibitors, either by themselves, such as the hydrogenated tallow amine (such as Armeen® 2HT), or combined with the carboxylic acids as salts, such as that of a fat-derived carboxylic acid with N-alkyl-propane-1,3-diamine (Duomeen®) (product DCI-11 [70], Figure 5.8a) or the structurally similar 3-alkoxy-N-(3-aminopropyl)-propylamine (Tolad® 3222 [64]) (Figure 5.8b).

Amide derivatives are also used as corrosion inhibitors, such as that of TPSA reacted with one mole equivalent of N-octyl-bis(2-hydroxypropyl)amine [100], or oleic acid reacted with N-methyl-glycine ('Sarkosyl® O', Figure 5.9a [101]).

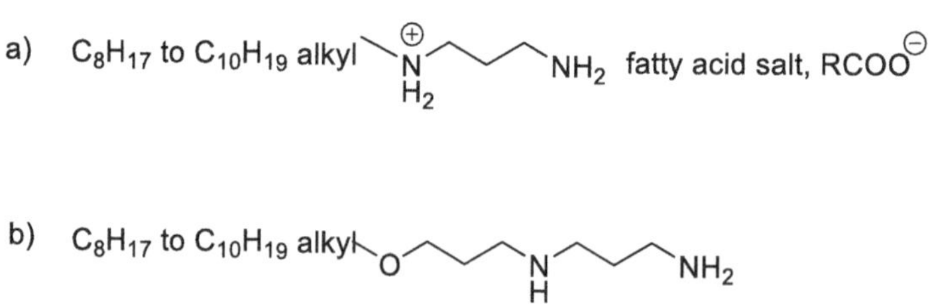

Figure 5.8 Corrosion Inhibitors DCI-11 and Tolad® 3222.

a)

b)

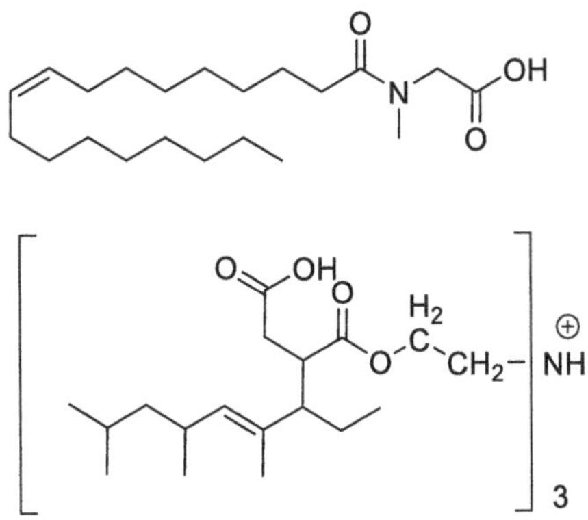

Figure 5.9 N-Acyl-Sarcosine and TPSA-Diethanolamine Corrosion Inhibitors.

Another process linking the amine to the acid is used in Tolad 357, for which three molar equivalents of TPSA are reacted with one molar equivalent of triethanolamine to give 'an amide and/or amine salt' (quoted [64, 70]); the chemistry tells us that this would be a tri-ester zwitterion (Figure 5.9b),[26] as described in the original patent [102], since a tertiary amine does not easily produce an amide. A molecule that carries three carboxylic acid groups and three ester groups adsorbs strongly onto metal surfaces and the three tetrapropenyl groups produce a tightly bound, hydrophobic surface.

Simple amines, however, are not mentioned in a review of copper corrosion inhibitors in the aqueous environment – it is clear that the much more effective nitrogen heterocycles (below), which may also bear amine groups, are of greater commercial interest [103].

5.7.2 Corrosion by Sulphur

While acid attack and rusting are major problems for steel, sulphur tends to produce a hard-protective layer of iron sulphide on the surface of this metal but with group I-B metals, copper and silver, sulphur compounds are quite corrosive. Jet fuel usually contains higher proportions of sulphur than automotive fuels and this can lead to copper corrosion as black copper sulphide is formed (up to 0.25% in the Russian standard TS-1 fuel used in this reference [104]). Copper is also attacked by acids, while silver is not, which is why silver alloys are used in automotive fuel tank-level sensors. However, this application is sensitive to any silver corrosion and, although the levels of sulphur in automotive fuels are now very low, silver sulphide deposits, caused by the trace amounts of sulphur compounds in gasoline (particularly mercaptans) can build up over years causing the tank-level indicator to malfunction. Significant amounts of copper corrosion, as measured by the copper strip corrosion test (ASTM D-130), have been demonstrated in diesel with quite low levels of sulphur (four fuels with 9–19 ppm S were used) or gasoline (gasoline containing 36 ppm of sulphur was used); silver was badly tarnished by the gasoline [105]. In any fuel containing higher levels of sulphur, such as 0.05–0.5% in distillate marine fuels, furnace fuels and fuel for off-road applications, the corrosion of copper in brass or bronze fittings is a significant problem.

A range of heterocyclic compounds have been identified that inhibit the corrosion of copper and other metals – many of these have a wide range of applications. These compounds have universally understood modes of action: 'inhibiting corrosion of copper, steel or copper alloys in fuel systems is using corrosion inhibitors ... these are either sulphur scavengers or metal deactivators' [27] [106]. Initially, many of these developments were for the protection of copper in aqueous applications, as it is widely used in the electrical and plumbing industries [103]. Heterocyclic corrosion inhibitors, particularly for copper and silver, are thiadiazoles and benzothiazoles, imidazoles and benzimidazoles, triazoles and benzotriazoles and triazines (Figure 5.10). The efficacies of such inhibitors have been reviewed for copper corrosion in aqueous applications [103] and triazoles and thiadiazoles are commonly used today in gearbox oils [107]. The basic heterocyclic compounds and several their derivatives are generally available, some are commodities, because of their value as intermediates for the production of other chemicals such as dyestuffs, surfactants and epoxy resins. Alkyl groups added to these chemistries make them oil soluble and applicable in fuels for the inhibition of both acid and sulphur corrosion of copper and silver.

One group of effective copper corrosion inhibitors are 2,5-dimercapto-thiadiazoles such as dioctyl-dithio-1,3,4-thiadiazole which appears in blend patents in the 1990s as an existing, commercial inhibitor (Elco®461) [105, 106] for use

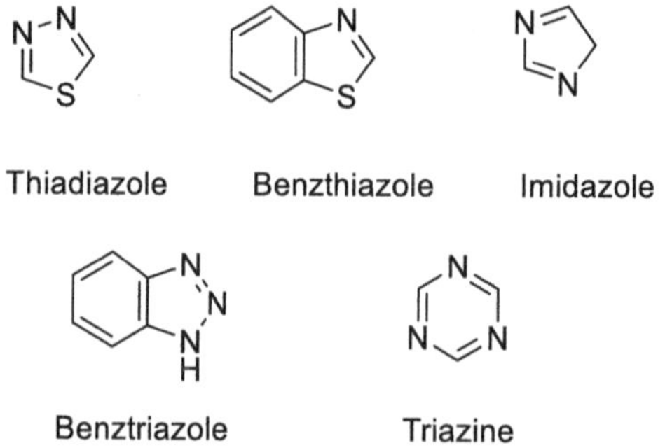

Figure 5.10 Heterocycles on Which Many Copper Corrosion Inhibitors Are Based.

in automotive fuels as copper or silver corrosion inhibitors, effective at 20–30 ppm in low sulphur diesel fuel. In a recent patent, an alkyl dithiothiadiazole, in combination with a polyether-amine, was tested in an automobile as a silver corrosion inhibitor to protect the fuel gauge sender unit and it was found to be effective – both in the car test and in the silver strip corrosion test [108]. This test used low sulphur (6 ppm S) unleaded gasoline, illustrating the sensitivity of silver even to such extremely low levels of sulphur. However, one third of this corrosion inhibitor, by weight, is sulphur; so 36 ppm treat that was used brings with it 12 ppm of sulphur. Such chemistry, with such high proportions of sulphur, is clearly not applicable in low sulphur automotive fuels, though they may be used in the remaining higher sulphur fuels – marine and some off-road diesel fuels.

Non-sulphur nitrogen heterocyclic corrosion inhibitors are clearly preferred for low sulphur fuels: foremost among these are the triazoles, triazines and imidazoles. Benzotriazole has long been recognised to be an effective copper corrosion inhibitor in aqueous systems [103] and it is soluble in ethanol and methanol, so it has been proposed for use in gasohol, which is more corrosive than straight petroleum gasoline [109]. As with other potential fuel additives, benzotriazole and triazoles have low hydrocarbon solubilities, so some oil-soluble derivatives have been developed for use in fuels. Triazoles react readily with oil soluble amines, such as di-(2-ethylhexyl)-amine, to give such established corrosion inhibitors as N,N-bis-(2-ethylhexyl)-4-methyl-1H-benztriazole-1-methanamine (also known commercially as Irgamet®39 [110] or Reomat® [106], Figure 5.11a) and the amino-triazole, Irgamet 30 (N,N- bis-(2-ethylhexyl)-1,2,4-triazol-1-ylmethanamine, Figure 5.11b [111]).

To further enhance the anti-corrosion effectiveness of tolyltriazole in fuels, other adsorbing polar groups may be added as in the 'adduct' (a salt) with N,N-bis(2-hydroxyethyl)-oleylamine[28], which is fuel soluble and, at a concentration of less than 10 ppm, prevented corrosion in the copper strip test applied to a

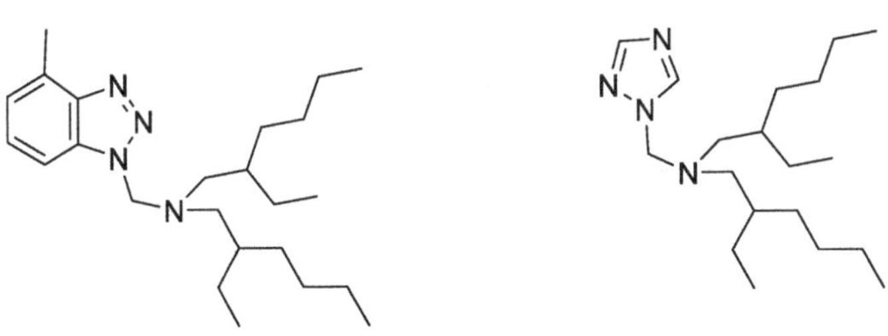

(a) (b)

Figure 5.11 Triazole Corrosion Inhibitors.

kerosene containing 20 ppm of sulphur [112]. Acylation of benzotriazole or 5-amino-benztriazole by hydrocarbon rich anhydrides, such as DDSA or poly-isobutylene-succinic anhydride (PIBSA), is also reported to give effective corrosion inhibitors for fuels or lube oils [113]. Triazole itself is prepared from cyanamide and hydrazine followed by reaction with a carboxylic acid, which may be used to attach an oil-solubilising alkyl group [114]; benzotriazole is prepared by diazotising 1,2-diamino-benzene (Figure 5.12).

Other additives designed to inhibit sulphur corrosion of copper and silver are benzimidazoles, which are also used to protect steels against corrosion by aqueous acids, as in some steel finishing or scale-cleaning operations [115, 116]. Although there is little sign benzimidazoles have found use in distillate fuels, the imidazoles themselves or, preferably, the thermally stable imidazolines are used as corrosion inhibitors in refineries where the raw, hot distillate streams are more corrosive than finished fuels [117, 118]. In commercial imidazolines, their corrosion activities are enhanced by the inclusion of an extra polar substituent. There are at least three commercial processes for the production of imidazolines [119, 120]; one is simply to heat a carboxylic acid, such as oleic acid, with a 1,2-diaminoethane or, preferably, a polyamine such as diethylene-triamine (DETA), to give, first, an amidoamine intermediate then the fuel soluble aminoethyl-substituted imidazoline (Figure 5.13 [117]) – an effective inhibitor of steel corrosion in diesel fuel [121]. The use of di-propylenediamine in place of DETA contributes to the water demulsifying efficiency of this derivative [118]. Another imidazoline enhancement is to ethoxylate the

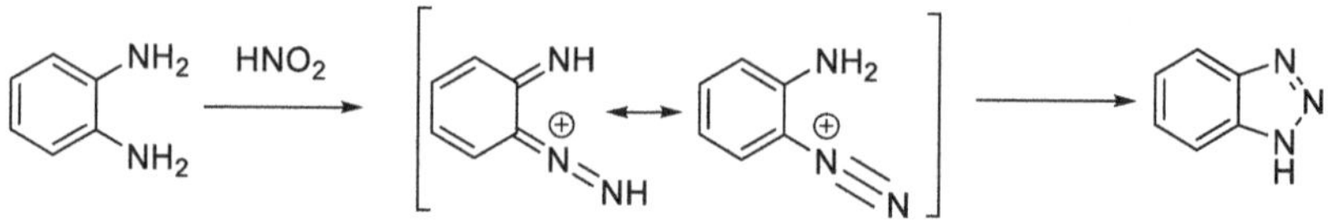

Figure 5.12 Synthesis of BenzoTriazole.

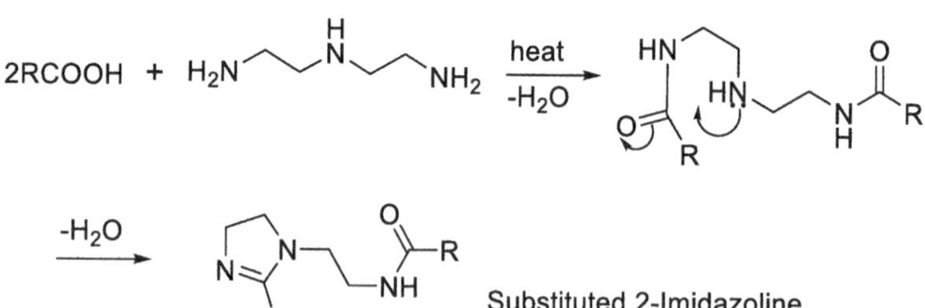

Figure 5.13 Imidazoline Formed from Reaction of Carboxylic Acid with DETA.

1-nitrogen to carry a 2-hydroxyethyl group, again with the oil solubilising oleyl-group on the 2-position – a derivative which is also commercial and is reported to be a very effective corrosion inhibitor in diesel fuel [121] and in gasohol [70].

Triazines can be prepared from the readily available cyanuric chloride (2,4,6-trichloro-1,3,5-triazine), itself made by heating cyanogen chloride over charcoal impregnated with an alkali earth chloride (Figure 5.14 [114, 122, 123]). Cyanuric chloride can be sequentially aminated with different alkylamines to give oil soluble melamines (Figure 5.14 [114, 124, 125]). Another process is to heat nitriles to give alkyl-substituted, oil-soluble triazines. Other widely available triazines are cyanuric acid (s-triazine triol), made by heating urea and removing the resultant ammonia, and melamine (2,4,6-triamino-s-triazine), also made by heating urea but keeping the ammonia in the reaction [123]; both of these can be further functionalised by reacting the amines with reagents bearing oil-solubilising groups, such as fatty acid derivatives.

Triazines are very susceptible to nucleophilic attack, which makes them effective hydrogen sulphide scavengers [126] – of relevance to heavy fuels in which, combined with their film forming ability on metal surfaces, they give good protection against sulphur corrosion [122, 127]. Hexahydro-triazines, which are easily prepared from the reactions of substituted amines with aldehydes [128], are also effective hydrogen sulphide scavengers. These are relevant because hydrogen sulphide and mercaptans are not only poisonous and obnoxious impurities found in a high sulphur fuels, they are very corrosive to copper and steel. Fuel oils and heavy fuels, then, are more corrosive than modern automotive fuels so the same corrosion inhibitors are used at much higher treat rates – for example, up to 86 ppm [39].

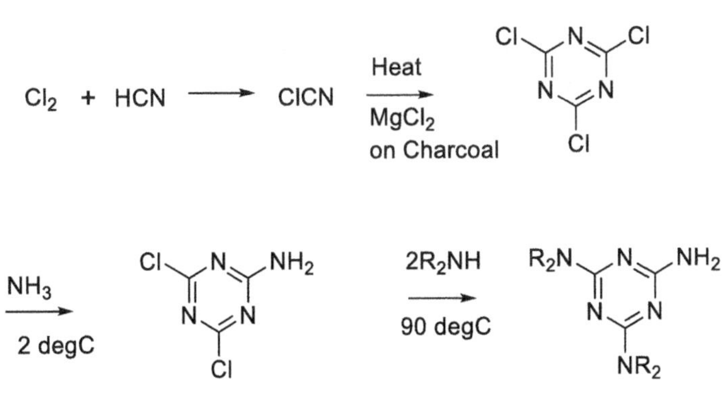

Figure 5.14 Formation of Melamine Analogues.

5.7.3 Corrosion by Vanadium Pentoxide

As already described (section 5.6.5), the exhausts from residual fuels also bring the problem of vanadium and sodium mixed oxides, which have relatively low melting points, adsorbing onto and causing corrosion to turbine blades, exhaust valves and other parts exposed to products of the burnt fuel. This problem may be controlled by the use of magnesium, calcium or zinc oxides, with which the vanadium pentoxide reacts, in preference to the engine-system metal surfaces, producing a higher melting ash in place of the low melting vanadium/sodium containing flux [87]. Calcium derivatives tend to react more readily with the sulphur oxides to produce calcium sulphate, which is unreactive in the exhaust environment; magnesium and zinc do not suffer this problem and so are preferred antidotes to vanadium oxide. Inhibition of vanadium oxide corrosion in a model gas turbine plant, burning residual fuel, was demonstrated for fuel containing the oil soluble zinc naphthenate [87].

The addition of slurries of magnesium or calcium oxides or hydroxides to residual fuels was widely used in power plants to neutralise sulphur acids and control vanadium corrosion [129], but these had to be added to the fuel just before injection as they rapidly separate on standing. Patents refers to this practice as current in 2006 [130], suggesting that older processes of more finely dispersing the metal oxide [131] did not solve the problem. Clearly a solubilised form of magnesium or zinc oxide or carbonate is preferred. The oil soluble metal salts, such as fatty acid carboxylates or alkylbenzene sulphonates, carry quite low levels of metal (converting to the oxide in the combustion chamber) so are relatively expensive additives for such low-cost fuels. Oil soluble additives that carry large proportions of metal carbonate are used in crankcase lubricant additives, where they are known as overbased[29] metal detergent inhibitors. One such additive, overbased magnesium sulphonate, is supplied to overcome the vanadium corrosion problem [129]. Overbased metal detergent inhibitors are colloidal dispersions of calcium or magnesium carbonate, stabilised and solubilised by a surfactant such as an alkyl-benzene-sulphonate, alkyl-salicylate or alkylphenate of the same metal – such additives contain a substantial proportion of magnesium carbonate relative to the surfactant. The structures of such additives are similar to the particle stabilised by adsorbed PIBSA-PAM (Figure 3.23) with magnesium carbonate cores in place of the deposit core of the diagram and, for example, an alkyl benzene sulphonate in place of the PIBSA-PAM. Overbased magnesium sulphonates are well established as components in lube oil formulations [132] and their preparations are also well established [133, 134].

5.8 Mechanisms of Corrosion and Its Inhibition

5.8.1 Corrosion by Water/Oxygen and by Carboxylic Acids

The action of acids on all but the noble metals (such as silver) can be simply the formation of hydrogen while the metal dissolves into adventitious water. With acetic acid – an example of a carboxylic acid that may be formed by the oxidative degradation of fuel and which may be officially present in gasohol at up to 56 ppm [64] – such a reaction would be:

$$Fe + 2CH_3.COOH + nH_2O \rightarrow Fe^{++}aq. + 2CH_3.COO^{-}aq. + H_2$$

However, the corrosion of iron and steel to produce rust – hydrated ferric oxide/hydroxide – is an oxidative process that does not need the presence of acids. The microstructure of an iron surface has three layers on the iron itself that are progressively electron rich: next to the metal is ferrous oxide (FeO), then magnetite[30] (Fe_3O_4) and, on the external surface, ferric oxide (Fe_2O_3) [135]. The chemical process is one that needs the presence of water and oxygen as follows:

In the lattice: $Fe^{0} + H_2O \rightarrow Fe^{++}\ aq. + 2e^{-}$,

while $O_2 + 2H_2O + 4e^{-} \rightarrow 4HO^{-}$

then: $2Fe^{++}\ aq. + 4HO^{-} \rightarrow 2Fe(OH)_2,\ H_2O + \frac{1}{2}O_2 \rightarrow 2Fe(OH)_3 \rightarrow Fe_2O_3.H_2O$

Corrosion in this way is described as anodic/cathodic as there is an electron transfer involved. The metal acts as the anode, providing electrons into the surrounding medium, which must be conducting – such as brine in the bottom of ships bunkers or wet earth underneath storage tanks [61]. There the electrons are picked up by oxygen plus water to form hydroxy anions that react with ferrous ions in the region of corrosion, the cathode, completing the circuit and forming ferrous oxide. Further oxidation, by the same mechanism, raises the oxidation state of iron to three as hydrated ferric oxide – rust. Rust is not a tightly bound layer, wherever it is formed, it tends to fall off into the fuel as particles that can cause problems by blocking filters or narrow pipes and causing wear in fuel pumps and injectors [22]. This loss of the oxide layer enables access of oxygen to the metal itself, enabling further oxidative corrosion. Steel is such an integral part of engines, fuel lines, pipelines and storage tanks that its corrosion can lead to very high cost problems, hence corrosion inhibitors are widely used in fuels.

Other metals that are used in engine systems, zinc in brass fittings and aluminium in alloys in aircraft, may also undergo oxidation plus dissolution by any acids present, leading to leakages or gradual failures in joints, valves and critical parts of an engine. Acids may be sulphuric, from the sulphur in high sulphur fuels or biofuel processes[31], or organic, particularly carboxylic acids that are products of oxidative degradation of hydrocarbon fuels. Biofuels are more susceptible to free radical oxidation, having hydrogen atoms attached to carbons adjacent to oxygen atoms or carbonyl groups. FAME-based biodiesel is itself more corrosive than 100% petroleum diesel[32] [72] and is susceptible to hydrolysis by adventitious water to form carboxylic acids (which make the hydrolysis autocatalytic). For example, biodiesel has been found to take up zinc from brass fittings in fuel distribution systems [136]. ULSD, to which long chain carboxylic acid lubricity improvers have been added, also leach of zinc from brass fittings [137].

Copper, however, is resistant to attack by non-oxidising acidic water because it has a positive electrode potential relative to hydrogen (+0.34 volts), while iron and zinc have negative potentials relative to hydrogen (-0.44 and -0.76 volts respectively). In air, copper forms a thin, passivating layer of copper oxide that prevents further oxidation – unlike iron's fragile layer of rust.

Corrosion inhibitors need to form cohesive films on metal surfaces that protect them from attack by the corrosive agents [90]. Corrosion inhibitors adsorb by the interaction of their polar groups with the metal or metal oxide surface and the attached oleophilic group then produces a water-resistant (hydrophobic) layer [138]. Clearly, the surface coverage must be thorough, otherwise unprotected areas would be open to corrosion in the form of pitting. To provide this full coverage, corrosion inhibitors are selected to adsorb strongly and carry bulky hydrocarbon groups. With strong adsorption, only extremely low concentrations, such as 5–15 ppm, are needed in normal circumstances [22]. Military jet fuel requires not less than 6 ppm and not more than 36 ppm of corrosion inhibitor (using DDSA or dimer acid [99]). In a perfect world, the metal surface might only need to be treated once but there is always attrition or desorption of adsorbed additive, needing the continuous availability of replacement additive in the fuel. To illustrate the degree of attrition of adsorbed corrosion inhibitor, it has been reported that treat rates of 5–20 ppm may be selected to ensure that a fuel, such as gasoline, emerges from a pipeline carrying only 1 ppm of the additive [138].

In support of the supposition that corrosion inhibitors provide protective coverage of the metal surface, experiments with some corrosion inhibitors showed that the corrosion rate of steel in aqueous hydrochloric acid was inversely proportional to the degree of surface coverage by the corrosion inhibitor, going from a corrosion rate of 0.84 mg/cm^2 /day for unprotected steel to 0.05 mg/cm^2 /day when the surface is 95% covered [89]. Indeed, effective corrosion inhibitors that form films on the metal surface are often referred to as metal passivators [139].

As stated, when considering the structures of corrosion inhibitors, it is generally accepted that in order to provide adequate, protective metal surface coverage, corrosion inhibitors need three structural features [90]:

- An anchoring group that can adsorb strongly to the metal surface.
- Multiple adsorbing groups to improve the surface adhesion.
- A hydrocarbon group large enough to provide hydrophobic coverage of the metal surface (sometimes referred to as the backbone).

The anchoring group is always a polar group and, to be most effective, it should have strong affinity for the metal involved. Iron has a high affinity for oxygen based ligands and readily forms complexes such as $[Fe_3O(O_2CR)_6L_3]^+$ [140] with many carboxylic acid ligands. It is not surprising, then, that various carboxylic acids are effective as corrosion inhibitors for steel – coordinating with the iron oxide structure on the surface. Carboxylic acids can associate with many metals and are generally inexpensive, so are widely used to provide corrosion inhibition in automotive fuels, heating oils and jet fuel – with other metals also in mind, such as aluminium, copper and magnesium for jet fuel [70, 141]. As with lubricity, a favoured fatty acid corrosion inhibitor is refined tall oil fatty acid ('TOFA'[33]), which has a relatively low melting point and high oil solubility.

However, mono-carboxylic acids are not very strongly adsorbed so they are improved upon by using multiple carboxylic acid groups in the same molecule. Multiple adsorbing polar groups greatly increase the strength of adhesion of these molecules in two ways. The obvious one is that two groups on the same molecules provide twice the strength of adhesion. In addition, there is the entropic advantage for one molecule with two adsorbing groups compared with two molecules, each with one polar group, adsorbing onto the metal surface. Another way of looking at this is to say that while an anchor group may desorb at any time, the molecule with one adsorbing group will be lost, while one anchor group desorbing in a molecule with two such groups is most likely to be held in place by the second group; the desorbed group being held in place is well positioned to re-adsorb. So, as carboxylic acid corrosion inhibitors, molecules such as DDSA and dimer acid, which have two carboxylic groups, are much preferred. Surface protection potency is often further enhanced by adding additional polar groups such as an aminoethyl or hydroxyethyl substituent -often added onto an imidazoline [121, 142] – or a multiplicity of acid, amine, amide, and ester groups [143]. Maleic anhydride can be added to the carbon–carbon double bonds of the unsaturated acids in TOFA to combine the carboxylic acid and anhydride functions which, after part esterification, can give multiple carboxylic and ester functions and carry more alkyl groups [97].

Branched alkyl groups provide bulky hydrophilicity, such as 2-ethyl-hexyl in the toluiltriazole Irgamet 30® [111] or tetrapropenyl in TPSA [71]. With strong adhesion and good metal surface coverage by the anchor group, the attached alkyl provides not only oil solubility but also provides a hydrophobic outer surface. Keeping water away from the surface itself prevents a major source of corrosion – oxidation to metal ions and their dissolution as salts or formation of hydrated oxides that do not adhere to the metal surface, as in rusting. Too large an alkyl group will make the molecule too oil soluble so it will have a lower tendency to adsorb, in the adsorbed/solution equilibrium; also, the potency of a molecule with a larger-than-necessary alkyl group is reduced directly in proportion to the excess molecular weight.

5.8.2 Corrosion by Sulphur

Copper and silver have a strong affinity for sulphur – the source of their susceptibility to corrosion by sulphur and its compounds. The affinity for sulphur is stronger than for oxygen, so compounds other than carboxylic acids are needed that are more attracted to copper and silver to prevent sulphur corrosion. Sulphur corrosion of copper requires an oxidative step which may be controlled, to some extent, by the presence of an antioxidant:

$$2Cu^0 + 2RSH + [O] \rightarrow 2Cu^+RS^- + H_2O$$

Experiments on hydrogenated (desulphurised) fuel with added sulphur compounds, at 150°C, identified the corrosive sulphur compounds as disulphides and mercaptans, but not sulphides [104]. It is generally accepted that sulphur-based corrosion of copper is the worst with elemental sulphur, followed by hydrogen sulphide and mercaptans, then disulphides and sulphides are the least bad [144].

BHT is a free radical trap antioxidant (see 2.5.6) so would be expected to interfere with the disulphide corrosion, which proceeds by radical chain homolysis to give thiolate radicals (RS ·) that give cuprous thiolate. Mercaptans (thiols) are the most aggressive sulphurous corroders of copper, even at one tenth the concentration of the disulphide, and they are not significantly affected by the presence of an anti-oxidant; in corrosion tests on jet fuel, corrosion was, therefore, assigned to the 0.003% of mercaptans that it contained and which gave a sediment that analysed as cuprous sulphide (from proportions given, probably a mixture of Cu_2S, cuprous sulphide and Cu_2S_2, cuprous disulphide) [104].

Given the affinity of copper and silver for sulphur, it is not surprising to find that stable compounds containing sulphur, such as thiazoles and thiadiazoles (Figure 5.10), are effective corrosion inhibitors for these metals. Again, they cannot be used in ultra-low sulphur fuels because of the sulphur that they would introduce into both the fuel and the additive blend[34]. However, there is a link with other additive applications in that 'copper corrosion inhibitors ... are either sulphur scavengers or metal deactivators' [108]. Triazine and triazoles derivatives are effective both as sulphur scavengers and as corrosion inhibitors for copper and silver.

Triazoles and benzotriazoles form tightly bound, non-permeable layers on copper metal surfaces [111]. An SEM study[35] has identified the formation of a layer of the tolyltriazole dialkylamine on the surface of copper conductors, with little copper sulphide, when oil in which the copper was immersed was treated with this corrosion inhibitor; by comparison, in the untreated oil the copper conductor surface was covered with cuprous sulphide [144].

The protective film forming activity of these nitrogen rich heterocycles is ascribed to the overlap of the empty d orbitals on the copper atom with the electron rich π orbitals and lone pairs on the nitrogen atoms [103]. 3-Aminotriazole, benzotriazole, tolyltriazole and imidazoles – with or without various substituent groups – adsorb strongly and form stable complexes with copper on the metal surface [103]. Benzotriazole has been found to form a strongly bonded, chemisorbed two-dimensional barrier film less than 5 nanometres thick [145].

5.8.3 Corrosion by Vanadium Pentoxide

Maximum levels allowed for vanadium contents in residual fuels, in various specifications, can vary up to 600 ppm (mg/Kg), though actual levels are mostly less than 200 ppm with less than 15% of fuels containing more than 200 ppm [88]. Sodium levels are mostly (about 95%) less than 100 ppm and are usually associated with water contents (less than 1%). During combustion and followed by further oxidation in the cooler area of the combustion chamber or exhaust duct, the vanadium in a fuel is converted to vanadium pentoxide, V_2O_5, which has a relatively low melting point of 675°C. The sodium ends up as sodium sulphate by reaction with the sulphuric acid, from sulphur in the fuel, which condenses out at temps below 890°C. Together, the sodium sulphate and vanadium pentoxide form a range of complexes that can have melting points as low as 300–400°C. These complexes condense onto and adhere to metal surfaces, where they dissolve the iron oxide surface layers leaving the hot, bare metal open to further attack by oxygen and acids in the exhaust stream.

The solution to the problem is to add basic magnesium compounds to the fuel (section 5.7.3). Not only does this neutralise some of the sulphur acids but the magnesium oxide also combines with the vanadium/sodium combustion products to produce a much higher melting ash that is non-corrosive, voluminous and friable, which does not adhere to the oxide surface of the metal. The effectiveness of this chemical approach to vanadium corrosion of metals was demonstrated in a gas turbine simulation apparatus, which measured the amount of oxidation occurring on blades made from corrosion-resistant nickel-steel alloy [87]. Initially, mock blades were coated with pastes of vanadium pentoxide alone and pastes of vanadium pentoxide that had been roasted with magnesium oxide or with zinc oxide (1:3 molar equivalents). The blades were heated in a furnace at 700°C for 100 hours. With no vanadium pentoxide present, 0.5 mg of metal oxide was formed on the blade; with the vanadium pentoxide alone, 6.7 mg of oxide had formed; with the combination of vanadium pentoxide and magnesium oxide, 0.4 mg of oxide was formed; and with vanadium pentoxide and zinc oxide the measurement of oxide was a loss of 0.3 mg[36]. Both the combined oxides had given, effectively, no corrosion, as in the absence of vanadium pentoxide. Following this up in the gas turbine apparatus, burning a fuel oil that contained vanadium and sodium, corrosion of the blades was followed by the weight of alloy oxide being formed over a period of 50 hours at 750°C. With the untreated fuel oil, the oxide accumulated at 0.1755 $mg/cm^2/hour$, while fuel containing zinc naphthenate gave an oxide accumulation rate of only 0.0125 $mg/cm^2/hour$. Availability of overbased magnesium sulphonates from the lube additive industry ensures the preference for the magnesium-based additive for preventing vanadium oxide corrosion – metal detergent inhibitors for lubricating oil are either calcium or magnesium derivatives [146, 147].

Notes

1 Hydrofining, hydrotreating, hydroprocessing and hydrodesulphurisation are all forms of hydrogenation at elevated temperatures and pressures over a complex catalyst.

2 Jet fuel is a well specified kerosene cut which is also known as aviation kerosene or turbine oil.

3 A steel ball is held, under pressure, against a steel ring, rotating at 240 rpm, the lower part of which is immersed in a reservoir of the fuel under test. After the test, the wear scar on the ball is measured under a microscope; the more wear, the larger the scar (ASTM D5001 test method, as describe [14]).

4 Use of ethanol in gasoline has been mandated since 1976 in Brazil with proportions varying between 10% and 25% [19].

5 In the USA, gasoline contained 250 to 350 ppm of sulphur, 1978–98, and Europe started to be reformulated to a maximum 150 ppm sulphur in 2000 [148].

6 Bosch also specifies its gasoline injectors for GDI for use at pressures up to 20 MPa.

7 The Worldwide Infineum Winter Diesel Fuel Quality Survey appears every two years, such as WDFQS 2014 [25].

8 The Worldwide Fuel Charter is an organisation set up in 1998 by automotive manufacturers' organisations in Europe, the USA and Japan. Its purpose is to increase understanding of the fuel quality needs of motor vehicle and engine technologies and to promote fuel quality harmonisation worldwide.

9 The two improved HFRR results given in this patent in Table 1 [31] are of ULSD with 20 or 5% of vegetable oil mixed with n-octanol (4/1 ratio).

10 Ideal solubility depends upon the inverse of melting point – Standard Physical Chemistry [149].

11 Descriptions of lubricity improvers or friction modifiers are sometimes found in their MSDSs (Material Safety Data Sheets), such as TOFA, dimer acid or poly-ethoxylated alkylphenol-formaldehyde-condensates (APFCs).

12 Standard esterifications of the alkyl salicylic acid gave samples that contained about 35% of the ester. Treating the lithium salt of the alkyl salicylic acid with glycidol gave the 80% ester product [44].

13 While the European specification is WSD maximum of 460 microns, an unofficial target is around 350 microns to allow for variability in HFRR test results.

14 In this patent, such esters were being considered for use as thickening agents with 'excellent lubricity' in various lubricating oils, gear oils and transmission fluids.

15 To quote 'It is well known to make fuel lubricity additives by reacting alkylene glycols with monomeric and dimeric carboxylic acids using sulphur-containing catalysts such as toluene sulfonic acid' leaving the average product with 11 ppm of sulphur. Some batches may have more sulphur than allowed for blending into ULSD in the USA, which specifies that additives themselves should also contain a maximum of 10 ppm sulphur; the improved product [150] contains almost no sulphur so allows higher active ingredients in the product.

16 A nitrogen compound within the typical boiling range of gas oil would have a molecular weight of about ten times the atomic weight of nitrogen, so 15 ppm nitrogen resides in 150 ppm of its parent compound.

17 A description of metal surfaces after atmospheric exposure is one of layers of metal, metal oxides of lower oxidation levels, then higher oxidation levels, with bound layers of water and organic molecules [151].

18 Air is expelled when a tank is warm and drawn in when cold, either day by day or day/night variations.

19 Also known as furnace oils in some countries.

20 Listed as also being used for off-road and marine diesel.

21 Marine fuels may be distillate fuels or heavy fuels; heavy fuels contain residual fuels and, usually, also contain distillate fuels for stability and viscosity reasons; marine fuels are kept in bunkers on ships so are often referred to as bunker fuels or just bunkers.

22 In the large Wartsila engine (Chapter 3), for example, there are 160 g of fuel per injection per cylinder, at 100 rpm this amounts to 16 kg per minute. At 0.5% of sulphur that is 80 g sulphur per minute or 240 g H_2SO_4 per minute per cylinder, or 14.4 kg per hour, and running continuously.

23 In the large Wartsila engine (Chapter 3), for example, there are 160 g of fuel per injection per cylinder, at 100 rpm this amounts to 16 kg per minute. 0.02% V is 3 g vanadium per minute or 5 g V_2O_5 per minute per cylinder, or 300 g per hour, and running continuously.

24 These have been identified by commercial name and chemistry in a patent concerned with novel applications to gasohol [70].

25 Published by the Technical Committee of Petroleum Additive Manufacturers in Europe (ATC).

26 One of the acid groups will have given its proton to the tertiary nitrogen to become a carboxylate.

27 See Chapter 2 for the role of metal deactivators.

28 The adduct is believed to be a Petrolite® product.

29 The term is an extension of the idea that an acidic surfactant, such as an alkyl benzene sulphonic acid, is neutralised by a base, such as magnesium oxide or carbonate. When more the base than is needed for neutralisation is incorporated as colloidal particles, then the system is termed over-based.

30 Magnetite is a mixed Fe^{2+} and Fe^{3+} oxide, also known as ferrous ferric oxide.

31 Toluene sulphonic acid is used as a catalyst in some transesterification processes used to produce FAME.

32 The specification for 100% FAME biodiesel, B100, EN 14214, allows up to 0.5 mg KOH/g total acid, i.e. 500 ppm KOH equivalent. This needs 2,500 ppm of a 280-mol. wt. amine (such as hydrogenated-tallow-amine) for neutralisation; when a B10, 250 ppm of amine would be needed. Acids catalysts used, such as toluene sulphonic acid, are limited at max. 10 ppm S, so most of the acids present are probably carboxylic (EN 14214 allows 3.5% of non-ester).

33 TOFA contains linoleic and oleic acids (36 and 48% in one analysis [91]).

34 Some such inhibitors contain 50% sulphur, even low treats of 10ppm could make a ULSD off-spec. In the USA, fuel additives are also bound by the low sulphur specifications of automotive fuels.

35 This test was in a transformer-oil at 140°C for 1400 hours.

36 The patent says that the ash with zinc oxide was particularly friable so, presumably, was either carried off the naturally occurring oxide coating or these low results (-0.3 to + 0.6. mg) were within the margin of error.

References

1 Wei, D.P., Spikes, H.A., and Korcek, S. (1999). The lubricity of gasoline. Society of Tribology & Lubrication, *Tribology Transactions* 42: 813–823.

2 Fuel Additives: Use and Benefits, ATC Doc (2013). 113. atc-europe.org. http://www.atc-europe.org/public/Doc113%202013-11-20.pdf (Cited: 2 February 2014).

3 ATC Europe (2008). Lube additives and the environment. atc-europe.org. http://www.atc-europe.org/publications.asp (Cited: 1 October 2013).

4 Moore, C.C. and Fischer, P.W. (1946). Diesel engine fuel. US Patent 2,527,889. Filed USA, 19 August.

5 Owen, K., Coley, T.R., and Weaver, S.R. (1995). *Automotive Fuels Reference Book*, 2nd edn. Society of Automotive Engineers, Warrendale, P.A. ISBN 1560915897.

6 Booth, M. and Wolveridge, P.E. (1993). Severe hydrotreating of diesel can cause fuel injector pump failure. August, *Oil & Gas Journal* 91 (33): 71.

7 Chevron Global Aviation – Aviation Fuels Technical Review (2006). cgabusinessdesk.com. https://www.chevron.com/-/media/chevron/operations/documents/aviation-tech-review.pdf (Cited: 16 April 2020).

8 Lois, E. and Arkoudeas, P. (2012). Lubricating aspects of automotive fuels. In: *New Advances in Vehicular Technology and Automotive Engineering* (eds. J.P. Carmo and J.E. Ribeiro), IntechOpen Ltd London, Ch.3, pp. 91–118.

9 Beltzer, M., Craig, R.C., Panzer, J. et al. (1986). Alkanol fuel compositions. EP 0,257,149. Filed Europe, 21 August.

10 Cooke, J.A. and Robinson, I.C.H. (1983). A diesel fuel based on methanol and/or ethanol. GB Patent 2,143,846. Filed UK, 25 July.

11 Blau, P.J. (2007). *Scuffing: From Basic Understanding to Engine Materials Testing*. Detroit: Oak Ridge National Laboratory, U.S. Department of Energy. DEER Conference, Detroit.

12 Dukek, W. G. (1988). Ball-on-Cylinder Testing for Aviation Fuel Lubricity. *SAE Transactions*, 97. 475–482. Society of Automotive Engineers, Warrendale, P.A.

13 ExxonMobil (2008). World jet fuel specifications with avgas supplement. http://large.stanford.edu/courses/2017/ph240/chhoa1/docs/exxon-2008.pdf (Cited: 11 June 2020).

14 Biddle, T.B. and Edwards, W.H. (1998). Evaluation of corrosion inhibitors as lubricity improvers. United Technologies Corp., US evaluation AD-A198 743, Report nmber AFWAL-TR-88-2036.

15 Cooper, D. (1995). Laboratory screening tests for low sulphur diesel fuel lubricity. January, *Lubrication Science* 7: 133–148.

16 Washecheck, P.H., Liu, A.T.C., and Kennedy, E.F. (1981). Methanol fuel and methanol fuel additives. US Patent 4,375,360. Filed USA, 12 January.

17 PEW Environment Group (2011). History of fuel economy. PewEnvironment.org/CleanEnergy. April http://www.pewtrusts.org/~/media/assets/2011/04/history-of-fuel-economy-clean-energy-factsheet.pdf (Cited: 15 March 2017).

18 Shaub, H. and Waddy, W.E. (1984). Method for improving the fuel economy of an internal combustion engine using fuel having hydroxyl-containing ester additive. US Patent 4,617,026. Filed USA, 15 August.

19 Wikipedia Contributors. (2017). Ethanol fuel in brazil. Wikipedia.org. March (Cited: 16 March 2017). Page Version ID: 769090870.

20 Wikipedia contributors (2017). Ethanol fuel in the united states. Wikipedia.org. March (Cited: 16 March 2017). Page Version ID: 770087082.

21 Motor Gasolines Technical Review. (2009). chevron.com. https://www.chevron.com/-/media/chevron/operations/documents/motor-gas-tech-review.pdf (Cited: 26 November 2012).

22 Chevron corp (2007). Diesel fuels technical review. https://www.chevron.com/-/media/chevron/operations/documents/diesel-fuel-tech-review.pdf (Cited: 26 November 2012).

23 Robert Bosch GmbH (2013). Gasoline direct injection – Key technology for greater efficiency and dynamics. bosch-di.com. http://www.bosch-mobility-solutions.us/media/documents/e-product-brochure-gasoline-direct-injection.pdf?WT.mc_id=272_DE_OEM_OEM (Cited: 13 June 2017).

24 Gasoline Direct Injection was the most widely adopted emerging fuel saving technology in 2018: 51% (2019). Green car congress. 16 April https://www.greencarcongress.com/2019/04/20190416-gdi.html (Cited: 4 June 2020).

25 Infineum International Ltd (2014). Infineum worldwide 2014 winter diesel fuel survey. Infineum.com. http://www.infineum.com/media/80722/wdfs-2014-animated.pdf (Cited: 20 June 2015).

26 Infineum Worldwide Winter Diesel Fuel Survey 2018 (2018). Infineum.com. https://www.infineuminsight.com/en-gb/resources/winter-diesel-fuel-quality-survey/2018-winter-diesel-fuel-survey (Cited: 9 May 2020).

27 Bovington, C., Caprotti, R., Meyer, K. et al. (1995). Development of laboratory tests to predict the lubricity properties of diesel fuels and their application to the development of highly refined diesel fuels. December, *Tribotest Journal* 2 (2): 93–112.

28 Hadley, J.W., Owen, G.C., and Mills, B. (1993). Evaluation of a high frequency reciprocating wear test for measuring diesel fuel lubricity. Society of Automotive Engineers, Warrendale, P.A. SAE Technical Paper 932692.

29 Hardy, W.B. and Doubleday, I. (1922). Boundary lubrication, the paraffin series. *Proceedings of the Royal Society* 100: 550–574.

30 Bergazli, C. and Henry, C.P. (2006). Synergistic additive composition for petroleum fuels. WO 2008,054,368. Filed USA, 12 September.

31 Snower, G.M. (2007). Fuel oil composition. US Patent 2008,222,946. Filed USA, 15 March.

32 Knothe, G. (2005). The history of vegetable oil-based diesel fuels. In: *The Biodiesel Handbook* (eds. J. vanGerpen, J. Krahl, and G. Knothe), Ch. 2. AOCS, academia.edu.

33 Knothe, G. and Steidley, K.R. (2005). Lubricity of components of biodiesel and petro-diesel. Origin of Biodiesel Lubricity. American Chemical Society, *Energy & Fuels* 19: 1192–1200.

34 Wadumethsrige, K., Ara, M., Salley, S.O. et al. (2009). Investigation of lubricity characteristics of biodiesel in petroleum and synthetic fuel. American Chemical Society, *Energy & Fuels* 23: 2229–2234.

35 Munoz, M., Moreno, F., Monne, C. et al. (2011). Biodiesel improves lubricity of new low sulfur diesel fuels. Elsevier, *Renewable Energy* 36: 2918–2924.

36 Afton Chemical. Lubricity Improvers. http://www.aftonchemical.com/Solutions/FuelAdd/Specification_Distribution/Diesel/Pages/LubricityImprovers.aspx (Cited: 9 March 2015).

37 Ansell, C. and Wardle, R.M.W. (2007). Method of formulating a fuel composition. US Patent 2008,110,080. Filed USA, 18 October.

38 Hong, J. (2014). New lubricity challenges, infineum trends 2014. Infineuminsight.com. September http://www.infineuminsight.com/insight/sep-2014/new-lubricity-challenges#top (Cited: 17 February 2015).

39 Innospec Fuel Specialties (2020). Fuel additives product summary. https://www.innospecinc.com/our-markets/fuel-specialties/refinery-specialties (Cited: 14 May 2020).

40 Caprotti, R. (1994). Bovington C.H. and macrae C.D.J. Fuel composition. EP 680,506. Filed 20 January.

41 Shibata, K., Kase, M., Fukuhara, S. et al. (2008). Method for producing mono-glyceride containing composition for gas oil additive. JP 2010,121,088. Filed Japan, 21 November.

42 Packet, D. (2000). Fuel composition. EP 1,216,288. Filed Europe, 7 September.

43 Clayton, C.W. and Miller, D. (2007). Formulating fuel compositions. US Patent 2007,240,361. Filed USA, 18 April.

44 LeDeore, C. and Caprotti, R. (1997). Lubricity additives for fuel oil compositions. US Patent 6,258,135. Filed USA, 15 September.

45 Malec, R.E. (1979). Anti-Wear additives in diesel fuels. US Patent 4,204,481. Filed USA, 2 February.

46 Ball, K.F., Bostick, J.G., and Brennan, T.J. 1998. US Patent 5,891,203. Filed USA, 20 January.

47 Kenbeek, D., Verboom, C., and Van der Waal, G. (2000). Complex esters, formulations comprising these esters and use thereof. US Patent 6,462,001. Filed USA, 31 March.

48 Furey, M.J., Appeldoorn, J.K., and Turbak, A.F. (1965). Lubricity additive – Hydrogenated dicarboxylic acid and a glycol. US Patent 3,287,273. Filed USA, 9 September.

49 Ludema, K.C. (1996). *Friction, Wear, Lubrication: A Textbook in Tribology.* CRC Press. ISBN 0849326850.

50 Mortier, R.M., Fox, M.F., and Orszulik, S.T. (2010). *Preface of Chemistry and Technology of Lubricants,* 3rd edition.Springer Dordrecht Heidelberg London New York .pp v-x. ISBN 9781402086618.

51 Matzke, M., Litzow, U., Jess, A. et al. (2009). Diesel lubricity requirements of future fuel injection equipment. SAE Paper 2009-01-0848. Society of Automotive Engineers, Warrendale, P.A., ISSN 01487191.

52 Wei, D. and Spikes, H.A. (1986). The lubricity of diesel fuels. Elsevier Sequoia, *Wear* 111: 217–235.

53 Sanders, J.V. and Tabor, D. (1951). Structure of thin films of aliphatic esters and alcohols on metals. The Royal Society, *Proceedings of the Royal Society* 204: 525–533.

54 McGraw-Hill (2007). *Encyclopaedia of Science and Technology*, 10th edition, Volume 4, p 775. McGraw-Hill Education.

55 EU: Fuels: Diesel and Gasoline. (2018). Transportpolicy.net. https://www.transportpolicy.net/standard/eu-fuels-diesel-and-gasoline (Cited: 8 June 2020).

56 United States Environmental Protection Agency (2016). Overview of EPA's tier 3 gasoline sulfur regulations. September https://www.epa.gov/regulations-emissions-vehicles-and-engines/final-rule-amendments-related-tier-3-motor-vehicle (Cited: 4 January 2017).

57 EU Fuel Quality Status, Today and Tomorrow (2009). Asian clean fuels association news. December Vol. 7(8).

58 Department for Transport (2012). Fuel quality directive gas oil requirements. January http://webarchive.nationalarchives.gov.uk/20101007153548/http:/www.dft.-gov.uk/pgr/roads/environment/fuel-quality-directive/pdf/fuelquality.pdf (Cited: 5 January 2017).

59 Infineum International Ltd (2016). Time ticking for marine fuels. December http://www.infineuminsight.com/insight/dec-2016/time-ticking-for-marine-

fuels?utm_campaign=Infineum+Insight+Dec16+email&utm_medium=email&utm_source=Infineum+Insight (Cited: 14 December 2016).
60 Corradi, M. and Hughes, E. (2019). *IMO 2020 and Beyond, Industry Preparedness for Imminent Sulphur Cuts and Other Emissions Regulations that Lie Ahead*. Insight from Infineum International Ltd. 28 October.
61 Mohitpour, M., Yoon, M.S., and Russell, J.H. (2012). *Hydrocarbon Transmission Pipeline and Storage Systems*. Design and Operation. ASME.
62 Knothe, G., Van Gerpen, J., and Krahl, J. (2005). *The Biodiesel Handbook*. AOCS Press. ISBN 9781893997622.
63 Abboud, Y., Abourriche, A., Saffaj, T. et al. (2006). Corrosion inhibition of steel in acidic media by 2,2′-bis-benzimidazole derivatives. Elsevier Science B.V., *Applied Surface Science* 252: 8178–8184.
64 Cunningham, L.J. and Dumont, R.J. (2007). Method and compositions for reducing deposits in engines combusting ethanol-containing fuels and a corrosion inhibitor. US Patent 2008,168,708. Filed USA, 11 January.
65 Silver Corrosion Inhibitors (2016). Ecolab.com. http://en-uk.ecolab.com/program/silver-corrosion-inhibitors (Cited: 10 June 2020).
66 Duissenov, D. (2013). Production and processing of sour crude and natural gas – Challenges due to increasing stringent regulations, p96. In: *Department of Petroleum Engineering and Applied Geophysics* . Supervisor Gudmundsson, J.S. Trondheim: Norwegian University of Science and Technology. M.Sc. thesis submitted June.
67 Guéniat, M., Harjono, M., Missbach, A. et al. (2016). Dirty diesel – How Swiss traders flood Africa with toxic fuels. Publiceye.ch. September https://www.publiceye.ch/fileadmin/files/documents/Rohstoffe/DirtyDiesel/PublicEye2016_DirtyDiesel_A-Public-Eye-Investigation.pdf (Cited: 10 June 2020).
68 Fuel Regulations India (2016). IS 1460: 1995Specifications (2000). DieselNet.com. March https://www.dieselnet.com/standards/in/fuel.php (Cited: 10 June 2020).
69 ATC Europe. (2004). Fuel additives and the environment. 1 January http://www.atc-europe.org/public/doc52.pdf (Cited: 10 June 2020).
70 Wolf, L.R. and Baustian, J.J. (2012). Corrosion inhibitor compositions for oxygenated gasolines. US Patent 10,131,859. Filed USA, 28 December.
71 DuMont, R.J., Kulinowski, A.M., and Cunningham, L.J. (2007). Methods and compositions for reducing deposits in engines combusting alcohol-containing fuels. US Patent 2008,202,561. Filed USA, 1 March.
72 Jin, D., Zhou, X., Wu, P. et al. (2015). Corrosion behaviour of ASTM 1045 mild steel in palm biodiesel. *Renewable Energy* 81: 457–463.
73 ASTM Committee D02 (2002). Standard specification for fuel oils, D396-02. annual book of ASTM standards. ASTM international. Vol. 5.

74 Sulphur Content of Heating Oil to be Reduced in North Eastern States. (2012). U.S. energy information administration – EIA – Independent statistics and analysis, eia.gov. 18 April www.eia.gov/todayinenergy/detail.cfm?id=5890 (Cited: 27 November 2014).

75 (2015). BS 2869:2010+A1:2011, BSI Standards Publication, Fuel oils for agricultural, domestic and industrial engines and boilers – Specification. 28 September http://home.aktor.qa/External%20Documents/Intenational%20Specifications/British%20Standards/BS/BS%2002869-2010%20+%20A1-2011.pdf (Cited: 12 June 2020). ISBN 978 0 580 76376 2.

76 Specifications guide, Japan domestic oil products (2020). https://www.spglobal.com/platts/plattscontent/_assets/_files/en/our-methodology/methodology-specifications/japan-domestic-oil-products-methodology.pdf (Cited: 12 June 2020).

77 Pippel, E., Woltersdorf, J., and Schneider, R. (1998). Micromechanisms of metal dusting on fe-base and ni-base alloys. *Materials and Corrosion* 49: 309–316.

78 Grabke, H.J. (2007). Metal dusting. In: *Corrosion by Carbon and Nitrogen: Metal Dusting, Carburisation and Nitridation* (ed. H.J. Grabke and M. Schutze), Ch. 1, pp 1-23 CRC Press-Woodhead Cambridge England.

79 El-Meshari, A.I. (2008). *Metal Dusting of Heat Resistant Alloys, Final Thesis.* Cambridge: University of Cambridge – Department of Materials Science and Metallurgy. Ph.D. thesis, October.

80 Bouwel, E.V. (2016). *Sulphur Countdown, Impact of Global Sulphur Limit on the Marine Fuels Market.* Insight from Infineum International Ltd. 13 July.

81 Vermeire, M.B. (2012). Everything you need to know about marine fuels. PDF4PRO.com. June (Cited: 12 June 2020).

82 Power Fuel Specialties, Power Corrosion Inhibitors. Innospec. (2020). https://www.innospecinc.com/our-markets/fuel-specialties/power-fuel-specialties/power-corrosion-inhibitors (Cited: 12 June 2020).

83 CONCAWE Marine Fuels Ad Hoc Group (reporting to FE/STF-24) (2013). Marine fuels and hydrogen sulphide. ©CONCAWE, Brussels, Report no. 8/13.

84 Amosa, M.K., Mohammed, I.A., and Yaro, S.A. (2010). Sulphide scavengers in oil and gas industry – A review. *NAFTA* 61: 85–92.

85 Smythe, J. (2015). *Trimming Costs Come with Hidden Expense: Cost Saving Measures Can Result in Cold Corrosion of Marine Cylinder Liners.* Insight from Infineum International Ltd. 15 April.

86 The Learning Resource For Marine Engineers (2001). marinediesels.co.uk. http://www.marinediesels.info (Cited: 13 June 2020).

87 Taylor, T.A., MacFarlane, J.J., and Stephenson, N. (1951). Improvements in or relating to fuel oils and the combustion of fuel oils. GB Patent 731,242. Filed UK, 20 June.

88 Leigh-Jones, C. (2014). *Vanadium and Sulphur in Marine Fuels*. The International Bunker Industry Association. June http://ibia.net/wp-content/uploads/2014/06/Vanadium-Sulphur-in-Marine-Fuels.pdf (Cited: 13 June 2020).

89 Migahed, M.A. and Beneficial, A.-S.M. (2009). Role of surfactants as corrosion inhibitors in petroleum industry: A review article. Taylor & Francis group, 1 May *Chemical Engineering Communications* 196: 1054–1075.

90 Papavinasam, S. (2014). *Corrosion and Control in The Oil and Gas Industry*. CH.7 Mitigation of Internal Corrosion, section 7.4 Corrosion Inhibitors. Pp 374-397 Elsevier N.Y. and London. ISBN 9780123970220.

91 McGuire, J.M. and Powis, P.J. (1998). Gas chromatographic analysis of tall oil fractionation products after methylation. 1 February *Journal of Chromatographic Science* 36: 104–108.

92 Oosterhout, J.C.D. and Banks, W.D. (1956). Manufacture of alkenyl succinic acid. US Patent 2,741,597. Filed USA, 8 March.

93 MacMillan, J.A. and Brewer, M.L. (1997). Fuel additives. US Patent 6,156,082. Filed USA, 30 May.

94 Landis, P.S., Backensto, M., and Backensto, E.B. (1951). Rust inhibitor for light petroleum products. US Patent 2,632,695. Filed USA, 20 September.

95 Sublett, M.D. and Blewitt, C.W. (1999). Process for producing light color dimer acids from the dimerisation of unsaturated fatty acids. WO Patent 00/75252. Filed USA, 4 June.

96 Barrett, F.O., Goebel, C.G., and Peters, R.M. (1954). Process of dimerizing monounsaturated fatty acids. US Patent 2,793,219. Filed USA, 13 December.

97 Dougherty, J.A., OudeAlink, B.A., and Dear, G.W. (1994). Corrosion inhibition by formation of iron carboxylate. EP 667,402. Filed Europe, 7 September.

98 Moser, F.R. (1935). Process for manufacturing lubricants. US Patent 2,124,628. Filed USA, 21 March.

99 Larkin, C. (2013). Quantification of corrosion inhibitor/lubricity improver in military fuels using infrared spectroscopy. Defence Technical Information Centre. 1 January https://apps.dtic.mil/sti/citations/ADA580288 (Cited: 14 June 2020).

100 Moss, P.H. and Yeakey, E.L. (1978). Detergent and corrosion inhibited motor fuel composition. US Patent 4,144,035. Filed USA, 27 March.

101 Hovemann, F. and Nottes, I.G. (1967). Corrosion inhibitors for fuel oils. GB Patent 1,212,935. Filed UK, 18 March.

102 Godar, R.L., Hendricks, C.C., and Roux, K.R. (1980). Corrosion inhibitors. US Patent 4,253,876. Filed USA, 19 February.

103 Antonijevic, M.M. and Petrovic, M.B. (2007). Copper Corrosion Inhibitors. A review. www.electrochemsci.org, 30 November, *International Journal of Electrochemical Science* 3: 1–28.

104 Chertkov, Y.B. and Ignatov, V.M. (1966). Comparative evaluation of the effect of jet fuel additives. June Khimiya i Tekhnologiya Topliv i Masel, Vol. No.6, pp. 53–56. Translation in Chemistry and Technology of Fuels and Oils. UDC 665.521.3.

105 Poirier, M.A. and Gervais, B.L. (1996). Distillate fuel composition containing combination of silver corrosion inhibitors. CA Patent 2,174,171. Filed Canada, 15 April.

106 Poirier, M.A. (1994). Copper corrosion inhibitors in or for distillate fuels. EP 618,284. Filed Europe, 29 March.

107 Caines, A. and Haycock, R. (1996). *Automotive Lubricants Reference Book.* Society of Automotive Engineers, Warrendale, P.A.

108 Poirier, M.A. (1995). Distillate fuel composition containing combination of silver corrosion inhibitors. US Patent 5,653,787. Filed USA, 6 June.

109 Knepper, J.I., Garecht, R.J., and Dear, G.W. (1983). Corrosion inhibition in engine fuel systems. US Patent 4,595,523. Filed USA, 1 July.

110 Reyes-Gavilan, J.L., Lichtenberg, F.T., Oshode, A.O. et al. (2000). Stabilised hydrotreated and hydrodewaxed lubricant compositions. EP 1,054,052. Filed Europe, 10 May.

111 Schaut, A. and Eeckhoudt, S. (2012). Effects of triazole additives in transformer oils. CIGRE. Paper D1-102_2012.

112 Weers, J.J. (1989). Composition for inhibition of corrosion in fuel systems and methods for preparation and use thereof. EP 330,416. Filed Europe, 21 February.

113 Andress, H. and Piotrowsky, H. (1972). Fuel and lubricant compositions. GB Patent 1,387,134. Filed UK, 19 May.

114 Dyestuffs Industry (2005). *Kirk-Othmer Encyclopedia of Chemical Technology*, 5th edition. Vol. 9. John Wiley and sons inc.

115 Wang, X., Wan, Y., Zeng, Y. et al. (2012). Investigation of benzimidazole compound as a novel corrosion inhibitor for mild steel in hydrochloric acid solution. (editors A. Weissberger. Library Congress Catalogue Card 53–7158) *International Journal of Electrochemical Science* 7: 2403–2415.

116 Khaled, K.F. (2003). The inhibition of benzimidazole derivatives on corrosion of iron in 1 M HCl solutions. Pergamon, *Electrochimica Acta* 48: 2493–2503.

117 Dajani, M. (1963). Inhibiting corrosion in petroleum processing equipment. GB Patent 1,011,314. Filed UK, 10 October.

118 Seffens, W.E. (1972). Corrosion Inhibitors for refining & petrochemical processing equipment. US Patent 3,766,053. Filed USA, 29 June.

119 Hofmann, K. (1953). Imidazoline and its Derivatives, pp 213-245, Part 1, and 367-373 Part 2. In: *The Chemistry of Heterocyclic Compounds* Vol. 6. Interscience PublishersInc. Chichester Susse.

120 Ebel, K., Koehler, H., Gamer, A.O. et al. (2014). Imidazole and derivatives. In: *Ullman's Encyclopedia of Industrial Chemistry* vols. 2. (ed. B. Elvers), Fine Chemicals.

121 Dwivedi, D., Lepkova, K., and Becker, T. (2017). Carbon steel corrosion: A review of key surface properties and characterization methods. Royal Society of Chemistry, January *RSC Advances* 7: 4580–4610. 10.1039/c6ra2509g

122 Gomez de la Oliva, C., Goya-Laza, P., and Ochoa de Ocariz, C. (2011). Six-membered heterocycles: Triazines, tetrazines and other polyaza systems. In: *Modern Heterocyclic Chemistry* Vol. 3, Chapter 20, pp 1777-1864. (eds. J.J. Vaquero, J. Barluenga, and J. Alvarez-Builla), Wiley-VCH Weinheim.

123 Huthmacher, K. and Most, D. (2014). Cyanuric Acid and Cyanuric Chloride. In: *Ullman's Encyclopedia of Industrial Chemistry* vols. 2, pp543-563. (ed. B. Elvers), Fine Chemicals. Wiley-VCH Weinheim. ISBN 9783527334775.

124 Kaiser, D.W., Thurston, J.T., Dudley, J.R. et al. (1951). Cyanuric chloride derivatives. I. Aminochloro-s-triazines. July *Journal of the American Chemical Society* 73: 2981–2983.

125 Kaiser, D.W., Thurston, J.T., Dudley, J.R. et al. (1951). Cyanuric chloride derivatives. II. substituted melamines. July *Journal of the American Chemical Society* 73: 2984–2986.

126 Compton, D.R. and Strickland, J.B. (2010). Method of removing hydrogen sulphide. US Patent 2012/0,012,506. Filed USA, 14 July.

127 Kelland, M.A. (2014). *Production Chemicals for the Oil and Gas Industry*, 2nd edition. CRC press.

128 Rivers, G.T. and Rybacki, R.L. (1992). Mixtures of hexahydrotriazines useful as H2S scavengers. US Patent 5,347,004. Filed USA, 9 October.

129 Performance, B. (2019). Marine fuel oil: neutralizing corrosion in large ships. 27 February https://www.bellperformance.com/fuel-treatment-for-ships (Cited: 20 June 2020).

130 Meffert, M.W. and Quinn, T.G. (2006). Additizing heavy fuel oil at terminals. US Patent 2008/0021713 USA, 18 July.

131 Sanders, H.R., Morway, A.J., and Clark, N.J. (1958). Residual fuels containing insoluble additives. US Patent 3,036,901. Filed US, 24 November.

132 Bloch, R.A., Lapinas, A.T., Outten, E.F. et al. (1995). Crankcase lubricant for modern heavy-duty diesel and gasoline fuelled engines. US Patent 6,004,910. Filed USA, 7 March.

133 Marsh, J.F. and Cleverley, J.A. (1979). A process for the production of basic magnesium sulphonates and the products obtained by this process. EP 015,341. Filed Europe, 7 December.

134 Marsh, J.F., Cleverley, J.A., and Simpkin, D.J. (1983). Basic alkaline earth metal sulphonates. US Patent 4,557,872. Filed USA, 11 October.

135 Jarman, R.A., Burstein, G.T., and Sheir, L.L. (1994). *Corrosion*, 3rd Edition. Butterworth Heinemann. vols. 1. Ch. 1.

136 Caprotti, R., Bhatti, N., and Balfour, G. (2010). Deposit control in modern diesel fuel injection systems. s.l.: Society of Automotive Engineers, Warrendale, P.A. SAE Technical Paper 2010-01-2250.

137 Caprotti, R., Breakspear, A., Graupner, O. et al. (2004). *Impact of Fuel Additives on Diesel Injector Deposits*. Tampa: SAE. 2004-01-2935, Powertrain & Fluid Systems Conference & Exhibition. Society of Automotive Engineers, Warrendale, P.A.

138 Richards, P. (2014). Gasoline additives, corrosion inhibitors. In: *Automotive Fuels Reference Book*, 3rd edition. Ch.11, pp 317-347. Society of Automotive Engineers, Warrendale, P.A.

139 Mang, T. and Dresel, W. (2007). *Lubricants and Lubrication*, 2nd edition. Wiley-VCH.

140 Cotton, F.A. and Wilkinson, G. (1988). *Advanced Inorganic Chemistry*, Fifth edition. John Wiley and Sons.

141 Martel, C.R., Bradley, R.P., McCoy, J.R. et al. (1974). *Aircraft Turbine Engine Fuels Corrosion Inhibitors and Their Effects on Fuel Properties*. National Technical Information Service. Report # AFAPL-TR-74-20.

142 Udelhofen, J.H. (1961). new imidazoline salts and hydrocarbon fuels containing same. US Patent 3,269,811. Filed USA, 28 April.

143 Diery, H., Grossmann, M., and Boldt, M. (1966). Corrosion-Inhibiting Compositions. GB Patent 1,165,024. Filed UK, 7 September.

144 Venkata Lakshmi, R.A. and Murthy, T.S.R. (2012). Studies on effect of copper corrosion passivators on the paper insulated copper conductor by XRD, SEM and EDX analysis. September-October, *International Journal of Engineering, Research and Applications* 2: 1295–1299.

145 Copper Development Association (2009). A1349 XX/09. *Working with Copper*. Benzotriazole: An effective corrosion inhibitor for copper alloys. https://www.copper.org/publications/pub_list/pdf/a1349.pdf (Cited: 21 June 2020).

146 Lubricant Additives: Use and Benefits (2016). atc-europe.org. August http://www.atc-europe.org/public/Document%20118%20-%20Lubricant%20Additives%20Use%20and%20Benefits.pdf (Cited: 21 June 2020).

147 Ahmed, N.S. and Nassar, A.M. (2011). Lubricating oil additives. (ed. C.-H. Kuo), Tribology – Lubricants and Lubrication. Intechopen.com, Ch. 10, 249–268.

148 The Impact of Gasoline Fuel Sulfur on Catalytic Emission Control Systems (1968). Manufacturers of emission controls association. September http://www.meca.org/galleries/files/sulfur.pdf (Cited: 16 March 2017).

149 Atkins, P.W. (1982). *Physical Chemistry*, 2nd edition. Oxford University Press. ISBN 0198551509.

150 Hoang, V.Q., Leung, P.L., and Rivers, G.T. (2012). No-sulphur fuel lubricity additive. US Patent 2012,317,876. Filed USA, 29 August.

151 Castle, J.E. (2008). The composition of metal surfaces after atmospheric exposure: An historical perspective. Taylor & Francis, April *Journal of Adhesion* 84: 368–388. ISSN: 0021-8464.

6

Combustion Improvers

6.1 The Need for Combustion Improvers

Combustion improvers for petroleum-derived fuels are used to treat three different problems: gasoline engine knock caused by unwanted auto-ignition, diesel engine knock caused by auto-ignition occurring too slowly and the emission of carbonaceous soot particles in the exhausts of all engines and fuel burners.

In gasoline engines, ignition is provided by a carefully timed electric spark while auto-ignition may cause a further, explosive ignition, out-of-time, heard as engine knock. Ignition in diesel engines depends upon the auto-ignition of fuel that is injected as the piston is almost at the top of its stroke – when the air in the cylinder has been heated up by compression – ignition must occur reliably at this point or the engine will suffer from diesel knock. In both engines, knocking causes damage to the engine. While the bulk of a fuel's anti-knock properties are provided for in the refining and blending of the fuel, for many years the fine-tuning of this property was aided by the use of anti-knock additives to control ignition: octane improvers to prevent auto-ignition of gasoline and cetane improvers to aid the auto-ignition of diesel fuel. While cetane improvers are still in use in diesel fuel, octane improvers have now mostly been displaced by blending of selected refinery components.

Another group of combustion improvers are those used to ensure complete combustion in systems using fuel spray burners. These are needed to minimise the production of particles, caused by incomplete combustion, that build up as sooty deposits in boiler and exhaust systems, or are emitted into the atmosphere where they are a hazard to health. Such particles are also produced by internal combustion engines, mainly diesel engines, but also gasoline engines after the move to direct injection. In the exceptionally large diesel engines that burn

Fuel Additives: Chemistry and Technology, First Edition. Robert D. Tack.

high boiling or residual fuels, as used in ships, combustion improvers are a necessity to prevent the blocking of turbochargers and heat exchangers. In the smaller diesel engines, used in motor vehicles, particulate traps are fitted to exhausts to trap soot particles. These traps would become blocked by the particles quite quickly if they were not burned off at some point – additives are needed to ensure that this occurs efficiently.

6.2 Combustion Improver Specific Problems

6.2.1 Gasoline Engine Knock and Octane Boosters

The four strokes of the gasoline engine[1] cycle (Figure 6.1) are:

- On the first piston down-stroke, the intake stroke, air and gasoline vapour are drawn in through the intake valve, into the cylinder. Nowdays, the gasoline may have been injected into the air stream or directly injected into the cylinder (see Chapter3).
- Second is the compression stroke as the piston moves up the cylinder, compressing the air/gasoline mixture
- Third, at the top of the compression stroke, an electric spark ignites the air/gasoline mixture; the burning and expansion of hot gases drives the piston down in the power stroke
- Fourth, the exhaust stroke, the piston moves up again, forcing the combustion products out through the exhaust valve

Besides gasoline engine knock, other out-of-time events that may occur are pre-ignition and run-on ignition [1]. Pre-ignition is the initiation of combustion

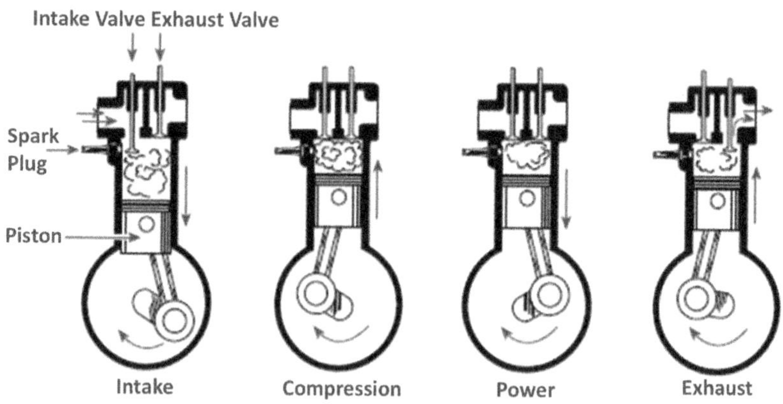

Figure 6.1 The Four Strokes of a Gasoline Engine. © 2019 Chevron. All rights reserved.

by the hot spark plug or hot particles on the piston that occurs before the timed spark, during the compression stroke. This effect can occur after the engine has been switched off, without the spark, causing carburetted engines[2] to run-on for a period. Improved engine design and better quality gasoline that resists auto-ignition have made such problems uncommon until recently, as turbocharged gasoline engines have been found to suffer from low-speed pre-ignition (LSPI), which may result from hot-spots or auto-ignition of lubricating oil vapour and droplets, or both [2]. Such problems as these, that are caused by deposits that build up in the combustion chamber and around the inlet valve or injector, are treated using effective detergents (Chapter 3).

Gasoline engine knock usually occurs as the flame front, initiated by the spark, moves through the engine cylinder, causing the remaining air/fuel mixture, lower in the cylinder, to auto-ignite under the growing temperature and pressure [3]. It then detonates, causing a spike in the pressure accompanied by a sharp sound (knock or ping). This out-of-time pressure spike causes a loss of power and prolonged knocking causes engine damage.

The tendency of a gasoline fuel to auto-ignite and cause knocking is measured in a one-cylinder gasoline engine that has a variable compression ratio; the compression ratio is increased until knocking is just detected. This is then compared with the mixture of iso-octane (2,2,4-trimthylpentane) (octane number 100) and n-heptane (octane number 0) that gives detectable knocking at the same compression ratio – the percentage of iso-octane in this mixture is the octane number of the test gasoline. When the test engine is run under low speed conditions (600 rpm), the result is called the research octane number (RON); when run at high speed and hot conditions, the result is called the motor octane number (MON). RON and MON usually differ by several numbers and sometimes the average of the two is used as the anti-knock index, AKI [4].

Knocking occurs when an automotive engine is running on gasoline of too low an octane number for its specification. In many countries, regular gasoline has a RON of 95 (AKI about 90) and in the USA, Australia, and New Zealand it is 92 (87 AKI); super or premium grades have higher RONs in the range 97 to 100. Motor manufacturers ensure that most vehicles under production will operate well using regular grade gasoline and such knocking should not occur unless the engine is not in correct adjustment.

In the early part of the twentieth century, the problem of knocking inhibited the development of the gasoline engine because the fuels were straight run distillates, which have RONs of less than 70 (Table 1.4). However, in the 1920s, it was found that tetraethyl lead (TEL) is a very effective octane booster, needing only small amounts, at low cost, to give gasoline the higher octane that it needed to allow for higher compression ratios and more efficient engines [4]. The advantages to the refiner of using TEL were such that its use gradually increased up to 0.66 mg/litre in the late 1960s. TEL is effective because it decomposes in the cylinder, with the burning fuel, into lead and lead oxide which effectively scavenge free radicals, killing radical chain reactions that cause detonation [5].

The effectiveness of TEL as an anti-knock additive[3] for gasoline was discovered by Thomas Midgley, working in a General Motors subsidiary in Dayton, Ohio. Midgley developed the manufacturing process to produce TEL using the reaction of a sodium-lead alloy with ethyl chloride. He also used ethylene bromide as a fuel co-additive, to enable the lead to be emitted with the exhaust as relatively volatile lead bromide compounds [6,7]. In 1924, a General Motors patent claimed the use of TEL in motor fuels in combination with various alkyl chlorides [8] while, in 1928, the Standard Oil Development Company patented an improved process for the manufacture of TEL [9]. Since the 1920s, there have been many patents on how to make and use the sodium-lead alloy, the use of catalysts and other process improvements to produce TEL. The overall chemistry is quite straightforward:

$$4\ NaPb + 4\ CH_3CH_2Cl \rightarrow (CH_3CH_2)_4Pb + 4\ NaCl + 3\ Pb$$

However, from 1970, the use of TEL was gradually diminished to being completely banned for use in automotive applications by the year 2000, in most countries [10]; in 2017, it was only legally used in Afghanistan [11]. The most important reason for the demise of TEL is that exhaust emissions containing lead compounds are poisonous, causing brain damage, particularly to young people in cities. Lead also poisons the emission control catalysts that were introduced in 1975 [4]. Since the 1960s, refinery technology had improved making it possible to produce the volumes of high-octane, unleaded gasoline required without the use of TEL.

The question arises as to how the refiner produces a gasoline that meets the expected octane requirement without TEL. Auto-ignition is a free radical process, so compounds that form relatively stable radicals stop or slow down free radical chain reactions, thus inhibiting auto-ignition. Aromatic compounds, such as toluene, can form relatively stable radicals by either adding radicals or losing a hydrogen atom. Highly branched alkanes, such as iso-octane, can, at their tertiary carbon atoms, also form relatively stable radicals. Both aromatic and highly branched alkanes have high octane numbers (Table 1.4).

Control of a fuel's octane number is achieved by the octane-blending of various components available in a refinery, some of which are produced especially for their high octane numbers, such as the aromatics from the fluid catalytic cracker (FCC) or the catalytic reformer, and branched alkanes and alkenes from the alkylation unit [4] (Table 1.4). The FCC [12] and alkylation processes [13] were only developed to commercialisation in the late 1930s and were rapidly developed with the onset of the Second World War for use in aeroplane engines, which required high octane gasoline. Catalytic reformers only appeared in 1949 [14].

While TEL has all but vanished from use in motor vehicles on the road, it is still in use in most current grades of aviation gasoline [15]. The reason for this is that many aircraft gasoline engines require a 100-octane number fuel and the metallurgy of many rely upon protective features of lead in the fuel. There are several grades of TEL-containing aviation gasoline which are differentiated by their octane values, as indicated by colour [16] (Table 6.1). There are also unleaded grades and engines that can use these grades, but the complete removal of TEL from aviation gasoline has met with difficulty and current testing of alternatives is expected to be complete in 2021. It is expected that the need for leaded aviation gasoline will eventually disappear along with the older models of aircraft that use such fuels.

While TEL was being phased out, other metallic octane improvers were brought into play. In the 1960s, ferrocene had been proposed as a gasoline anti-knock agent [17] and methyl-cyclopendadienyl manganese tricarbonyl (MMT) had been developed for use in motor fuels in the 1950s [18]: both of these were in limited use as TEL replacements during the early years of unleaded gasoline but such additives are no longer in use in commercial motor gasoline.

6.2.2 Diesel Knock and Cetane Improvers

To understand the phenomenon of diesel knock and the role played by the cetane rating of a fuel and cetane improvers, it is necessary to understand the diesel auto-ignition process [10,19]. The four strokes of the diesel engine cycle are (Figure 6.2):

- On the first stroke, with the intake valve open, air is drawn into the cylinder[4], as the piston moves down.
- Second stroke, with the intake valve closed, the piston moves up compressing and heating the air to over 550°C.
- Third stroke, at the top of the compression stroke, fuel is injected into the compressed, hot air where it auto-ignites and burns quickly, pushing the piston down as the hot, gaseous products of combustion expand.
- Fourth, the piston moves up again, forcing the combustion products out through the exhaust valve.

Table 6.1 Leaded Aviation Gasoline Grades.

Grade:	G80	G91	G100	G100LL
Octane	80	91	100	100
Max. lead, g/ litre	0.14	0.56	1.12	0.56
Colour	Red	Brown	Green	Blue

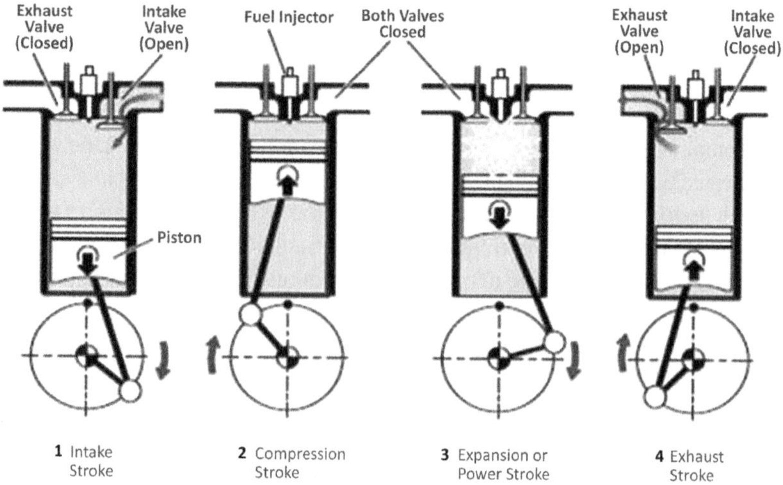

Figure 6.2 The Four Stokes of the Diesel Engine. © 2019 Chevron. All rights reserved.

It is essential for smooth expansion that the ignition starts as soon as possible after the start of fuel injection – preferably whilst fuel injection is still proceeding. The delay between the start of fuel injection and ignition, the ignition delay, is a critical parameter. If ignition is delayed until all the fuel is injected and mixed with air in the cylinder, there is too much fuel to burn evenly and it may detonate all at once, abruptly and violently, producing more noise – known as diesel knock [20]. This results in lower fuel economy, more polluting exhaust emissions and smoke, irregular operation and lower power. Such effects are usually most noticeable on cold starting; later, the hot engine block raises the temperature in the cylinder which aids auto-ignition of diesel fuel.

The degree of ignition delay depends upon the quality of the diesel fuel, as specified by its cetane number – a fuel with a high cetane number (≥50) has a short ignition delay, which is good, and fuel with a low cetane number (≤40) has a long ignition delay, which is bad. Cetane number was initially determined by finding the ignition delay of the fuel under test in a single cylinder test engine that has variable compression, running under a standard set of conditions. Comparison is made with the cetane number standard that gives the same ignition delay. Primary cetane number standards are mixtures of 1-methylnaphthalene[5], which is resistant to auto-ignition and has a cetane number of zero, and n-hexadecane (cetane), which readily auto-ignites in the engine and has a cetane number of 100. The percentage of n-hexadecane in the mixture that matches the ignition delay characteristics of the test fuel is its cetane number.

The original, primary method of determining the cetane number has, in everyday use, been replaced by a few alternatives, such as the ignition quality tester and the fuel ignition tester. But most used is the cetane index, a number which is calculated from the density and the distillation range of the fuel. In Europe, EN 590 requires diesel fuel to have a cetane index of not less than 46.0 and a cetane number of not less than 51.0; in North America, the EPA and ASTM D975 standards require a minimum cetane number of 40 but individual states are more demanding, for example California requires a minimum cetane number of 53 [21]. Premium grades usually have higher cetane numbers in the range 52 to 60.

In contrast to the case for gasoline, a diesel fuel is blended with minimum proportions of aromatic compounds and highly branched alkanes – both of which form relatively stable radicals so inhibit auto-ignition. Saturated, long-chain alkanes, preferably with minimal branching, have the best auto-ignition properties so are maximised in diesel fuel blends. Diesel fuel cetane is helped by the severe hydrogenation processes that are in place to remove sulphur that also saturate many aromatics and alkenes – compare the cetane number of 0 for 1 – methylnaphthalene with 48 for decalin (fully hydrogenated naphthalene) [19]. The ease of auto-ignition is also reflected in the lowest temperatures at which auto-ignition can take place for the various compounds in diesel fuel, such as 202°C for cetane compared with 472°C for iso-cetane and 579°C for 1-methylnaphthalene [20].

Cetane numbers are improved by the addition of cetane improvers, which are usually alkyl nitrates – especially 2-ethylhexyl-nitrate (EHN). For example, the addition of 0.4% of EHN can raise a fuel cetane number of 45 to 51 [22]. EHN decomposes to give alkoxy and alkyl free radicals above 130°C (Figure 6.3), below the temperatures needed for auto-ignition of cetane (202°C). On decomposition, EHN also gives nitrogen dioxide, a stable free radical having an odd number of valence electrons (Figure 6.3). Nitrogen dioxide is an oxidising agent, able to abstract hydrogen-atoms from hydrocarbons, or from the formaldehyde by-product of EHN's decomposition, initiating a free radical autoxidation chain [23] (see Chapter 2 for details of autoxidation). The nitrous acid produced decomposes at the high temperatures in the cylinder to give hydroxyl radicals and nitric oxide. Hydroxyl radicals are very reactive, so they abstract hydrogen atoms from nearby hydrocarbons, thus initiating another free radical chain. Nitric oxide also has an odd electron and reacts rapidly with oxygen to produce more nitrogen dioxide, thus completing a cyclic process that throws off free radical chains every cycle [24].

It has been shown in controlled experiments that the ignition delay reduction by EHN is the same as that produced by the same amount of nitrogen dioxide, nitric oxide or nitrous acid that would be produced by its decomposition. Using

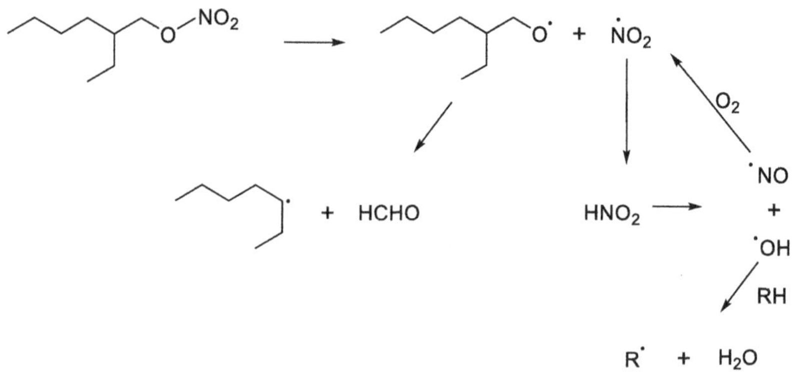

Figure 6.3 The Thermal Decomposition of Ethylhexyl Nitrate[6].

a single cylinder engine, regular diesel was found to give an ignition delay of 7.4 milliseconds (ms); the addition of 1% EHN reduced this to 4.9 ms as did adding the amount of NO, NO_2 or HNO_2 that would be produced by 1% of EHN. The formaldehyde that would be produced had little effect on the ignition delay.

Di-Tertiary Butyl Peroxide (DTBP), which has also been used as a cetane improver, was found to be a little less effective than EHN: the delay with 1% DTBP was 5.9 ms [24]. It seems that the mechanisms of the two cetane improvers are not the same but may be complementary. In the same work, a mixture of the two gave an ignition delay of 5.5 ms but with much less variability than found using either one cetane improver separately. However, in low sulphur kerosene, the DTBP was completely ineffective while the EHN was fully effective, further suggesting differences in the details of their activities. DTBP is a limited provider of free radicals: each molecule provides two free radicals as it decomposes (Figure 6.4) compared with many from the cyclic process of nitrogen dioxide production from EHN (Figure 6.3).

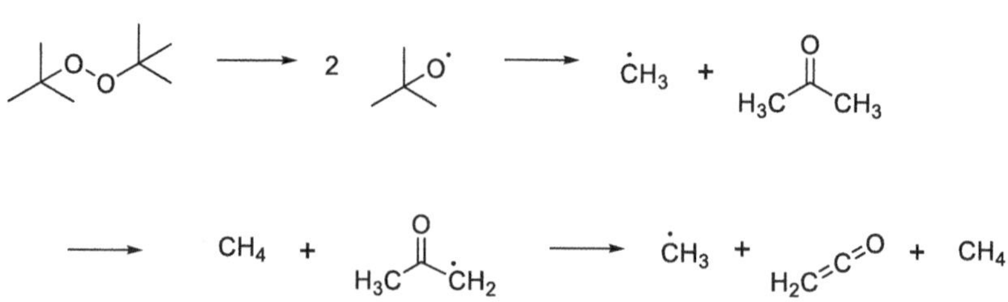

Figure 6.4 The Thermal Decomposition of DTBP.

The effectiveness of EHN at improving auto-ignition at lower temperatures is of particular help to diesel cold start performance: the addition of 0.1% EHN to diesel fuel gives reductions in time to first fire, time to stable idle and time to 50% reduced smoke, as well as reducing emissions of particulate matter [10]. EHN may be added at the refinery to adjust the cetane numbers of the blended diesel fuel streams or as part of an additive package, added to standard diesel to produce premium grade diesel fuels that have higher cetane numbers (up to 60). EHN would normally be used at 0.015 to 0.20% of the fuel and above [5]. DTBP is also used as a cetane improver to some extent – while it is less effective than EHN, it does not produce additional nitrogen oxides ('NOx') in the exhaust [19].

6.2.3 Combustion Improvers for Heating Oils and Heavy Fuels

Systems that burn oil to provide heat do so either directly, as in refinery still furnaces, or by boiling water to provide steam, such as for heating apartment blocks or for electricity generation in steam turbines. In such systems, the fuel is sprayed as fine droplets by a jet into the combustion zone and flame. Heating oils (furnace fuels) are the products into which are blended the aromatic middle distillate and kerosene fractions that are by-products of the FCC – these have low cetane numbers so are unsuitable for diesel fuel blending (Chapter 1). Heating oils may also contain higher boiling gas oils, which cannot be blended into diesel fuel because of distillation specifications. Aromatic and high boiling hydrocarbons tend to burn with sooty flames because of their high carbon contents.

Usually, in oil burners, the higher boiling components do not readily vaporise when the fuel is sprayed into the combustion zone. While the more volatile components are burning, the remaining, higher-boiling fuel droplets are starved of oxygen and are converted by the burning fuel into carbonaceous particles, soot, which leave in the exhaust [25]. This is a general process in all fuel injection systems, though perhaps more visible in oil burning heating systems and peak power electrical plants, where the boilers are used to produce steam to turn gas turbines [26]. Here, the relatively large sooty particles build up on boiler surfaces and in the exhaust systems, reducing heat exchanger efficiency and causing back pressures. Such problems have been growing in boiler systems that burn heavy fuels containing residual fuels, as these have been deteriorating in quality, containing more asphaltenes, while the demand for the better quality distillates[7] has grown [27]. Up to 1% of such fuels remains in the exhaust streams as unburnt hydrocarbons and carbon that form sticky deposits on heat exchanger surfaces, decreasing boiler efficiencies. The addition of 0.2% of a combustion improver additive ('VNII NP-200') brought about a reduction of 80% of the amount of unburnt carbon in the deposits on heat exchangers in a Russian power plant [27].

The addition of the naphthenate[8] salts of iron, cobalt or manganese to motor fuel, 'to improve the octane number of the fuels and to improve the burning qualities', and 'to add these salts to heavy residual fuels ... to reduce the large amount of carbonaceous materials ... when burned in marine type boilers' was already known well before the 1940s [28]. However, it appears to have been only then that they were applied to domestic heating fuels to reduce smoke and the build-up of carbonaceous deposits on boilers [28]. Ferrocene (or the more soluble methyl ferrocene) was known as a combustion improver for jet fuel, though the fuel containing ferrocene also needed antioxidants to stabilise it [29]. The gasoline anti-knock additive, MMT, was developed in the late 1950s to control soot and sooty deposits in oil burning furnaces [30,31] and, in combination with oil soluble calcium compounds, was proposed for smoke reduction in jet fuel [32]. MMT and ferrocene are still included in some fuel additive packages for use in heating oils and diesel fuels [33,34] and may be components in the Russian additive NP-200.

6.2.4 Combustion Improvers for Particulates in Diesel Engine Exhausts

In all diesel engines, liquid fuel is sprayed into the cylinder. In marine engines and large engines used for electricity generation, the distillate diesel fuels used contain blend components that are high boiling and rich in aromatics, when compared with motor diesel. These engines may also burn residual fuels that contain asphaltenes – compounds that already resemble carbon and are quite involatile. Such fuels have problems of incomplete combustion and the production of large amounts of sooty, carbonaceous exhausts. The carbonaceous soot particles in exhaust streams cause major problems as they build up on turbochargers and heat exchangers, in addition to their build upon valves and piston rings in the combustion chamber. The build-up of deposits on the heat exchangers and turbochargers necessitates various cleaning procedures. The advantages of ferrocene (preferably at 10 to 30 ppm) as a combustion improver was demonstrated in a 33,000 kW ship's engine: its use gave a great reduction, in places to zero, in the amount of deposits that would be removed in daily and monthly cleaning procedures [35].

In general, it is said that combustion improvers for this application contain various metals, such as magnesium, calcium, iron and manganese, and have long been in use in these fuels [5,36] – though the calcium and magnesium salts are also for control of corrosion by vanadium-related deposits (Chapter 5). Providers of commercial combustion improvers for use in heavy fuel oils, used in marine engines and boilers, claim that their additives (used at 100 to 1,000 ppm, dilution unknown) improve the combustion of carbon particles, thus improve combustion efficiency, reduce carbonaceous deposits in the combustion zone, in the exhaust valves and ports, and in piston grooves[9], and also reduce smoke emissions [25,37].

When motor diesel fuel contained moderately high boiling fractions, such as those boiling up to 380°C or higher, as recently as in the 1980s, ferrocene had been proposed to clean up and minimise carbonaceous deposits in diesel engines [38]. However, diesel fuels are now much cleaner, with limits on the final boiling points and limited amounts of alkyl benzenes, with no polyaromatics. Also, the injection of fuel into the cylinder is better controlled by the engine management system. Despite fuel improvements, carbonaceous particulates are still found in the exhausts of diesel engines and modern gasoline direct injection (GDI) engines, ranging from the black smoke of poorly maintained engines to invisible micron and nanometre-sized particles. In the past, attention was concentrated on diesel engines but the introduction of GDI and tighter specifications for allowable particulate emissions has meant that gasoline particulate filters may also be necessary; additive technology to aid the burning of gasoline engine particulates will probably follow that used for diesel engines.

Unfortunately, in diesel exhausts, the smallest particles that have carbon cores and are coated with organic compounds provide the worst threats to health because they lodge, irremovably, deep in the lung [39,40]. The distinction between particulates of different size ranges, in motor vehicle emissions, has been recognised as important: PM10 particulates are those less than 10 micron and PM2.5 are those less than 2.5 micron in diameter. Particulates larger than PM10 are trapped in the nose, those between PM10 and PM2.5 are deposited in the respiratory tract but PM2.5 particulates can penetrate into the bronchioles and alveoli, where they cause inflammatory responses, increasing the risks of asthma and COPD[10] [40,41]. In view of this problem, diesel vehicle emission standards have targeted particulate emissions, requiring them to be massively reduced. For example, Euro emissions standards for large goods vehicles allowed up to 0.40 g/kWh of particulate matter (PM) in 1995, in 2012 this maximum was reduced to 0.02 g/kWh [42].

In the 1980s, ferrocene was proposed to improve diesel vehicle efficiencies by removing carbon deposits inside the cylinders [38]. Then, diesel particulate emissions were also thought to be controllable by the use of ferrocene – it was claimed that a ferrocene additive, at 0.08 ppm of iron, reduced the amount of particulate emissions from a diesel fuel by 43% by catalysing the combustion of the particulate carbon [43]. However, such additives cannot reduce particulates down to modern emission standards, so diesel particulate filters (DPFs) are now fitted to the exhausts systems of most new diesel vehicles [44].

Particulate material trapped by DPFs, made of a ceramic material, gradually block the pores of the filter, causing a back-pressure that reduces the engine's efficiency and could lead to failure. Such systems rely upon the carbon particulates being burned off when the filter gets sufficiently heated by the exhaust stream. However, the temperatures required to ignite the carbon, > 600°C [45], are too high to occur in most everyday vehicle uses. Ignition temperatures may be reduced to more accessible temperatures either by including a catalytic metal, such as platinum, in the material used to construct the DPF, or by adding a fuel soluble form of a metal, such as iron, copper or cerium [46], to the fuel. The former system has the disadvantage that such metals as platinum are sensitive to poisoning by sulphur in the fuel – still a problem in many parts of the world. The latter, termed fuel borne catalysts or DPF additives, are insensitive to sulphur in the fuel and, as the metal becomes included in the particles, it reduces the temperature required to burn off the carbon by 150°C or more [45]. Such additives are used at concentrations in fuel of 10 to 50 ppm [47].

DPF fuel additives initially relied upon iron as an oil soluble salt or ferrocene itself. Solubility of the DPF additive in hydrocarbons is often a limitation to its use in a concentrate that would be used to blend additive packages or to be added to fuel. Such a concentrate needs to be stable at winter temperatures to enable widespread, year-round storage and not give deposits from the treated fuel over extended periods at low temperatures. Ferrocene itself has limited solubility, able to hold only 1.5% of iron at −15°C in solution in an aromatic solvent, compared with the proposed alternative 2,2-bisferrocenylpropane, that is soluble at up to 7.5% of iron [48]. This same patent also shows that fuel containing 20 ppm of iron in the form of iron tris(2-ethylhexanoate) produces deposits on storage, whilst fuels containing 20 ppm of iron as ferrocene or as the2,2′ – bisferrocenylpropane do not cause deposits on storage. Another patent claims the use of the ferric isodecanoate as a DPF additive, being soluble enough in fuel, at 10 or 40 ppm of iron, for storage over long periods [49].

An alternative approach to supplying the maximum amount of iron in the minimum amount of soluble additive, used by Rhodia, is to use a colloidal dispersion of extremely small particles of iron oxide (nanoparticles) supported by a surfactant. Such dispersions may be prepared by precipitating ferric hydroxide and dehydrating to ferric oxide in the presence of surfactants (iso-stearic acid used in the example), to yield particles of amorphous iron oxide, 3 nm across [50]. Tested in a diesel car, fitted with a particulate filter, it was claimed that this additive, at 7ppm of iron in the fuel, enabled the carbon to be burned off the DPF at 400 to 425°C, compared with 600°C in the absence of iron.

In its patent, Infineum was concerned with the effective delivery of colloidally dispersed nanoparticles of crystalline iron oxide (magnetite form) [51]: the use of a PIBSA (such as 1000 molecular weight) was demonstrated to give clear advantages as the stabiliser for iron oxide particle dispersions (crystalline or amorphous) – over mono-carboxylic acids and when compared with iron neodecanoate – providing minimum additive viscosity, colloid stability in fuel and lack of injector fouling (DW-10 test).

Recent joint patent activity of Lubrizol and Rhodia is also concerned with the no-harms application of dispersions of crystalline iron oxide nanoparticles (size preferably less than 10 nm), stabilised by both iso-stearic acid and the Lubrizol Mannich and PIBSA-DMAPA detergents (described in section 3.4.2), in particular the quaternised versions of these detergents [52]. The advantages provided by these detergents to the iron oxide dispersion, demonstrated by test data in a B10 biodiesel, are:

- improved iron oxide dispersion stability in the fuel
- inhibition of injector deposit formation
- improved oxidative stability of biodiesel blends (stated that FBC additives can reduce oxidation stability)

In addition, the effective catalysis of the DPF regeneration, by improved burn-off of captured exhaust particulates, was also demonstrated for these combinations.

Cerium oxide, CeO_2, has also been deployed as a diesel fuel additive that reduces particulate emissions, in the form of a dispersion of particles with a mean diameter of less than 100 nm. Examples in a patent disclose particles of cerium oxide of 10 or 12 nm in diameter, made by precipitation of cerium hydroxide ($Ce[OH]_3$) from an aqueous solution of cerium nitrate by ammonium hydroxide. The cerium hydroxide was then oxidised to cerium oxide with hydrogen peroxide, in the presence of a colloid stabiliser, such as 2-[2-(2-methoxyethoxy)ethoxy] acetic acid [53]. When making the concentrate, for application, the aqueous dispersion was further dispersed in diesel fuel by combining with the surfactant Aerosol AOT and 1-hexanol. The product was evaluated in diesel fuel, at a concentration of 42 ppm of cerium oxide, in a diesel engine driven electricity generator. When the generator engine was running at a steady state, the exhaust was passed through a filter for 150 seconds to collect the particulates. The carbon collected on the filter turned it grey, measured as a percentage greyscale[11]. Untreated diesel fuel gave the filter a grey scale rating of 70% (dark, due to much being trapped), while the fuel containing cerium oxide gave, under the same conditions, a grey rating of 40% (much lighter), indicating less particulate matter in the exhaust due to their more complete oxidation. In addition, the diesel generator provided 3% more power output with the treated fuel, compared with the untreated fuel.

Lubrizol have claimed the use of copper as the metal to reduce soot formation, initially to improve combustion properties of furnace fuels [54]. For this purpose, the copper was solubilised by an oxime and a Mannich detergent (see Chapter 3). In a later patent, a number of combinations of copper naphthenate with amine-derived complexing agents and an anti-oxidant are described and claimed for use '...in diesel fuels for operating diesel engines equipped with exhaust system particulate traps ... useful in lowering the ignition temperatures of exhaust particles collected in the trap' [55]. Support for the efficacy of a copper additive for this application initially relied upon an earlier patent by a German company, which describes equipment to be fitted to the exhaust of an internal combustion engine to filter off the soot and to occasionally burn-off the deposited soot [56]. Since the temperatures available at the filter were too low to burn-off the soot, the equipment contained a system to spray a cuprous chloride solution (from a storage tank) onto the soot, thereby reducing its ignition temperature to between 350°C and 375°C. The additive company has presented more detail in a paper which describes fuel additives for DPF regeneration as copper complexes with the oxime of 4-dodecyl-salicylaldehyde or with the structurally similar oxime of 5-dodecy-2-hydroxy-acetophenone (Figure 6.5) [57]. The reader may have noticed the similarity of this copper complex and that of the favoured MDA, when complexed with copper to inhibit its pro-oxidative effect to stabilise fuel on storage (Figure 2.4); however, its purpose here is to provide copper to any carbon particles formed during combustion. Using these derivatives to provide 30 ppm of copper to a diesel fuel, the DPF regeneration temperature was reduced from 510°C to 375°C. Lubrizol's copper FBC additive, OS96401, was used in a study of how combinations of metals aid DPF regeneration [46].

Marine diesel engines would be expected to produce more particulate emissions – especially in view of the sizes of these engines and the large amounts of polyaromatics, sulphur and inorganic compounds in the fuels that they burn. There have been many studies of particulate emissions from ships and this is indeed the case [58,59] though currently the focus on improving emissions from marine engines is the reduction of sulphur [60,61]. Fortunately, the large reductions in sulphur to be allowed in marine fuels will entail the use of cleaner fuels, probably those containing high proportion of desulphurised distillates, or the scrubbing of exhaust gases, both of which will reduce ships' particulate emissions.

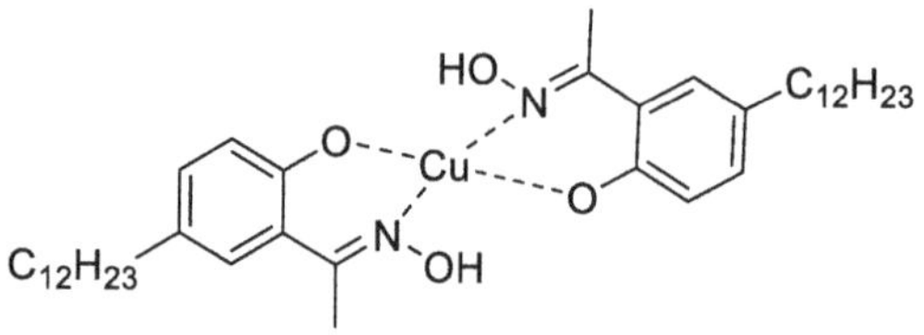

Figure 6.5 Copper Complex with 5-dodecy-2-hydroxy-acetophenone.

6.3 Mechanisms of Soot Formation and Its Removal

Soot formation occurs as fuel is sprayed into the flames in the combustion zones of burners in jet engines, oil fired heating systems, industrial furnaces and steam generators for electricity generating turbines. While the fuel in a diesel engine is burned in a closed cylinder, it too relies upon a flame that is formed in a jet of fuel that is injected into the cylinder. Here, the discussion will concentrate on the formation of soot in fuel injection engines – particularly diesel engines. In all cases of liquid or gaseous hydrocarbon combustion, the hydrocarbon molecules lose hydrogen, as water, and break down into smaller molecules, such as acetylenes, before ending up as oxides of carbon.

Soot is formed in a series of chemical reactions followed by agglomeration of particles, all of which depend upon the changes in the surrounding physical environment. First, the formation of soot in relation to the processes in the diesel engine cylinder and in the exhaust will be a discussed. Then follows a discussion of the possible chemical effects of metallic additives.

6.3.1 The Formation of Soot

The general picture of soot formation is that of fuel burning in a fuel-rich environment, with limited oxygen, which occurs close to a fuel injection point. The small, reactive molecules that are formed in this environment combine to form aromatic compounds or, for fuels containing aromatics such as kerosene, build on the aromatics already present [62]. Supported by studies sampling soot samples from various parts of an ethylene flame, and examined by Transmission Electron Microscopy (TEM), the following picture has emerged [63]. As small molecules are formed and combine with the initial benzene rings, they grow into Poly-Aromatic Hydrocarbon molecules (PAH) having molecular weights of about 200. These then grow into the soot micro-particles of 1.5 to 15 nm across (10^3 to 10^6 molecular weight) by a process which may be a combination of coagulation by surface reactions and PAH growth. As the micro-particles are exposed to the high temperatures in the flame, they are converted to graphite crystallites. Coagulation of the micro-particles, by numerous collisions, enables these to grow into primary soot particles, which then also coagulate to form the final soot particles of 10^4 to 10^8 mol. weight units in size. A TEM (Figure 17.3 of this paper [63]) shows a final soot particle, which extends for hundreds of nanometres, as a chain-like aggregate of many spherical primary particles, each 35 nm in diameter – TEM examination of soot from an acetylene flame gives the same picture [64].

Soot forms in and around flames at temperatures of 1000°C to 1400°C. In flames of ethylene/air or heptane/air mixtures, soot particles have been found to form in the hottest parts (1300°C to 1400°C) [65,66]. Another investigation, using a plug flow reactor and a propane gas flame, was able to plot soot formation against time, demonstrating it to be fast at about 1200°C and falling off at temperatures moving down to 1100°C [67]. The data at 1200°C and above covers soot formation over a timescale of up to 30 milliseconds (ms), while down to 1100°C, soot production is notably slower in a timescale of up to 70 ms. These timescales should be compared with the time for a down-stroke of a diesel engine, running at 1000 rpm, which is 28 ms [68].

Details of soot formation in the cylinder of a passenger car diesel engine have been monitored by a probe that was positioned inside the area of the fuel spray of the fuel injector [68]. The information in this paper has been gathered onto one diagram, covering the critical first 4 ms after fuel injection starts, to enable the various occurrences to be viewed together (Figure 6.6). From the point at which the piston approaches Top Dead Centre (TDC), soot formation during the combustion stroke unfolds as follows:

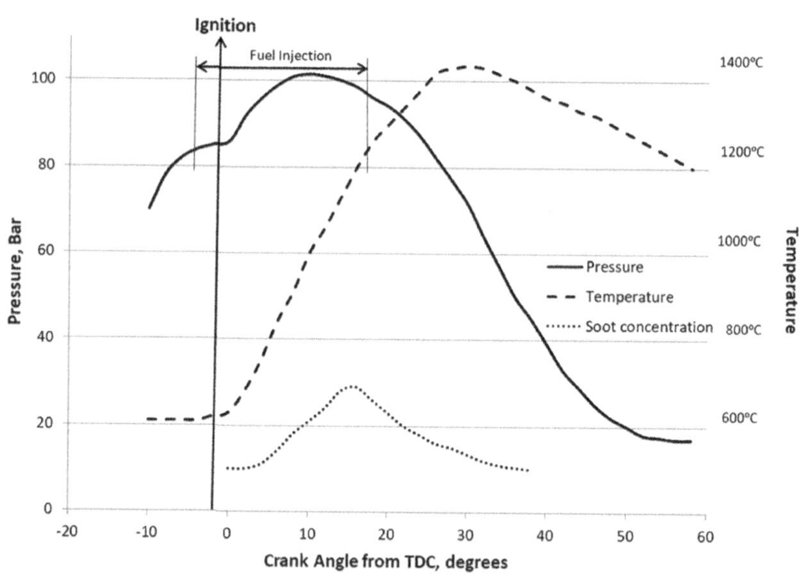

Figure 6.6 Soot Formation in a Passenger Car DI Diesel Engine, Running at 2,600 rpm.

- As the piston approaches TDC, the air in the cylinder is under maximum compression and reaches a temperature of just over 600°C.
- Fuel injection, which lasts about 1.2 ms, starts just before TDC and ignition occurs after a delay of about 0.2 ms, close to TDC. In this region, there is an excess of fuel – some vapour and some as liquid droplets – in an inhomogeneous mixture with air, so soot forms.
- Fuel continues to be injected into the flame and the environment moves towards burnt gases over the next 1 ms. This is still a fuel-rich environment so continues to produce soot leading to a maximum in local soot concentration. The average cylinder temperature and pressure both rise as the fuel is burning – though the temperature in the flame is much higher than this.
- The piston now moves down, pushed by the pressure that has built up as the fuel burns and the temperature rises. Soot concentration at the sampling point falls as it is diluted by the expanding gases and is burnt in the air with which the burning fuel becomes effectively mixed.
- This continues over the next 4 to 5 ms, during which the soot particles grow by coagulation of the primary particles, at an average temperature in the cylinder of over 1000°C. Soot particles are then less than 100 nm, mostly much smaller (95% 3 to 10 nm).
- In the remaining 12 ms of expansion, some soot particles continue to grow, while others are burned as they are mixed with the excess of air that a diesel engine uses.
- The piston then moves upwards, and the exhaust valve opens for the products of combustion – mainly carbon dioxide and water – with the nitrogen and unused oxygen from the air, plus the unwanted emissions of, mainly, NOx and soot. The exhaust gases pass through the particulate filter, which captures the soot particles.

The chemistry of the formation of soot has attracted much interest in most areas of fossil fuel combustion, in view of its negative economic and environmental consequences. Despite the difficulties of studying chemistry in flames, there is a generally accepted picture of the chemistry involved in the formation of the PAH molecules in the first step of soot formation. Hydrocarbons, especially in fuel-rich regions, at the high temperatures that they experience in flames, crack and are gradually oxidised, losing hydrogen to oxygen and hydroxy radicals (see Chapter 2 for oxidation mechanisms). They break down to small, reactive molecules, in particular acetylenes [69–71]. Of the reaction schemes that are possible, one that has been proposed and for which there is much evidence, starts from the acetylenes (Figure 6.7): acetylene oligomers and vinyl acetylenes form under free radical conditions and these cyclize to form benzene rings.

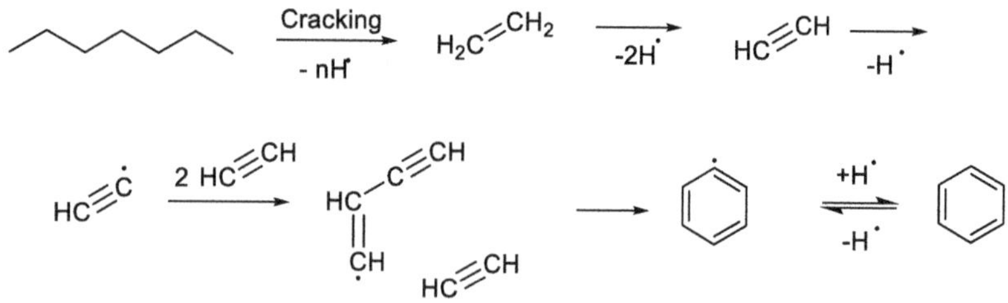

Figure 6.7 Formation of Benzene Rings from Acetylenes.

Acetylenes react with aromatics to form alkynyl – and alkenyl-substituted aromatics that further cyclise, losing hydrogen atoms[12], forming naphthalene then acenaphthalene (Figure 6.8).

Aromatic molecules themselves can combine under the free radical conditions in an alternative route to PAHs via phenanthrene (Figure 6.9).

Repetitions of such reactions lead to the growth of PAH molecules of increasing size that are the beginnings of the physical sequence described above (Figure 6.10).

It has been proposed that the free radical reactions could also be cationic reactions [72], though most authors use the free radical route. Inside the flame, the very reactive radicals can undertake any feasible reaction that leads to the formation of graphite-like structures – alongside the oxidation reactions in

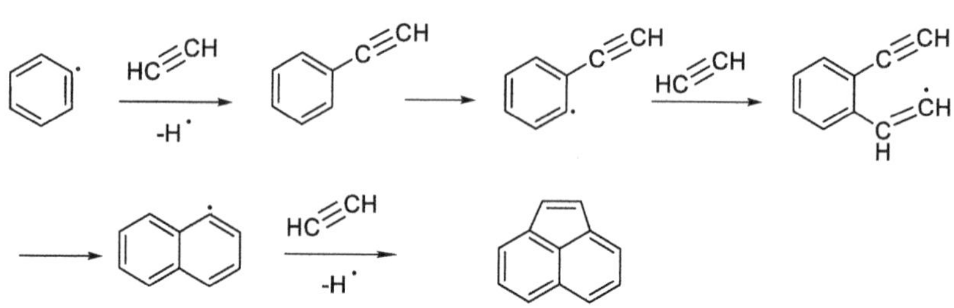

Figure 6.8 Formation of Acenaphthalene.

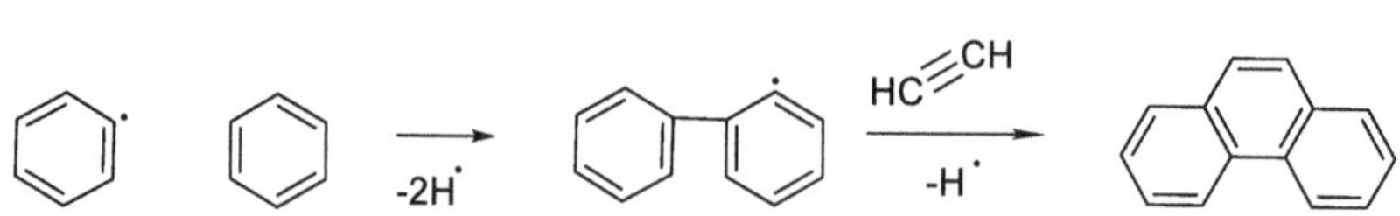

Figure 6.9 Formation of Phenthanthrene.

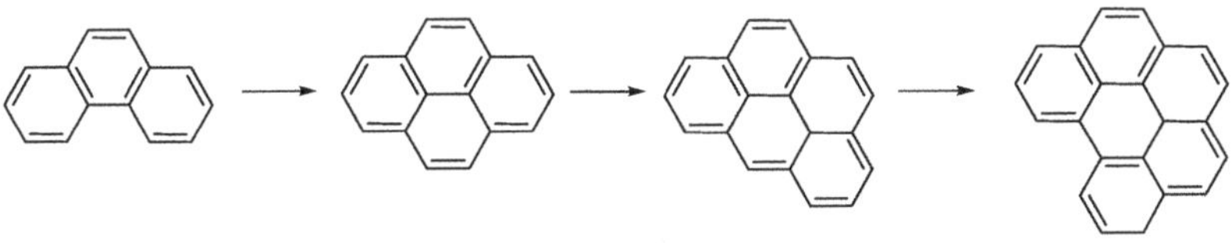

Figure 6.10 Growth of Large Poly Aromatic Hydrocarbons.

which the great majority of the hydrocarbon is converted to water and carbon dioxide.

Metallic additives are believed to reduce the amount of soot in exhausts by three mechanisms [66,73]. The first applies to metals including group IA and IIA metals, such as potassium and calcium. This is believed to be a charge stabilisation against agglomeration of the early, primary particles – a technique, relying upon potassium ions, that has long been in use by carbon black producers to prevent particle agglomeration. In this way, the soot particles are kept small, so have high surface areas enabling their oxidation, a surface reaction, whilst still in the hot, combustion-supporting zones.

By the second mechanism, group IIA metals form hydroxyl radicals, by interaction with water molecules in the flame, which rapidly remove soot and soot precursors. The third mechanism applies to metals with variable oxidation states, such as iron, manganese, copper and cerium, which become included in the soot particle where they catalyse soot oxidation by aiding oxygen transfer to carbon. It appears that this effect applies late in the flame in the oxygen-rich environment, which would apply to soot particles being burned in the later part of the piston down-stroke and as the exhaust gases leave the cylinder. The effects of metals are also seen as reductions of about 100°C to 200°C in the temperatures at which soot ignites and may be burned off a filter. That metals incorporated into soot particles aid their ignition before they reach the particulate filter follows from the reductions in diesel exhaust particulates or smoke [43], or reductions of soot in the flue gases from furnaces [73] by the use of metallic additives, that were well established before DPFs were introduced. In fact, the method used to demonstrate reduction of soot by cerium oxide, described above, was to collect the soot in a filter with no intervening DPF [53].

Notes

1 Also known as the spark-ignition engine.

2 In a carburetted engine, rotation of the engine draws fuel vapour and air into the engine, enabling the run-on effect after the ignition is switched off. Fuel

Injection is also switched off with the ignition, so run-on is not a problem in FI engines.

3 Also known as anti-knock agents, additives, or compounds.

4 Or forced in under pressure as modern diesel engines are usually turbo-charged – the air is pressurised before entering the cylinder to allow for more fuel, providing more energy, in the same cylinder.

5 1-methylnaphthalene is not stable in storage so has been replaced by a secondary standard 2,2,4,4,6,8,8-heptamethylnonane (iso-cetane) which has a cetane number of 15 [19].

6 Figure 6.3 and Figure 6.4 only show the immediate decompositions of EHN and DTBP, the full process of the free radical autoxidation and combustion is beyond the scope of this book.

7 High demand for motor fuels meeting modern specifications cause refiners to convert more of the high boiling distillates to motor fuel components, leaving less available to be blended with residual fuel (Chapter 1).

8 Naphthenic acids are alkylated cyclopentene and cyclohexane (naphthenes) that bear carboxylic acid groups on the alkyl chains [74]. Present in crudes at up to 3%, they are extracted from gasoline and middle distillate products with sodium hydroxide because they are corrosive [75]. As refinery by-products, they are cheap and have been widely used in many applications, including making metals salts for use in lube oils.

9 The piston rings sit in the grooves and flex, improving the seal between piston and cylinder walls; carbonaceous deposits in the grooves prevent this flexing, causing excessive wear of rings and cylinder and may cause rings to break.

10 COPD is Chronic Obstructive Pulmonary Disease such as emphysema and bronchitis.

11 Greyscale is a photographic term of the darkness of the object, here the filter. The patent used the online Adobe Photoshop system to make the measurement.

12 The H atoms may be taken by hydroxyl radicals, forming water, or by oxygen or by other free radicals in the vicinity.

References

1 Marshall, E.L. and Owen, K. (1995). *Motor Gasoline*. Society of Chemical Industry. ISBN 0854044094.

2 Crepeau, G. (2018). *PSA Looks for the Right Balance*. Infineum International Ltd. 9 July.

3 Wikipedia Contributors. Engine Knocking. 17 December 2017. (Cited: 6 March 2018). Page Version ID: 817048490.

4 Gibbs, L., Anderson, B., Barnes, K. et al. (2009). Motor gasolines technical review. https://www.southwestairsports.com/ppgtechinfo/top80/hrservicenotes/fueloilspecs/motor-gas-tech-review.pdf (Cited: 12 July 2020).

5 Fuel Additives: Use and Benefits, ATC Doc. 113. (2013). atc-europe.org. https://www.atc-europe.org/public/Doc113%202013-11-20.pdf (Cited: 12 July 2020).

6 Midgley, T., Jr. (2018). Editors of encyclopaedia britannica. 2 February https://www.britannica.com/biography/Thomas-Midgley-Jr (Cited: 8 March 2018).

7 Editors of Encyclopaedia Britannica. (2018). Tetraethyl lead. 18 January https://www.britannica.com/science/tetraethyl-lead (Cited: 8 March 2018).

8 General_Motors_Research_Corp. (1923). Improvements in or relating to fuels for internal combustion engines. GB Patent 224,102. Filed UK, 1 December.

9 Standard_Development_Company (1927). An improved manufacture of organometallic compounds. GB Patent 290,444. Filed UK, 21 May.

10 ATC Europe (2004). Fuel additives and the environment. 1 January http://www.atc-europe.org/public/doc52.pdf (Cited: 10 June 2020).

11 Johnston, I. (2017). UK company sells lead to last place on Earth where leaded petrol is legal. 22 August http://www.independent.co.uk/environment/leaded-petrol-algeria-still-legal-innospec-cheshire-uk-sale-export-tel-tetraethyl-lead-a7907196.html (Cited: 7 July 2020).

12 Wikipedia Contributors (2018). Fluid catalytic cracking. 16 February (Cited: 29 March 2018). Page Version ID: 825987299.

13 Wikipedia Contributors (2018). Alkylation unit design. 8 March (Cited: 29 March 2018). Page Version ID: 829396148.

14 Wikipedia Contributors. (2018). Gasoline. 26 March (Cited: 29 March 2018). Page Version ID: 832524110.

15 Wikipedia Contributors (2020). Avgas. 7 July (Cited: 12 July 2020). Page Version ID: 966518332.

16 ExxonMobil (2008). World jet fuel specifications with avgas supplement. http://large.stanford.edu/courses/2017/ph240/chhoa1/docs/exxon-2008.pdf (Cited: 11 June 2020).

17 Stevens, D.R. and Sweet, R.L. (1952). Organic compositions containing a metallo cyclopentadienyl. US Patent 3,294,685. Filed 21 April.

18 Brown, J.E. and Shapiro, H. (1952). Cyclomatic compounds. US Patent 2,818,416. Filed 10 December.

19 Bacha, J., Freel, J., Gibbs, A. et al. (2007). Diesel fuels technical review. https://www.chevron.com/-/media/chevron/operations/documents/diesel-fuel-tech-review.pdf (Cited: 12 July 2020).

20 Song, C. (2000). Introduction to the chemistry of diesel fuels. In: *Chemistry of Diesel Fuels* (eds. C. Song, C.S. Hsu, and I.,I. Mochida), 1–43. London, N.Y.: Taylor & Francis. Ch 1.

21 Smith, B.W. (2017). Cetane numbers: How diesel fuel's biggest unknown affects your truck. Motortrend. 11 April http://www.trucktrend.com/how-to/expert-advice/1704-cetane-numbers-how-diesel-fuels-biggest-unknown-affects-your-truck (Cited: 12 July 2020).

22 Danilov, A.M. (2012). Fuel additives, evolution and use in 2006–2010. Plenum Publishing Corp., *Chemisty and Technology of Fuels & Oils* 47 (6): 470–484.

23 Cotton, F.A. and Wilkinson, G. (1988). *Advanced Inorganic Chemistry*, Fifth edition. John Wiley and Sons. ISBN 0471849979.

24 Clothier, P.Q.E., Aguda, B.D., Moise, A. et al. (1993). How do diesel fuel ignition improvers work? Royal Soc. Chem., April *Chemical Society Reviews* 22 (2): 101–108.

25 Witzel, L., Moszkowicz, P., and Claus, G. (1995). Mechanism of particulate reduction in heavy fuel oil combustion. *Fuel* 74 (12): 1881–1886.

26 Vermeire, M.B. (2012). Everything you need to know about marine fuels. PDF4PRO.com. June. (Cited: 12 June 2020).

27 Danilov, A.M., Okina, N.G., Mitusova, T.N. et al. (2003). Increase in the Efficiency of use of Furnace Residual Fuel Oil. Plenum Publishing corp. N.Y., *Chemistry and Technology Fuels and Oils* 39: 6–9.

28 Standard Oil Development Company. (1942). Improvements relating to fuel oils. GB Patent 561,328. Filed UK, 24 June.

29 Griffing, M.E. and Burns, J.A. (1967). Improved jet fuels. GB Patent 1,171,648. Filed UK, 28 July.

30 Brown, J.E., DeWitt, E.G., and Shapiro, H. (1958). Hydrocarbon distillate fuel compositions. GB Patent 901,689. Filed UK, 5 December.

31 Percy, A.W. and Zollinger, W.J. (1957). Fuel oil and method of operating fuel oil furnaces. US Patent 3,112,789. Filed 26 December.

32 Strukl, J.S. (1972). Reducing deposits and smoke from jet fuels with additives incorporating an ammonium salt. US Patent 3,883,320. Filed 7 December.

33 McRobert, W.G. (1998). Hydrocarbonaceous fuel compositions and additives therefor. US Patent 5,944,858. Filed 17 February.

34 Fengwen-Chi (2010). Diesel oil smoke eliminating and oil saving additive. CN Patent 101,768,486. Filed China, 5 February.

35 Thunker, W., Lohmann, G., Marschewski, A. et al. (1997). Use of ferrocene. US Patent 5,746,784. Filed 14 April.

36 Kukin, I. (1964). Additive compositions to improve burning properties of liquid and solid fuels. US Patent 3,348,932. Filed 21 August.

37 EAZY Combustion Improver (2015). September https://www.eazychem.co.uk/eazy-combustion-improver (Cited: 13 July 2020).

38 Kracklauer, J.J. (1980). Method of conditioning diesel engines. US Patent 4,389,220. Filed 4 June.

39 Hesterberg, T.W., Long, C.M., Bunn, W.B. et al. (2012). Health effects research and regulation of diesel exhaust: An historical overview focused on lung cancer risk. Informa Healthcare, *Inhalation Toxicology* 24 (S1): 1–45.

40 Ristovsky, Z.R., Miljevic, B., Surawski, N.C. et al. (2011). Respiratory health effects of diesel particulate matter. 11 November *Asian Pacific Society of Respirology* 17: 201–212.

41 ACFA InFocus. (2013). Airborne particulates: Size does matter. Asian Clean Fuels Association. February http://www.acfa.org.sg/pdf/InFocus11%20 2013_02%20PM2_5.pdf (Cited: 14 November 2016).

42 Wikipedia Contributors (2018). European emission standards. Wikipedia. 23 February (Cited: 2018 March 2018). Page Version ID: 827196905.

43 Farrar, D.H. (1989). Engine cleaning additives for diesel fuel. EP 375,303. Filed Europe, 15 December.

44 Dallanegra, R. and Caprotti, R. (2014). Validation of fuel borne catalyst technology in advanced diesel applications. Society of Automotive Engineers, Warrendale, P.A. SAE Technical Paper 2014-01-1401. May doi:10.4271/2014-01-1401.

45 Caprotti, R. (2013). Meeting China's emissions legislation. Infineum International Ltd. Insight. March.

46 Jelles, S.J., Makkee, M., and Moulijn, J.A. (2001). Ultra low dosage of platinum and cerium fuel additives in diesel particulate control. *Topics in Catalysis* 16/17 (1–4): 269–273.

47 Ulricht, A. and Wichser, A. (2003). Analysis of additive metals in fuel and emission aerosols of diesel vehicles with and without particle traps. *Analytical and Bioanalytical Chemistry* 377: 71–81.

48 Cook, S.L., Kalischewski, W., and Marschewski, A. (2001). Compositions comprising dimeric or oligomeric ferrocenes. EP 1,313,748. Filed Europe, 30 August.

49 Caprotti, R. and Pilling, R.J. (2003). Iron Salt Diesel Fuel Additive Compositions For Improvement Of Particulate Traps. US Patent 7,306,634. Filed 19 May.

50 Blanchard, G., Chane-Ching, J.Y., and Tolla, B. (2002). Organic colloidal dispersion of iron particles, method for preparing same and use thereof as fuel additive for internal combustion engines. EP 1,458,471. Filed Europe, 19 December.

51 Thompson, R.M. (2008). Method of supplying iron to the particulate trap of a diesel engine exhaust. EP 2,129,750. Filed 7 March.

52 D'Alencon, L., Lallemand, M., Harle, V. et al. (2011). Fuel additive composition containing a dispersion of iron particels and a detergent. US Patent 2,014,007,494. Filed 20 December.

53 Reed, K.J. (2007). Cerium dioxide nanoparticle-containing fuel additive. EP 2,066,767. Filed Europe, 4 September.

54 Stoldt, S.H. and Walsh, R.H. (1985). Fuel additive comprising a metal compound and an oxime and fuel composition containing same. US Patent 4,673,412. Filed 24 September.

55 Huang, N.Z., Adams, P.E., Daly, D.T. et al. (1991). Organometallic complex-antioxidant combinations and concentrates and diesel fuels containing same. US Patent 5,344,467. Filed 13 May.

56 Erdmannsdorfer, H., Wagner, M., and Weyh, G. (1982). Method and apparatus for removing carbon material or soot from the exhaust of an internal combustion engine. GB Patent 2,098,086. Filed UK, 12 March.

57 Adams, P.E. and Kolp, C.J. (2000). *Diesel Particulate Emission Reduction with an After-treatment Filter and a Copper Based Fuel Additive.* SanFrancisco: Division of Petroleum Chemistry, ACS, ACS Preprints, Symposium,, New Chemistry of Fuel Additives.

58 Petzold, A., Hasselbach, J., Lauer, P. et al. (2008). Experimental studies on particle emissions from cruising ship, Their characteristic properties, transformation and atmospheric lifetime in the marine boundary layer. *Atmospheric Chemical Physics* 8: 2387–2403.

59 Kaspar, A., Aufdenblatten, S., Forss, A. et al. (2007). Particulate emissions from a low-speed marine diesel engine. February *Aerosol Science and Technology* 41: 24–32.

60 Dastillung, M., Flores, R., Gardzinski, W. et al. (2009). Impact of marine fuels quality legislation on EU refineries at the 2020 horizon. CONCAWE Refinery Technology Support Group. Report 3/09.

61 CONCAWE Refinery Technology (2017). Support group. Marine Fuel Facts. concawe.org. January https://www.concawe.eu/wp-content/uploads/2017/01/marine_factsheet_web.pdf (Cited: 15 July 2020).

62 Vovelle, C., Delfau, J.L., and Reuillon, M. (1994). Formation of aromatic hydrocarbons in decane and kerosene at reduced pressure. In: *Soot Formation in Combustion, Mechanisms and Models* (ed. H. Bockhorn). 50–65. Springer-Verlag. Paper 3.

63 Dobbins, R.A. and Subramaniasivam, H. (1994). Soot precursor particles in flames. In: *Soot Formation in Combustion, Mechanisms and Models* (ed. H. Bockhorn), 290–301. Springer-Verlag. Paper 17.

64 Haynes, B.S. and Wagner, H.G. (1981). Soot formation. *Progress in Energy and Combustion Science* 7 (4): 227–273.

65 Glassman, I., Nishida, O., and Sidebotham, G. (1994). Critical temperature of soot formation. In: *Soot Formation in Combustion, Mechanisms and Models* (ed. H. Bockhorn), 316–324. Springer-Verlag. Paper 20.

66 Wu, J. (2004). *Soot Abatement Using Fuel Additives*. Thesis in Mechanical Engineering. Pennsylvania State University. December.

67 Huth, M. and Leuckel, W. (1994). Soot formation from hydrocarbons in a plug flow reactor: Influence of temperature. In: *Soot Formation in Combustion, Mechanisms and Models* (ed. H. Bockhorn), 371–381. Springer-Verlag. Paper 23.

68 Pischinger, F., Lepperhoff, G., and Houben, M. (1994). Soot formation and oxidation in diesel engines. In: *Soot Formation in Combustion, Mechanisms and Models* (ed. H. Bockhorn), 382–395. Springer-Verlag. Paper 24.

69 Bockhorn, H. (1994). Overview. A short introduction to the problem – Structure of the following parts. In: *Soot Formation in Combustion, Mechanisms and Models* (ed. H. Bockhorn), 3–7. Springer-Verlag.

70 Richter, H. and Howard, J.B. (2000). Formation of polycyclic aromatic hydrocarbons and their growth to soot – AReview of chemical reaction pathways. Pergamon, *Progress in Energy and Combustion Science* 26: 565–608.

71 Frenklach, M. and Wang, H. (1994). Detailed mechanism and modelling of soot particle formation. In: *Soot Formation in Combustion, Mechanisms and Models* (ed. H. Bockhorn), 162–192. Springer-Verlag. Paper 10.

72 Calcote, H.F. and Gill, R.J. (1994). Comparison of the ionic mechanism of soot formation with a free radical mechanism. In: *Soot Formation in Combustion, Mechanisms and Models* (ed. H. Bockhorn), pp 471–484. Springer-Verlag Berlin Heidelberg New York London Paris Tokyo Hong Kong Barcelona Budapest. Paper 29.

73 Lahaye, J., Boehm, S., and Ehrburger, P. (1994). Metallic additives in soot formation and post-oxidation. In: *Soot Formation in Combustion, Mechanisms and Models* (ed. H. Bockhorn), 307–315. Springer-Verlag. Paper 19.

74 Wikipedia Contributors (2017). Naphthenic acid. 29 November (Cited: 12 July 2020). Page Version ID: 812721870.

75 Littmann, E.R. and Klotz, J.R.M. (1942). Naphthenic acids. II Manufacture and Uses, February *Chemical Reviews* 30: 97–111.

7

Additives to Treat Problems during the Movement and Storage of Fuels

7.1 Introduction

Fuels are either on the move or being stored in tanks while waiting to be used. Bulk movements across large distances are often through pipelines that criss-cross many parts of the world. When the pipelines were built, the volumes that they were designed to carry were adequate to meet demand. However, growth in demand has meant that their capacities are often no longer sufficient, being limited by resistance to flow and the energy needed to pump more fuel. Drag reducers reduce the resistance to flow and increase the capacities of existing pipelines – a much less expensive option than building a new pipeline.

Flow through pipes to an end user, such as on the filling of an automotive fuel tank or the wing tanks of an aeroplane, can raise high voltage static electricity. At the exit of the pipe, the charge in the fuel seeks to escape and run to earth, creating a spark that can ignite the fuel–vapour/air mixture in the receiving vessel. Such static sparks have been the source of disastrous explosions and fires in storage tanks and shipping tankers. Operating procedures and the use of anti-static additives have enabled the control of this problem.

Fuel being pumped into another tank also entrains air and may form voluminous foam. Such foam may break in a few tens of seconds, in the meanwhile it causes the tank to overflow and prevents continuous filling. This is annoying to someone filling the tank of his passenger car but a significant waste of time for a truck driver filling his 200-litre fuel tank. Anti-foam agents prevent such foam problems, meeting a market demand.

Water in fuel is avoided for several reasons. Refineries make wide use of steam in distillation processes so raw distillate streams contain water that refiners go to extra means to remove. However, more water is often picked up

Fuel Additives: Chemistry and Technology, First Edition. Robert D. Tack.

by a fuel in transit or in long-term storage, such as leaks of rain into tanks or absorption from a humid atmosphere. Water separating to the bottoms of storage tanks is kept below the level of the take-off point and is easily removed through a separate valve; unfortunately, some fuel components and some fuel additives may aid water-in-oil emulsions which holds the water in suspension. To prevent this problem, demulsifiers are routinely added to fuels.

In reality, water may still be absorbed after leaving a well-managed storage facility. At, or near to, sub-zero conditions, remaining water in automotive fuels may separate and freeze; the ice that forms then interferes with fuel flow to the engine. Jet fuel suffers temperatures down to around −40°C in the wing tanks, during flight, so is particularly at risk of icing. This problem calls for anti-icing additives, especially in jet fuel.

The water that has settled in long-term storage tanks, such as those of a fuel distributor, at a minor airport or heating oil stored between seasons, can itself cause problems. Fuel tank corrosion is part of a wider problem (Chapter 5), but there is a further problem of the growth of various organisms at the water/fuel interface. Clumps of biomass can form that move with the fuel and, if present, block fuel filters and injectors, such as in oil heating systems. So, for fuels expected to be held in long-term storage, the final fuel additives discussed are biocides.

7.2 Drag Reducing Agents

7.2.1 The Pipeline Problem

Crude oils and distillate fuels – gasoline, diesel fuel, heating oils, kerosene and jet fuel – are often transported long distances by pipeline. 'Drag' is the everyday term used for the frictional resistance to flow. At a given pumping pressure, drag reduces the capacity of a pipeline by reducing the flow rate. Alternatively, more energy is needed to increase pumping pressure to maintain a higher flow rate. The capacity of a pipeline can be increased by its overall replacement with a larger cross-section pipeline or by erecting another pipeline; pipelines can extend for thousands of kilometres, so such solutions are avoided, given the very high capital expenditure that they would require. An alternative is to use drag reducing agents (DRAs), which are high molecular weight, hydrocarbon-soluble polymers. DRAs added to a distillate fuel, such as diesel fuel, at 10–25 ppm, can bring about a substantial reduction in drag such that the capacity of the pipeline, at the same pumping energy, may be increased by 40%, or the energy required at the same pumping rate may be reduced by 40% [1]. DRAs are now widely used in pipelining petroleum liquids.

The use of a poly-1-alkene (also known as poly-alpha-olefin, PAO) DRA was demonstrated in diesel fuel in the laboratory and in crude oil in a 28-mile section of an Oklahoma pipeline in 1972 [2]. However, the catalyst for the widespread use of DRAs seems to have been their use in the trans-Alaska-pipeline-systems (TAPs) from 1979 [3].

It is worth mentioning that these same polymers are also applicable as anti-misting agents. In the event of a vehicle or aeroplane crash, fuel exiting a bursting tank is readily ignited because of the fine spray of small droplets that is formed. Such a spray is made less flammable by the fuel containing DRA polymers which cause the fuel droplets to be much larger, so less easily ignited [4].

7.2.2 Chemistries of DRAs

Early references to DRAs pointed to the use of the very oil-soluble, high molecular weight poly-iso-butylene (PIB[1]), but this was limited because PIB is unstable to shear: PIB is too easily broken down to ineffective, low molecular weights by pumping and passing through orifices [2]. PAOs are more shear stable and high molecular weights are readily obtained by Ziegler-Natta catalysed polymerisations of 1-alkene (Figure 7.1)). PAOs remain the most widely used DRAs [5–7]. Drag reduction requires high molecular weights as drag reduction increases with increasing molecular weight. PAOs made from mixtures of n-alkenes are preferable to those made from individual n-alkenes because they have lower crystallinity, so have better solubilities in fuels [4]. While PAOs from longer chain alkenes have better shear stability, 1-alkenes such as hexadecane give PAOs that have greater side-chain crystallinity, so lower solubility, and they are more difficult to polymerise to the high molecular weights because of steric congestion. Polymers with shorter or no side

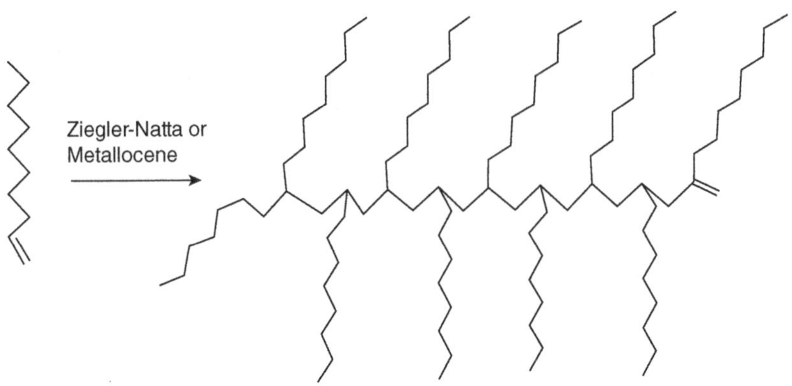

Figure 7.1 Poly-1-Decene.

chains, such as polyethylene or polypropylene, have higher backbone crystallinity and lower solubility. Minimum crystallinity for PAOs occurs with side chains C_6 to C_{10}, so preferred PAOs are made from n-alkene mixtures with chain lengths of C_8 to C_{12} or C_6–C_{10} [7,8]; such PAOs have faster polymerization reactions and better solubilities in oils [4,6].

Shear stability matters in two opposing ways. One is the need for moderate shear stability. Pipelines have multiple pumping stations along its length to keep the fuel moving, so a polymer of low shear stability would be mostly broken down to ineffective, low molecular weights at each pumping station. It would then need to be replenished after all pumping stations, which are often in isolated places such as the middle of Alaska or Siberia. A relatively, shear-stable polymer, added at a small over-treat, will survive sufficient DRA activity through perhaps two pumping stations, so would need to be replenished only at every third pumping station (in the TAPs, about one-third of the activity was lost at each pumping station so the DRA was added at up to 30 ppm, as 10 ppm is still effective [3]). At the end of the pipeline, a fuel should no longer contain any active DRA polymer as it can interfere with the fuel's quality [1], causing problems in fuel injection systems by disrupting the spray patterns through their anti-misting properties. Such problems in jet fuel, in which DRAs are not allowed, are avoided by running the pipeline with untreated fuel (gasoline or diesel fuel) at the end of that batch[2] of fuel to ensure that no DRA is left in the pipeline when a jet fuel batch is pipelined [3].

A problem with these polymers is that solutions of only 5%–10% polymer have high viscosities, even in the most suitable hydrocarbon solvents [4]. This may require heating of the solution in a volatile, flammable solvent at a remote pipeline pumping-station to enable it to be injected into the pipeline fuel. Dilute solutions have high transportation costs – carrying 90 tonnes of solvent for every tonne of polymer [7]. The solution approach to delivery of DRAs has been largely displaced by the use of slurries of small particles of the PAO-DRA in low flammability polar solvents such as higher alcohols, glycols or glycol ethers. This approach allows the additive to contain 25%–40% of DRA polymer. The process used to prepare such slurries is to freeze the polymer below its glass transition temperature (in liquid nitrogen) then grind the polymer down, in the presence of a particle stabiliser[3] such as ethylene-bis-stearamide, to a free flowing powder of particles 50–1000 nm across. The powder thus obtained is suspended at 30%–50% in a medium such as dipropylene glycol and iso-octyl alcohol to form a free-flowing latex, stabilised with a small amount of hydroxypropyl cellulose to prevent settling[4] [9]. Such a suspension was reported to have been stored in arctic condition with temperatures down to −20°C while remaining fluid and pumpable, so still able to provide drag reduction in crude oil.

Poly(alkylacrylates), poly(alkylmethacrylates) and poly-(N-alkylacrylamides) have also been used as DRAs (comb structure Figure 4.19). The high molecular weights needed for DRAs, such as 5 million, were obtained by free radical, emulsion polymerisation. The preferred monomers are 2-ethylhexylacrylate [10] or isodecyl-methacrylate [6]; however, such monomers are more expensive than are n-alkenes making such polymers less popular as DRAs.

7.2.3 The Process of Drag Reduction

The process of drag reduction goes back to a finding by B.A. Toms, in 1948, that the frictional coefficient of chlorobenzene, during turbulent flow through a pipe, was reduced by half by having 0.25% of poly-(methylmethacrylate) dissolved in the chlorobenzene – since called the 'Toms effect' [7]. Basically, during turbulent flow, a component of the motion is movement of elements of the liquid transverse to the main, axial flow down a pipe (Figure 7.2a). Mass transfers across the cross section, caused by the continuous stirring, moves elements of the liquid from the centre of flow, with maximum axial velocity, towards the pipe walls, at which there is zero flow. At the same time, elements of near zero flowrate are moved into the mainstream where they need to pick up momentum from the rest of the fluid. Such an interchange results in the loss of energy from the moving liquid to the pipe walls and a slowing down of the flow, such is viscous drag. DRA polymers which, in solution, take the form of a linear macromolecular coil, are extended in the direction of flow by the laminar flow near the pipe walls, thickening the laminar flow region (Figure 7.2b). This layer reduces the transfer of momentum and loss of energy to the walls. PAOs such as polydecene or poly-(2-ethylhexyl)-acrylate are more effective than polymers that have small or no sidechains (such as polypropylene or polyethylene) because the longer side chains, with solvating hydrocarbon fuel, produce a fatter, extended molecule which maximises this layer thickening [5].

7.3 Static Dissipaters

7.3.1 The Problem of Static Electricity in Fuels

When fuel is pumped rapidly through a hose, as in the filling of a vehicle fuel tank or aeroplane wing tanks, it builds up a static charge. This charge can have a high potential – tens of kilovolts – that can discharge as a spark when the fuel is close to a grounded object, such as the receiving fuel tank or the vehicle from which it is pumped (which then carries the opposite charge if not grounded). If the spark discharges in a flammable vapour/air mix, it can cause an explosive

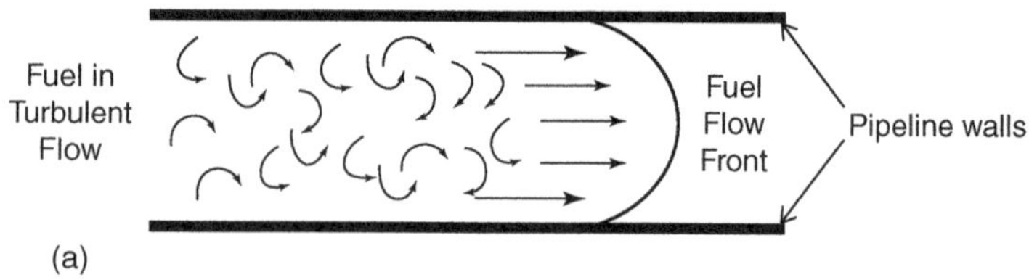

Figure 7.2a Fuel in Turbulent Flow with No DRA.

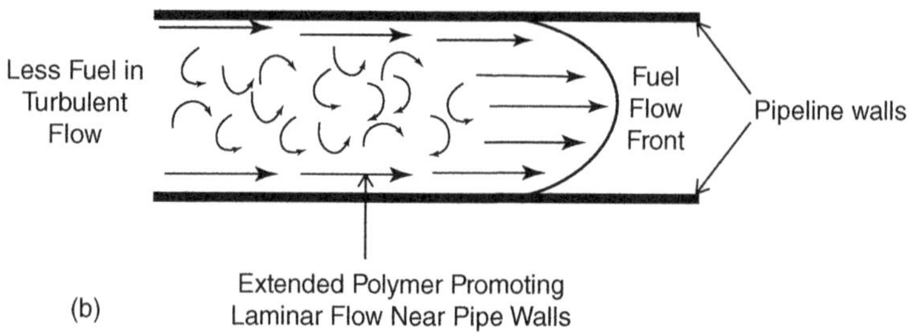

Figure 7.2b Fuel Containing DRA with Laminar Flow at the Pipeline Walls.

ignition. This problem has been long recognised in the petroleum industry as the cause of numerous explosions and fires in refineries, stores and tankers [11–13]. Procedures and physical means of grounding or neutralising static charges, such as electrically connecting a tanker to a receiving tank ('bonding') and ensuring electrical connection of tanks and fuel handling equipment with the ground, have been established to enable the dissipation of any charge build up and minimise the risk.

However, because hydrocarbon fuels are extremely poor conductors and fuel is often pumped at very high rates, grounding and bonding alone are not sufficient to completely prevent the build-up of static charges in petroleum fuels [14]. It is to prevent this possibility that static dissipaters[5] were introduced in the 1950s and are often added to fuels to increase their conductivities, thus allowing charges to dissipate rapidly [15–17]. The earliest patents in this area seem to be those of Shell, filed in the 1950s [12,18,19], which state that it was known to add electrical conductivity improvers to hydrocarbon liquids to prevent static build up, with early references to the problem in dry cleaning fluids on the 1940s [20].

The risk of any spark causing an ignition of the fuel vapour is highest for vapour/air mixtures in their explosive proportions. Gasoline is very volatile so the space above the liquid in a fuel tank usually contains little or no air, so cannot be ignited [21]. A general comment is that 'Conductivity is rarely a concern for automotive gasoline' [22]; in support of this comment, patent claims for static dissipaters rarely use gasoline as the test fuel – they are mostly concerned with jet fuel then, secondarily, with diesel fuel. However, there are occasions when filling an empty tank that is full of air, that a flammable vapour/air mixture may be produced; even filling an empty plastic or metal gasoline-can, resting on a rubber mat in the back of a pick-up truck, can produce both a flammable vapour/air mixture and a significant build-up of static [22]. Gasoline can have a moderate conductivity – older literature points to gasoline as usually having sufficient conductivity to dissipate charge [15] – but current processing to very low sulphur fuels removes most of the polar impurities that provided that conductivity [23]. Untreated gasoline now has a low conductivity (see section 7.3.3 and Table 7.4). The conductivity required for safe operations is specified for aviation gasoline and jet fuel at 50–450 picosiemens per metre (pS/m, sometimes also referred to as conductivity units) [24]. Motor gasoline specifications do not mention conductivity, and, in gasoline additive discussions, it is said that conductivity additives are more commonplace in fuels of intermediate volatility, such as kerosene, though low sulphur gasoline is increasingly treated [17].

At the other extreme, the middle distillates diesel fuel and heating oil have low volatilities so do not usually provide enough vapour above the liquid in a fuel tank to provide a flammable vapour/air mixture. But the real world is usually not so clear-cut. Switch loading is the practice of loading diesel fuel into a tanker that previously contained gasoline, which can create a highly hazardous situation [25]. A high static charge is produced by the rapid pumping of an untreated middle distillate – especially ULSD, in which hydrodesulphurisation has removed the conductive polar compounds [16,17,26]. A single static discharge from electrically charged diesel to the tank wall can ignite the residual gasoline–vapour/air mixture, causing an explosion. For switch loading, the European Petroleum Industry recommends that either fuel conductivity be increased, or pumping speed reduced, as grounding and bonding are not enough to prevent a fire from starting. In today's market, highly refined, low conductivity, low sulphur fuels are increasingly the norm [25], so Stadis® 450 or 425 are widely used in all ground fuels as 'best practice' to improve operational safety.

Jet fuel has an intermediate volatility so jet fuel tanks are particularly likely to contain ignitable fuel–vapour/air mixtures. Refuelling of aircraft is rapid and jet fuel is a poor conductor, a combination that has the potential for producing high static charges in the fuel [15]. For these reasons, static dissipaters have been added to jet fuel since the 1950s to augment extensive procedures that are in use to prevent the possibility of a static discharge initiating a jet fuel fire [27,28].

7.3.2 Chemistries of Static Dissipaters

The chemicals used as static dissipaters are generally described as fuel-soluble chromium materials, polymeric sulphur and nitrogen compounds and quaternary ammonium materials [16,17] but the detailed chemistries are now well known. Treat levels are extremely low, and typically only 2–20 ppm are needed.

Static dissipaters were originally metal salts. Shell's patent published in 1961 [12,18] for compounds to increase the conductivities of hydrocarbon liquids, particularly fuels, indicated that the need for conductivity improvers had already been recognised in the dry-cleaning industry [20]. A 1956 patent [18] points to the need for additives to increase the electrical conductivity of light hydrocarbon oils used to clean textiles and proposes their use in gasoline; previously, compounds such as magnesium or triethylammonium oleates or organic sulphonates had been proposed while this patent proposes the use of chromium diisopropylsalicylate along with sodium dioctylsulphosuccinate[6].

The first static dissipater that appears to be well established for the use in fuels is the Shell chromium salt of C_{14}-to-C_{18}-alkylsalicylic acid, a key component of the commercial static dissipater ASA-3, which is also described as a blend (Figure 7.3) of the chromium tris-alkylsalicylate, calcium-AOT[7],and a copolymer of 2-ethylhexylmethacrylate with 2-methyl-5-vinyl pyridine (10 mole%), in equal proportions, at a total of 50% in xylene [12,29]

These components and their conductivity-providing activities were demonstrated in gasoline. Untreated gasolines under test had conductivities of 0.01–10 pS/m which was raised to between 240 and 250 pS/m by the chromium alkylsalicylate at 11.35 ppm[8] alone and to 210 pS/m by only 0.385 ppm of the same chromium alkylsalicylate in combination with both the calcium-AOT

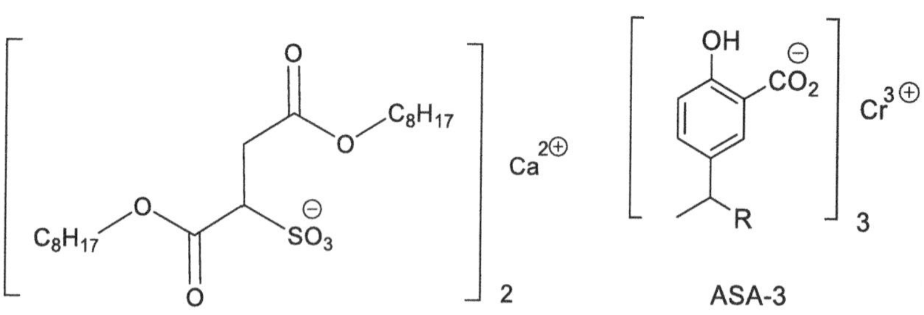

Figure 7.3 Calcium-AOT and ASA-3.

(0.44 ppm) and vinyl pyridine-alkyl methacrylate copolymer (10 ppm) – the advantage of the blend being the much lower level of metals required using the final formulation.

The key component of ASA-3, the chromium tris-alkylsalicylate, was made from an alkyl salicylate – a key part of Shell's lube oil detergent technology – by a route that utilises the Kolbe reaction (Figure 7.4)). Phenol is alkylated with a C_{14} to C_{18} 1-alkene mixture[9] by the Friedel-Crafts reaction – in the patent process, the catalyst used is zinc chloride with hydrochloric acid [12]. The mono-alkyl phenol (a mixture of ortho- and para-isomers) is converted to the sodium salt and then reacted with carbon dioxide, under pressure (the Kolbe reaction), which places a carboxylate group in the 2-position to give sodium alkylsalicylate. After acidifying and separating unreacted alkylphenols, the alkylsalicylic acid is neutralised with sodium hydroxide, dissolved in ethanol and a solution of chromium nitrate in ethanol is added. Benzene is added and the ethanol removed by azeotropic distillation, precipitating the by-product sodium nitrate, which is filtered off leaving a benzene solution of the chromium tris-alkylsalicylate [12].

ASA-3 was introduced outside the USA in 1962[10] and has been in use in jet fuel, gasoline and diesel fuel since then [17,21]. While ASA-3 was withdrawn

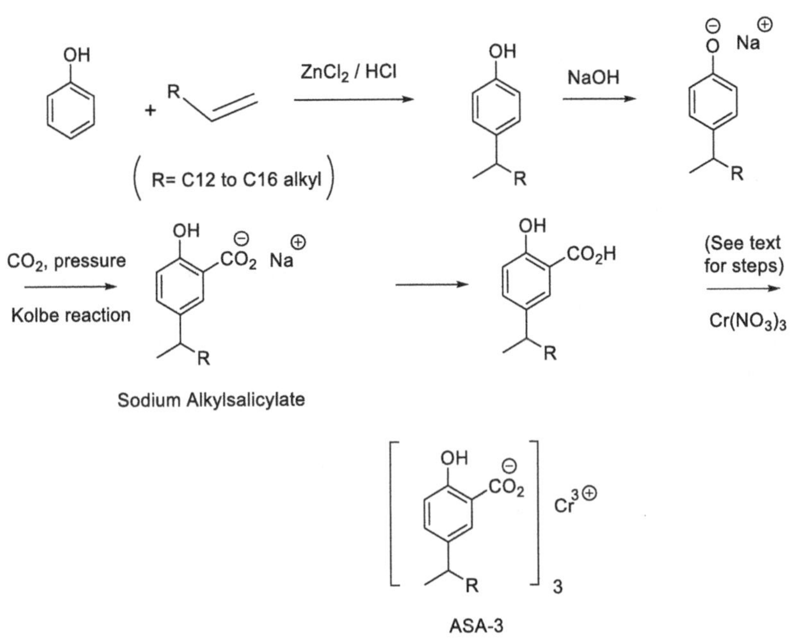

Figure 7.4 The Preparation of the Chromium Tris-alkylsalicylate, ASA-3.

by Shell in 1991, it may still be in use in eastern European countries [24] under the name SIGBOL. It is also reported as a current conductivity improver in the 2013 ATC document on fuel additives, which lists 'fuel soluble chromium materials' alongside 'polymeric sulphur and nitrogen compounds' and 'quaternary ammonium salts' for use in gasoline and in diesel fuel [17].

The most widely used static dissipater is Stadis 450. It was introduced first into the USA military aircraft in the late-1970s [30] and is the only one approved for use in many aviation fuels [23] – it has been in wide usage since 1983. Stadis 450 is a non-metallic static dissipater approved for use in jet fuel at 0.3–3.0 mg/l [24]. The related product Stadis 425 is recommended for ground use, especially for diesel fuel and gas oils [23]. Stadis 450 was developed by the Du Pont chemical company around the key component 1-decene polysulphone[11], made by free radical polymerisation of 1-decene with sulphur dioxide (Figure 7.5) [11].

Such polymers have relatively low ceiling temperatures – the temperature at which their rates of polymerisation and depolymerisation are equal, so they may form below that temperature and depolymerise above that temperature. For sulphur dioxide with butylene and higher homologues, the polysulphone ceiling temperatures are about 64°C[12] [31,32]. For Stadis 450, 1-decene and sulphur dioxide are copolymerised at 5°C–15°C in the presence of a small proportion of dodecyl mercaptan to control the molecular weight at, preferably, between 100,000 and 500,000.

The 1-decene polysulphone provides multiple advantages: in addition to conferring the necessary conductivity at a low concentration, it contains no metal, retains its activity over a prolonged time frame, is not removed by contact with water, and does not cause water emulsification [11]. For the most potent static dissipater activity, the polysulphone has been combined with a variety of amine derivatives, initially with quaternary amine salts such as dicoco-dimethylammonium nitrite[13]. Conductivities, in a purified kerosene, demonstrate the synergy between the two compounds and adequate conductivity was achieved at a total concentration of less than two parts per million (ppm) (Table 7.1).

The 1-decene polysulphone combined with the dicoco-dimethylammonium nitrite, at a ratio of 25 to 1, which has been demonstrated to give adequate

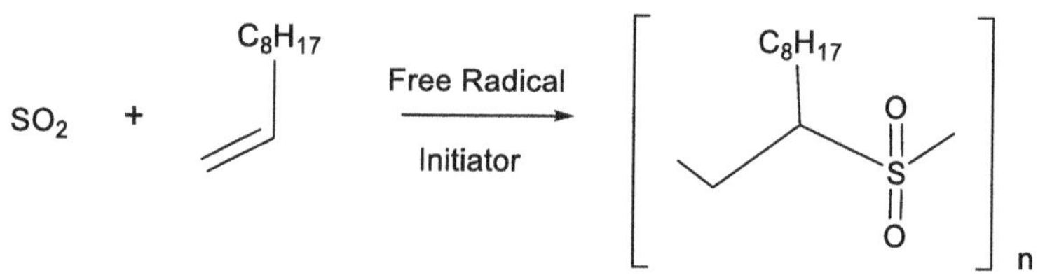

Figure 7.5 1-Decene Sulphone.

Table 7.1 Conductivities (pS/m) of Purified Kerosene with Static Dissipater Components.

1-Decene Polysulphone, ppm	Dicocodimethylammonium Nitrite, ppm	Conductivity, pS/m
0	0	≤20
1.2	0	20
0	2	20
0	4	40
1.2	0.25	300
1.2	1	450

conductivities (>150 pS/m) at concentrations below 8 ppm in a range jet fuels (Table 7.2) [11]. The same combination was demonstrated to provide adequate conductivities to diesel fuels and fuel oils, at concentrations of 1–10 ppm (variation over nine test fuels, which had untreated conductivities from 0 to 30 pS/m).

In the current era of ultra-low sulphur fuels, for which, fuel additives in the USA have the same limit on sulphur content as have the fuels (max. 10 ppm), the polysulphone has the drawback of containing sulphur . This point was made clear in claims for the use of a cationic copolymers of 2-ethylhexyl acrylate with 10 mole% 2-trimethylammonium-ethyl-methacrylate methyl sulphate (molecular weights 100,000–650,000), combined with various alkene/acrylonitrile/styrene (mole ratio 8/10/1) copolymers, as a zero halogen, very low sulphur additive [14]. These combinations gave adequate (about 200 pS/m) conductivities at 3–5 ppm in a 'high' sulphur diesel fuel (untreated conductivity 15 pS/m), while 7 ppm were needed for a ULSD (untreated conductivity 3 pS/m); in a kerosene (untreated conductivity 1 pS/m), 10 ppm gave conductivity 500 pS/m. The applicant for these patents is Petrolite, part of Baker Hughes, so it can be assumed that the cationic polymer, possibly with an alkene/acrylonitrile copolymer, is the active component in the static dissipater Tolad® 3514, which is described as a copolymer of an alkyl/vinyl monomer and a cationic vinyl monomer [33,34].

In a later patent [35], the same inventor combined these same polymers with the polysulphone to give high conductivities (up to over 1000 pS/m) when tested in kerosene at 5 ppm total combined treat. The minimum proportions of the polysulphone tested were 25% of the total and it might be expected that the 1.25% thus present would itself give adequate (>150 pS/m) conductivity to the fuels. However, the activity of the combination suggests that the concentration could be reduced by half and still provide adequate

Table 7.2 1-Decene Polysulphone with Dicocodimethylammonium Nitrite, Ratio 25/1, Tested as a Conductivity Improver in Various Jet Fuels.

Fuel	Untreated Fuel Conductivity pS/m	Active Ingredient, ppm	Treated Fuel Conductivity pS/m
JP-4 (A)	10	0.3	180
JP-4 (B)	0	3.2	215
Turbine fuel A	10	4.3	190
Turbine fuel B	10	7.6	230
Jet kerosene	10	6.5	210

conductivity. The advantage of this approach is also to provide the necessary conductivity at a reduced concentration of the sulphur-bearing polysulphone for use in ULSD and low sulphur gasoline – perhaps this is the approach used in Stadis 425.

Other companies have also proposed using static dissipater blends that contain lower levels of Stadis 450, whose activity is said to be boosted by synergy with other fuel additives, PIBSA/PAM fuel detergents [33] (see Chapter 3). The conductivities of jet fuel and diesel fuel were improved to more than adequate levels by the extremely low concentration of 0.25 ppm Stadis 450 when combined with detergents (Table 7.3). Since fuel detergents may be used at 100 ppm or more, this small quantity of Stadis 450 in ULSD, or low sulphur gasoline, can provide adequate conductivity at an acceptably low proportion of sulphur, when part of a fuel additive package. While these detergents alone provide more than adequate conductivity to the jet fuel alone, they are not approved for such use.

The advantage of using a polyamine was previously realised by Du Pont[14] [36] as is evident from the current formulation of Stadis 450, which is reported (from data sheet supplied) to be 66% toluene, 13.3% 1-decene polysulfone, 1.3% of a polyamine (the reaction product of N-tallow-1,3-propylenediamine and epichlorohydrin) and 7.4% dodecylbenzenesulfonic acid [33]. Stadis 450, however, was also reported in an earlier patent to contain a formulation of the 1-decene sulphone with a quaternary ammonium salt of dinonylnaphthyl sulphonic acid [37].

Given the effects of the fuel detergent on providing conductivity to diesel fuel, it may be that a specific static dissipater is not needed on top of a usual diesel additive package. Indeed, it has been shown that the conductivity of ULSD may be raised from 5 pS/m to an adequate level of 150–200 pS/m by the use of 740 ppm of a diesel additive package[16] [26].

Table 7.3 Effect on Conductivity of Combining Stadis 450 with Fuel Detergents[15].

Stadis 450, ppm	Detergent, ppm	Conductivity Jet Fuel, pS/m	Conductivity ULSD, pS/m
0.00	0	3	18
0.25	0	55	40
0.00	33	523	-
0.25	33	707	-
0.00	40	-	150
0.25	40	-	190

BASF have produced a static dissipater that contains no sulphur, by the trade-name Kerostat® 5009 [38], which is described as chromium tris-N-stearoyl-anthranilate (-2-aminobenzoate), example 1 of their patent [39]. Interestingly, BASF demonstrate that increasing the conductivity of diesel fuel by including 1–5 ppm of any one of the three commercial fuel static dissipaters also improves the performance of their wax anti-settling cold flow improver (section 4.2.7).

7.3.3 Understanding Static Dissipaters

The hazard of static electricity is that an electrostatic charge builds up in fuel when it is being pumped from tank to tank and a sudden discharge of this charge – as a spark – may be a source of ignition in the receiving tank [28,40]. The factors contributing to this risk are:

- A means of generating electrostatic charge – generation of separate charges in the contact layer between the hose and the fuel, and the fuel carries the charge downstream
- Sufficient charge to provide a spark of sufficient energy to cause ignition must accumulate, which depends upon its rate of generation less its rate of dissipation
- Electrostatic discharge (the spark) must occur at the same time, in the same space as the occurrence of an ignitable vapour/air mix
- The local ignition must propagate into the surroundings

It is necessary for all these factors to be in place together. There may be no problem with a system over a long period, giving the impression that it is safe, before an unexpected explosion occurs as these necessary items randomly come together one day.

When two surfaces of different materials are in contact, electrons flow from one to the other generating a charge difference. If the two surfaces are static and conducting, any charge difference is rapidly dissipated by the movement of electrons. If one or both surfaces are poor conductors, the charge difference dissipates slowly and if the surfaces are moving apart, the non-conductor (such as a hydrocarbon fuel) holds the charge as it moves away from the, now, oppositely charged surface [28][18]. Common examples of such separations are:

- Separation of fuel from a hose (rubber or plastic) or nozzle
- Fuel moving relative to other materials, such as flowing through a pipeline, being pumped or agitated
- Turbulent contact of immiscible liquids, such as a mixture of bulk water and fuels

Electrostatic charges leak away from a charged body as the electrons move and are conducted back to their now-positively-charged source, or to ground after reaching their conductive container – this is called dissipation. For charge to accumulate it must be generated more quickly than it is dissipated. Rate of dissipation depends upon the conductivity of the fuel, its container, and the ability of the container to lose its charge to ground or to the oppositely charged, bonded[17] body. If the fuel has a conductivity greater than 50 pS/m and its container is grounded, it should dissipate charge as fast as it is generated and static charge should not accumulate; rapid charge dissipation, prevents electrical discharge [16,17]. If the fuel container is also non-conductive or is isolated from the ground it will accumulate charge as it is being filled. Conductivities of untreated fuels vary (Table 7.4) but, as is apparent from those reported in patents for fuels used in testing, they are mostly in the lower end of these ranges.

EN 590 diesel specifies that it should have a minimum conductivity of 50 pS/m, though untreated ULSD usually has a much lower conductivity, such as a reported 4 pS/m (fuel with 10 ppm sulphur), while a higher sulphur gasoil

Table 7.4 Fuel Conductivities in pS/m.

Fuel	Quoted Values, pS/m	Patent Examples[18] pS/m
Gasoline	25 [41]	<24[19]
Jet Fuel	Usually < 5 [30], < 50 [28]	0.3, 3, 3.2, 4.3, 7.6, 6.5
Diesel	5 [41], 0.5–50 [28]	18, 2.5–10, 1, 3, 4, 2
Gas Oil	<50 [28]	na
Fuel Oil	50–1000 [28]	na

(36 ppm sulphur) was similarly non-conductive at 2 pS/m [33]. In this patent, a jet fuel used in the testing had a conductivity of 3 pS/m. As noted in the additives section, conductivity may be raised by some diesel additive packages, but the inclusion of a low treat rate of static dissipater removes any concerns.

Jet A and A-1 and aviation gasoline specifications state that conductivity should be 50–450 pS/m. The use of a static dissipater is optional in Jet A and mandatory in Jet A-1 (IATA standards [24]). The USA military specifies some differences, such as 150–450 for JP-8. Clearly, to be sure of such conductivities, a static dissipater is needed.

Static charge can discharge when the charge generated in the fuel accumulates to raise the voltage sufficiently to overcome the insulation of the atmosphere. Then the fuel with the accumulated charge, that is not grounded, can discharge to any conductive grounded object that is brought close to it, such as the side of a truck. For the spark to cause ignition of a flammable air/vapour mixture it must have enough energy, though it must be assumed that if an electrostatic spark is possible, it could well have enough energy to cause ignition.

Flammable air/vapour mixtures must be present in the space around the spark for ignition to take place. Low vapour pressure products, such as middle distillates, are usually handled below their flash points. For example, diesel fuel has a flash point of above 55°C (EN 590) so should not be a problem, unless it is at a temperature within 8.5°C of the flash point or higher [28]. However, there is a high risk of ignition if the high flash point fuel is contaminated with a lower flash point fuel, such as jet fuel or gasoline, or is being transferred into a container that still contains the vapours of a previous load of a lower flash point fuel (switch loading).

With the high vapour pressure fuel gasoline, the vapour space just above the surface of the liquid is too rich to support combustion – there is little or no air – and the space above the fuel in a gasoline tank is also likely to be too rich in vapours. When filling an empty container, though, some of the vapour space around the filler cap, open to the air, may create a flammable air/vapour mixture and it is here that a spark could occur – between the region of the filler cap and the fuel hose, unless the fuel is treated [28]. For the initial ignition to progress further, the flammable air/vapour mixture needs to extend somewhat or it will be extinguished for lack of either air or flammable vapour – but it is possible for a tank that has been emptied of gasoline to contain a large volume of the correct air/vapour mixture to produce an explosion. There are many reported examples of how this can happen: in 2007, during the filling of a 12,700 m^3 storage tank, previously used to store gasoline, with diesel fuel, after only 1,000 m^3 had been loaded, the too rapid flow of diesel and lack of full electrical precautions generated a static electrical spark which caused an explosion and fire [42]; a similar cause and effect was reported in 2011 during switch loading a road tanker with diesel fuel [43].

Because of the low concentrations of static dissipater used, there is potential for loss of activity by interactions with impurities or other additives in a fuel. It was found that anions of strong acids, such as alkyl benzene-sulphonic acid (a refinery introduced impurity[20]) and phosphoric acid (from some lubricity improvers such as Hitec 515 and 580) are antagonistic to the action of static dissipaters (ASA-3 and Stadis 450). Other fuel additives that might be present, but were not found to be significantly antagonistic, are carboxylic acids (lubricity improvers and corrosion inhibitors), naturally occurring compounds carrying O, N or S atoms (such as cresol, indole and benzothiophene), antioxidants (phenolic type) and anti-icing additive (FSII) 2-methoxyethanol [29].

Experiments on the conductivity of hydrocarbons containing static dissipaters ASA-3 or Stadis 450 showed that conductivity in a range of hydrocarbons had a strong dependence on viscosity, which is consistent with an ion conductance mechanism [29]. However, the authors of this reference have difficulty explaining the production of conductive ions in Stadis 450, suggesting a process of reversible proton transfer between the co-additives for both Stadis 450 and ASA-3. The alternative is one of proton transfer itself, which may occur through chains of hydrogen-bonded species, especially as we must include the presence of water[21] which would be strongly associated with the very polar species such as the polysulphone, the salicylate salts and any amine-based co-additives. Proton transfer conductivity can be high if both the anion and cation bear protons [44]. The presence of a high potential electrostatic field can align hydrated polar species and bring them into contact, such as in the electrostatic water separation process used to separate salt water from crude oil [45,46].

7.4 Antifoam Additives

7.4.1 The Problem of Foaming

Foaming of a vessel's liquid contents can interfere with the mixing of reactants in a reactor and reduce its capacity – a foaming mass can cause a reactor to overflow, much like milk boiling in a small saucepan. In fuels, the problem is that foaming can lead to severe limitations on the speed with which fuel is pumped from one vessel to another: from refinery or terminal to delivery tanker, from tanker to service station storage tanks and in the filling of the consumers' tanks. This applies particularly to diesel fuel and heating oil which have a natural tendency to produce foam [16,47]. At the service station, customers would experience the repeated triggering the automatic shutoff while filing their vehicle fuel tanks with diesel – otherwise foaming fuel would flow out onto the car body, shoes and forecourt.

Antifoam additives, also known simply as antifoams, are not needed in gasoline because of its low viscosity – foams collapse either because the bubbles burst in the presence of a surface active impurity or drain and burst as a low viscosity allows fast drainage. In jet fuel, antifoams are not permitted additives [24]. The tendency to foam is related to oil fraction viscosities, so becomes a problem in gas oils and diesel fuel and is a greater problem in the refining of lubricating oils [48].

7.4.2 What Antifoams Do and Their Chemistries

There were some organic antifoams, such as certain poly-acrylates and poly-alkylene glycols, that were claimed to be effective in lube oils, but only at relatively high treat rates (100–200 ppm [47]). However, it was later discovered that Poly-Di-Methyl Siloxane (PDMS) and its derivatives are the most effective antifoams; so, these became the most widely used in lube oils [49] and in petroleum products in general [48].

Simple PDMS and related polymers (such as poly-diethyl-siloxane) were known as antifoams for use in lube oils in the 1940s, being effective at only 10 ppm when finely dispersed into the oil (droplet sizes down to 2 micron were claimed to be easily obtained) [50]. Silicones are expensive and may be harmful to engines, clogging fuel filters and injectors, if used at concentrations providing more than a few ppm of silicon, so even at 10–20 ppm [51]. As a result, it has been stated that silicone antifoams should be effective at 2–4 ppm [51] in diesel fuel and, most preferably, at less than 3 ppm, as improvements to the silicone antifoam were realised [52].

Even at these low concentrations, optimised silicone structures quickly reduce foam in diesel almost completely – foaming is now not noticeable by a vehicle driver refilling his fuel tank at a service station. Tests used to measure foaming rely upon a repeatable method of producing the foam in a measuring cylinder, part filled with fuel, then measuring the proportion of foam first formed and noting its collapse over time. The foam may be produced by injecting diesel fuel into a measuring cylinder (for example 100 ml into a 250 ml measuring cylinder [51]), by shaking the fuel in a part-filled measuring cylinder for a fixed period of time [53], or by bubbling nitrogen or air, at a fixed flow rate (such as 50 ml/min), into the test oil through a sintered gas dispersion tube, for a fixed time period [54,55] (ASTM test D 892). Immediately after the foam is produced, measurements are taken of the volume of foam (using the gradations of the measuring cylinder) and the speed with which the foam is dispersed – either the time to complete disappearance or amount remaining after various, short time intervals.

To illustrate, 100 ml of fuel, treated with a diesel additive package, injected into an empty 250 ml measuring cylinder, produced 130 ml of foam (total volume 230 ml foam plus liquid) that took almost a minute to dissipate. With only 2 or 3 ppm of optimised silicone antifoams (well dispersed into the fuel), the initial foam was only about 5 ml and took only seven seconds to dissipate [52]. In other words, there would be no noticeable foaming at all during the filling of a vehicle fuel tank with diesel fuel containing such an antifoam additive.

7.4.3 Syntheses of Silicone Antifoams

A commercial synthetic route to these compounds starts with the direct reaction of silicon with methyl chloride, at high temperatures, in the presence of a copper catalyst (Figure 7.6) (products are separated by fractional distillation) [56,57].

Reaction of dimethyldichlorosilane with a controlled, minimal amount of water and acid or base catalyst, gives linear polymeric dimethyl siloxane dichlorides or glycols (Figure 7.7).

More water leads to the preference for glycols and, also, the production of cyclic siloxane oligomers, in competition with the linear oligomers/polymers. The cyclic siloxanes are mainly 6- or 8-membered rings of alternating silicon and oxygen atoms (Figure 7.8).

The cyclic siloxane oligomers are used for the more controllable ring-opening polymerisations (acid or base catalysed) which, in the presence of the

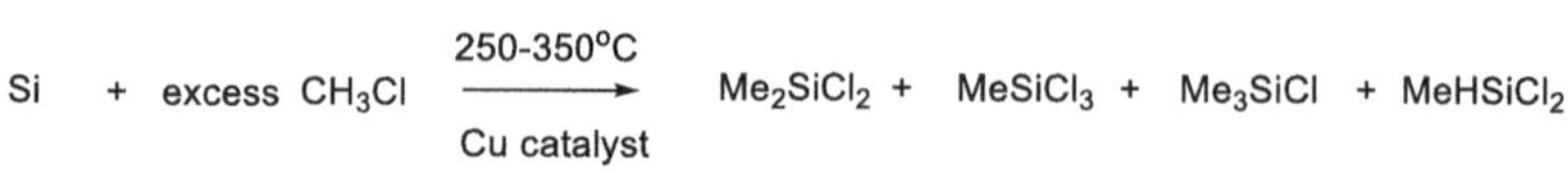

Figure 7.6 Formation of Methyl Chlorosilanes.

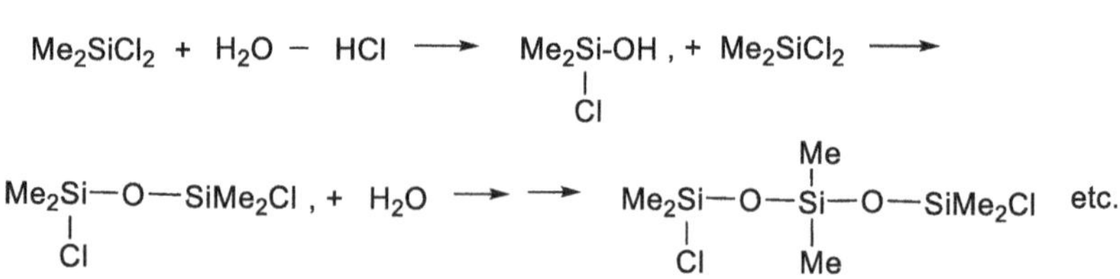

Figure 7.7 Reaction of Dimethyldichlorosilane with Water.

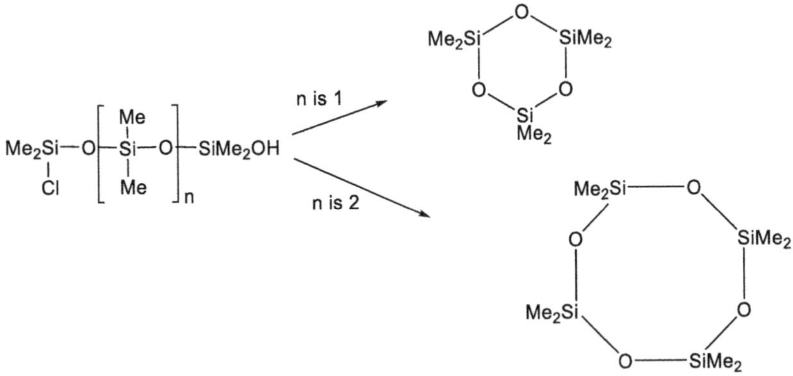

Figure 7.8 Cyclic Silicones.

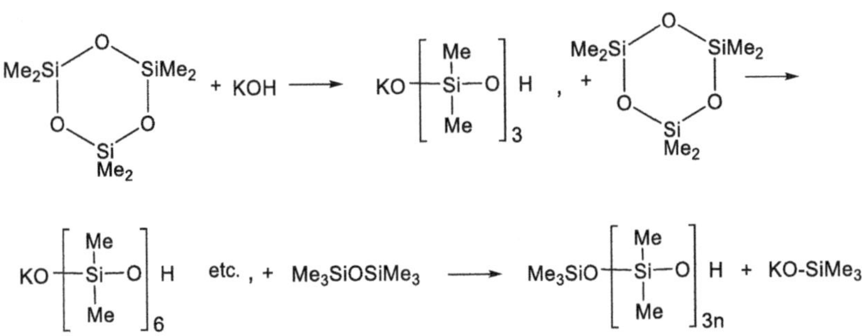

Figure 7.9 Ring Opening Polymerisation of Cyclic Silicones.

dimer hexamethyl-siloxane, will take on trimethyl-silyl end-groups, thus controlling the molecular weight (Figure 7.9).

Copolymerising the 8-ring polydimethylsiloxane with trimethylsilyl-terminated methyl-hydrogenosiloxane oligomers, or a cyclic siloxane bearing hydrogen as well as methyl groups, leads to a PDMS polymer with incorporated methylhydrogenosilyl groups, while PDMS formed in the presence of HMe_2Si-O-$SiHMe_2$ has hydrogenosilyl terminal groups. These Si-H groups will add across carbon-carbon double bonds in the presence of chloroplatinic acid catalyst (about 15 ppm), which enables further functional groups to be added to the PDMS and vary and optimise its properties. PDMS derivatives bearing Si-H groups, when reacted with allyl ethers of polyols, such as trimethylolpropane [47,54] or polyethylene-glycols [55], increase its polarity – hence oleophobic character – which improves their effectiveness as antifoams for diesel fuel and heating oil (Figure 7.10) [54]. Other alkenes or alkenyl aromatic compounds may also be added to balance the polarity to aid potency and ease of

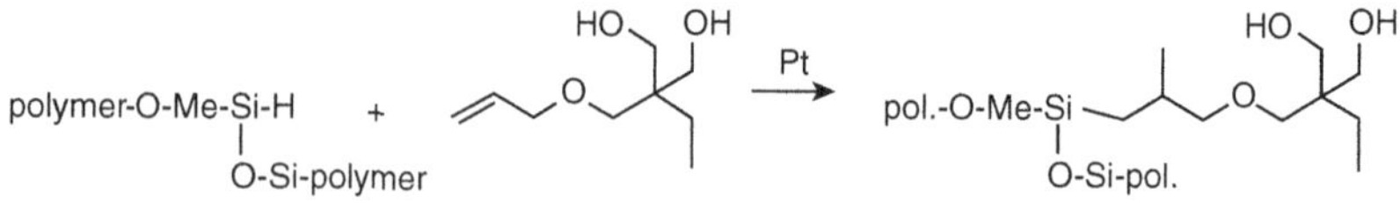

Figure 7.10 Adding a Polyol to a Silicone by Addition of a Hydrosilyl Group to the Alkenyl Group of an Allyl Ether.

dispersion, while enabling the silicone to be included in a stable additive package [47,52].

7.4.4 How Antifoam Additives Work[22]

The wall of a foam bubble is a sandwich of air/liquid/air surfaces. If surfactants are present, the air/liquid interfaces are air/surfactant/liquid interfaces. In fuels, the oleophilic (hydrocarbon) groups of any surfactant (such as added fuel detergent) are on the inside of the film in the fuel, and the polar groups are on the outside of the bubble surface, in contact with air. Bubble walls (films) may drain and the junctions between bubbles act as drains to carry the liquid away. A more viscous liquid drains more slowly, so foam stability increases linearly with kinematic viscosity, which is why foaming is more of a problem with diesel fuel than with gasoline. Surfactants inhibit the draining and the collapse of the film.

Antifoams must displace enough surfactant, if present[23], to create a hole and burst the bubble. In general, an effective antifoam needs to enter the oil/air interface and spread quickly on the oil surface. To do this, antifoams should [58]:

1) Be insoluble in the foaming medium.
2) Have lower surface tension than the foaming medium: poly-dimethyl siloxane has a surface tension of 20 mN/m at 20°C [48] compared with 26 mN/m for a North American diesel fuel [59].
3) Rapidly disperse in the foaming medium.
4) Be chemically inert.

PDMSs have low solubility in hydrocarbon liquids, low interfacial tensions with air and with hydrocarbons, and low viscosities so have great surface spreading ability on hydrocarbon liquids – they separate from and are not wetted by fuels. The PDMS methyl groups are hydrophobic (oleophilic) and oleophobic polar groups, such as polyols, are added to aid migration to the air/fuel interface, where they can interact with and displace any surfactant molecules that are stabilising the foam. This creates patches of low surface tension which leads to film weakness, causing the bubbles to burst more quickly than otherwise. The physical process is pictured as a globule of antifoam, in the liquid film of a bubble, migrating to the surface (top of Figure 7.11). After some draining of the film and through the process of de-wetting the antifoam globule, it is left as a temporary bridge in the film which rapidly breaks as fuel and antifoam separate (bottom of Figure 7.11) [60].

7.5 Demulsifiers and Dehazers

Demulsifiers and drag reducers are widely used in crude oils – they are generally considered to be oil-field chemicals. Suitable versions of these are also used in crude-oil-derived fuels. Demulsifiers break emulsions of water in oils and fuels, enabling the separation of water; they are also called dehazers when applied to the removal of fuel haze or cloudiness caused by a fine dispersion of small droplets of water. Demulsifiers are not always used in fuels and many distillate fuels do not, now, have water shedding problems.

7.5.1 The Problem of Water-in-Fuel Emulsions or Haze

As it comes out of the ground, crude is intimately mixed with water, either from natural groundwater, seawater or from its use in oil recovery processes[24]. By contrast, processes are in place to remove water from distillate fuels[25] before they leave the refinery. However, fuels pick up water during transportation and storage; agitation of wet fuel during transportation and pumping cause water-in-oil emulsions to form. Small amounts of water may separate from solution in a fuel simply because the fuel cools, causing a haze of fine water droplets [61]. Without an emulsion stabiliser, water dispersed in these ways may settle to the bottom of a fuel container, from which it can be tapped off. Fuel may contain natural emulsifiers – compounds containing oxygen or nitrogen[26] – and many fuel additives are surfactants, both of which can cause

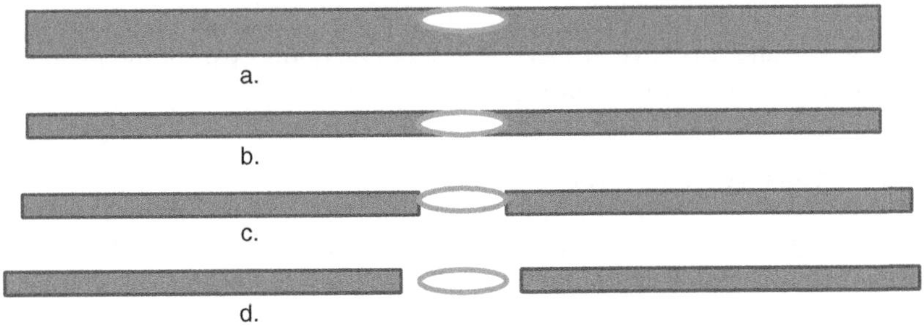

Figure 7.11 Schematic Representation of Antifoam Breaking a Bubble Film.

stable water haze or water-in-oil emulsions. In all fuels, water in fuel causes corrosion in pipes and tanks and allows the growth of micro-organisms [17]. In cold weather, water held up in a fuel can produce ice particles that block fuel filters [62]. Emulsified water may separate from static fuel as a sludge which, in view of its lower density than water alone, is readily re-dispersed into the fuel [63].

The lighter, low viscosity fuels, gasoline and jet fuel, did not always form stable emulsions with water until additives, such as fuel detergents and corrosion inhibitors, were incorporated in them [63,64]. Fuel additive manufacturers have been aware of this problem since the 1950s and have included this possibility in their development of fuel additives – as demonstrated by the case of a compound developed to stabilise fuel oil, act as a corrosion inhibitor and provide demulsifying activity all in one compound [65]. Fuel detergents may promote water emulsions so some claims for the use of fuel detergents also describe their use in combination with an alkoxylated-alkylphenol-formaldehyde resin demulsifier [66,67]. If a water haze or emulsion problem is present – whether a result of the fuel or additive used – a dehazer or demulsifier is often included in the fuel additive package [17].

Heavy fuels are particularly susceptible to emulsifying water as they frequently come into contact with water and they contain the same emulsifying compounds – asphaltenes and naphthenic acids – that are responsible for water emulsion problems in crude oil [17,68,69]. Gravity separation of water is, then, only partially effective and ships usually use centrifugation to speed up its separation, with the aid of demulsifiers which are included in additive packages designed for such fuels [70,71].

Jet fuel is an exception to the use of demulsifiers – they are not among the additives allowed for use in jet fuel [24]. As described in the discussion on anti-icing (section 7.6), jet fuel passes through a dewatering system before it is charged into an aeroplane's fuel tanks. In general, additives used in jet fuel have to pass a test that ensures that they do not interfere with the operation of the dewatering equipment. Water dissolved into jet fuel on the ground separates when the fuel is cold, in flight, and can build up in wing tanks; this usually settles and it is advised that water is drained off as part of the frequent, ongoing maintenance schedules (section 7.6).

7.5.2 The Chemistry of Demulsifiers

Separating water from water-in-oil emulsions has been, for more than a century, an important crude oil production step and this has been the driving force for the development of many water separation processes and chemical demulsifiers that speed up the separation of water, when used at low concentrations. 'Effective separation of oil and water is vital in ensuring the crude oil quality and low cost of oil production' and the use of chemical demulsifiers is the major process for effecting this separation [72,73] . The oil production industry has supported the development demulsifiers and research into the understanding of the processes involved. It is no surprise then, that demulsifiers used in fuels are selected from those already available in the heavily serviced oil-production industry.

Demulsifiers were first commercialised for separating water-in-oil emulsions that form in crude petroleum in the 1920s; these early demulsifiers were sodium soaps, fatty acids, and combinations of both [74,75]. The early inventors were using the knowledge that adding a detergent that promotes emulsions in the opposite case of oil-in-water destabilises a water-in-oil emulsion. So adding the sodium soap, well-known to promote oil-in-water emulsions, to a water-in-oil emulsion breaks that emulsion. Proportions needed were claimed to be from 0.1% to 1.0% for fatty acid derivatives [75], the amount used to demonstrate the effect in one patent was 1% of a mixture: 25% sodium soap, 65% oleic acid and 10% water [74]. Other compounds in use in this period were sodium alkaryl sulphonates, probably those used in the making of lubricant detergents. Developments of chemical demulsifiers have continued through the twentieth century (Table 7.5) [74,76].

This summary of the evolution of demulsifiers suggests that distillate fuels needed about 0.2% of demulsifiers available in the late 1920s while only 10 ppm are needed of modern demulsifiers, in distillate fuels, and 50–500 ppm are needed in heavy fuels [17].

Table 7.5 Chemical Demulsifiers for Water-in-Oil in the Twentieth Century.

Years	Chemistry	Treat Rates
1920–1930	Sodium fatty acid soap plus the fatty acid and olive oil.	0.5%–1%
1925–1935	Petroleum sulphonates or dioctyl-sulphosuccinate.	0.1%
1935–1948	Fatty acids, fatty alcohols and alkylphenols.	200 ppm
1948–1963	Ethylene oxide/propylene oxide copolymers and alkoxylated alkylphenol-formaldehyde resins.	100 ppm
1963–1975	Amine alkoxylates.	<100 ppm
1975–1985	Alkoxylated cyclic alkylphenol-formaldehyde resins.	<50 ppm
1985–1985	Polyesteramines and blends.	20 ppm

The source of the list (Table 7.5) also describes a more recent development which is the use of high molecular weight (10,000–20,000) cross-linked polyurethanes, made by reacting toluene diisocyanate with polypropylene glycol[27], as published in a 1974 patent [77]. These are probably the compounds referred to as cross-linked polyols in the following list of demulsifiers supplied by NALCO Champion, the company[28] that supported the research of this thesis [76]. 'Cross-linked resin/polyols' of this list derive from the inclusion of an alkylphenol-formaldehyde (APF) resin in the toluene diisocyanate/polypropylene glycol reaction.

The NALCO Champion list of demulsifiers was provided in a conference presentation [78]:

- Polyol esters
- Cross-linked polyols
- Cross-linked resin/polyols
- Polyamine oxyalkylates
- Sulphonates
- Oxyalkylated phenols
- Resins (low oxide)
- Resins (high oxide)

Polyol esters are polyols, such as ethylene glycol or trimethylolpropane, which are poly-alkoxylated with various proportions of ethylene oxide (EO) and propylene oxide (PO) [79] and are esterified with fatty acids or short chain diacids (such as succinic) [80,81]. Such conventional demulsifiers can be traced back to the 1930s when fatty acids, alcohols and alkylphenols were alkoxylated [82]. For polyamine oxyalkylates, the polyol core is replaced by a polyamine such as ethylene diamine, N-alkyl-1,3-propane diamines or a poly-(ethylenediamine) [81].

Cross-linked polyols are the polyols, such as EO/PO block copolymers, cross-linked by reaction with toluene diisocyanate, which increases their molecular weights and improves their demulsifying activities (shown at 30–150 ppm in a water/crude oil emulsion) [77]. The cross-linked resin/polyols are made by the same process with the inclusion of an 'isoalkyl-phenyl/formaldehyde polyoxyethylate' [77] (an ethoxylated alkylphenol/formaldehyde condensate, APFC).

The resins themselves are also listed as low and high oxide, which presumably refers to the proportion of alkoxylation ('oxyalkylate') and whose overall molecular weights would be 3,000–10,000 [76]. Polyamine oxyalkylates have higher molecular weights (>10,000), while sulphonates are amine or metal salts of long chain alkylbenzene-sulphonic acids (probably those produced for lubricating oil detergents[29] [67]).

Alkoxylated-APFCs ('resins') are the most widely used demulsifiers that appear to have been first introduced in the 1950s. An early patent [83] described the production of a wide range of APFCs, with alkyls from iso- or tert-butyl up to tetracosyl ($C_{24}H_{49}$), alkoxylated with ethylene oxide, propylene oxide or butylene oxide; their demulsifying abilities were demonstrated on a stable water/crude oil emulsion by breaking the emulsion in minutes when added at 40 ppm of a 50% solution. In another patent on such a demulsifier for water-in-oil emulsions [84], the same inventor is more specific on the need for certain 'hydrophile-hydrophobe' balance (now known as Hydrophile-Lipophile Balance or HLB). This balance is obtained by using suitable proportions of a nonyl-phenol or tetradecyl-phenol/formaldehyde resin (the hydrophobe, preferably para substituted), ethylene oxide (hydrophile) and propylene oxide (also hydrophobic). APFCs are readily prepared by acid- or base-catalysed reaction of an alkyl phenol with, usually, formaldehyde (Figure 7.12) though other aldehydes or ketones may be used.

The alkoxylation of the APFCs is usually a base catalysed, ring-opening anionic process – demonstrated here on a single alkylphenol molecule (Figure 7.13). As such, it is a living polymerisation that enables the production of block copolymers by, for example, adding propylene oxide (n molecules per alkylphenol) to form a block of poly-propylene oxide, followed by adding ethylene oxide (m + 1 molecules per alkyl phenol) to form a block of poly-ethylene oxide (Figure 7.14).

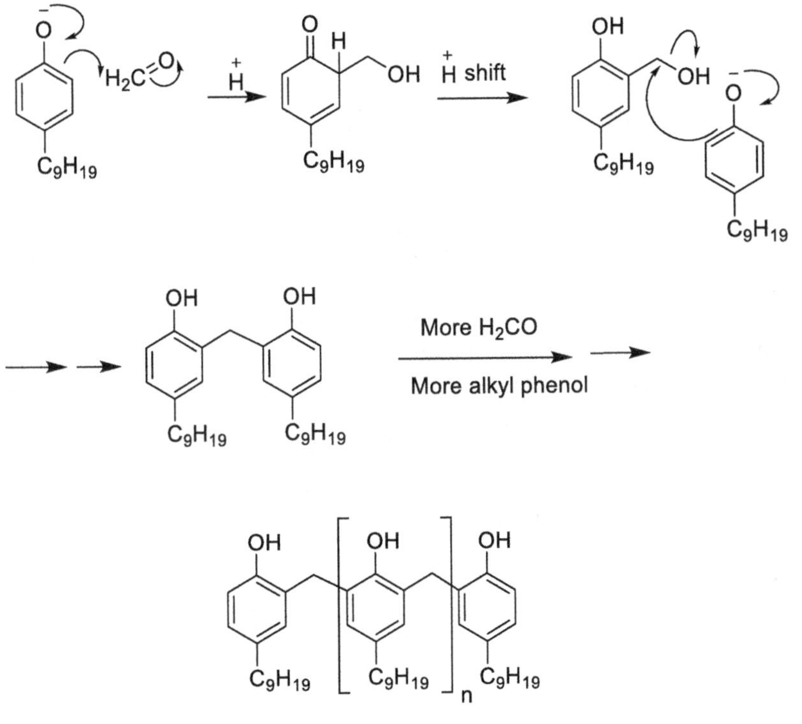

Figure 7.12 Base Catalysed Formation of Alkylphenol-Formaldehyde Resins[30].

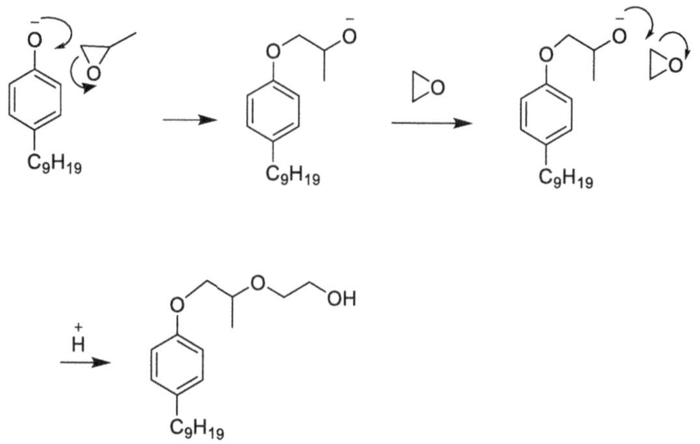

Figure 7.13 Propoxylation of an Alkylphenol Followed by Ethoxylation.

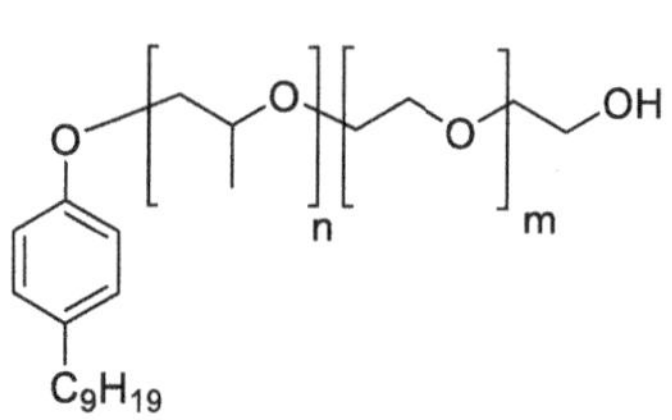

Figure 7.14 Alkylphenol with Block Copolymer of Propylene Oxide and Ethylene Oxide.

Such compounds as these dominate the demulsifier applications in gasoline and diesel fuel [17,61,68]; they are often added specifically to prevent water-in–oil emulsions in gasoline or diesel fuel containing a detergent (at 1–10 ppm) [67,85] or in a general fuel additive formulation as a dehazer [86]. The alkoxylated-APF resins may be also combined with polyols of an EO/PO polyalkoxylated amine, in such fuel additive formulations [85]. While demulsifiers are not now added to jet fuel, the alkoxylated APF resins have been claimed for use in preventing detergent-promoted water-in–oil emulsions in jet fuel as well as in gasoline [63]. The early inventors of the alkoxylated-APFs as demulsifiers, De Groote and Keiser, provided many examples of how to make the many variations that they claimed in many patent applications (48 were published in 1951 alone): variations of molecular weight, alkyl group and aldehyde used for the APF resin and alkoxylation by varied proportions of EO and PO, adding EO first, PO first or adding both together [87]. Such a range of variables renders this system the advantage of being able to fine-tune the HLB to match the needs of any water-in-oil system. Unfortunately, while these patents give much preparative and descriptive data of the alkoxylated resins and claims to their use, they give no quantitative measure of their relative efficacies as demulsifiers, which leaves us unable to deduce structural preferences.

Another major supplier of demulsifiers for the oil industry (Dow Chemical) also lists a wide range of variations of five classes of chemicals to enable the user to match their own particular application, because crude oils and the water phases (containing inorganic salts) vary widely. These demulsifiers are also recommended for use in heavy fuels [88].

Dow Chemical Demulsifier Types

- Polyol block copolymers
- Alkoxylated alkylphenol-formaldehyde resins
- Epoxy resin alkoxylates
- Amine-initiated polyol block copolymers
- Modified specialty resin alkoxylates

Four of these are similar compounds to those listed by NALCO Champion and epoxy resin alkoxylates probably refer to those patented by Dow [89] as a derivative of Dow's extensive expertise in epoxy resins. The chemistry starts with a commercial bisphenol-A diepoxy resin reacted with dibutylamine, followed by alkoxylation with EO and PO to give the demulsifier (Figure 7.15).

7.5.3 The Process of Demulsification

Polar materials, such as particles from the incomplete oxidation of fuel or water droplets, are dispersed into the fuel by surfactants that form inverse micelles. Such surfactants are molecules with large hydrocarbon groups and concentrated polar groups take an arrangement that keeps the polar groups together, out of contact with the fuel, surrounded by the hydrocarbon groups that are miscible with fuel (Chapter 3 and Figure 3.21). Such surfactants were, in the twentieth century, naturally present in fuels as molecules bearing oxygen, sulphur and nitrogen atoms. The severe hydrogenation that is used to remove sulphur from gasoline and diesel also removes most of the oxygen and nitrogen compounds so modern automotive fuels are usually less likely to form stable emulsions with water. It has already been pointed out, however, that some fuel additives, such as the detergents, are surfactants that may produce problems of stable water emulsions. Higher boiling heating oils and marine fuels still contain naturally occurring polar molecules that stabilise water emulsions. The emulsifying molecules in fuels have low HLBs (less than 5)[31], having high proportions of hydrocarbon in their structures so are most effective in stabilising water-in-oil emulsions.

Demulsifiers must destabilise these emulsions and enable the water to separate from the fuel. So they need to bring together the dispersed water droplets

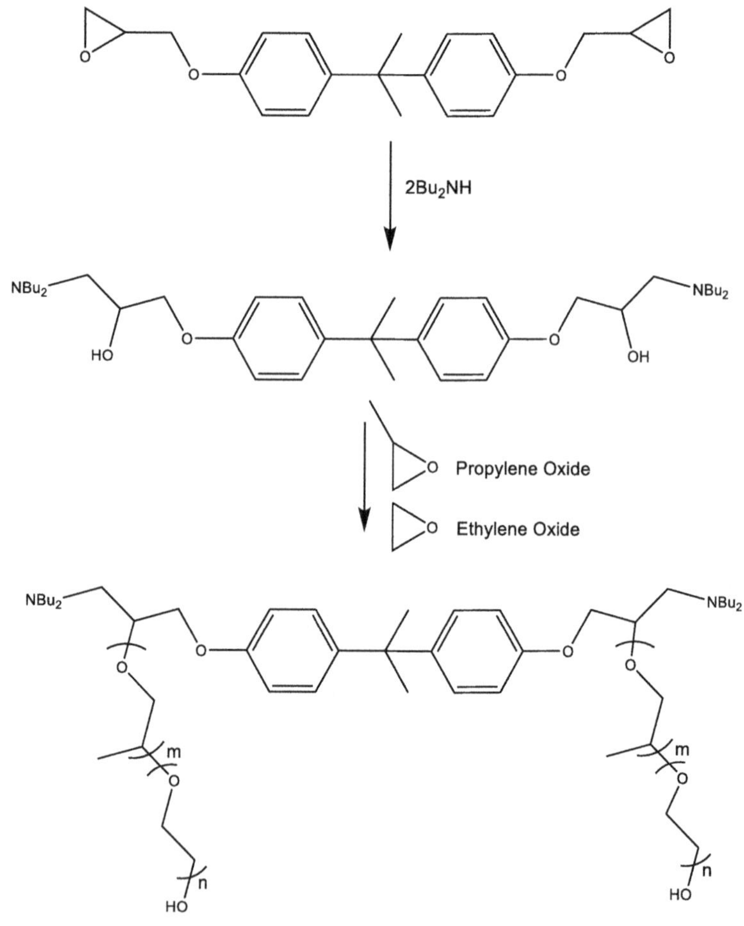

Figure 7.15 Demulsifier Derived from Bisphenol-A Diepoxy Resin.

(aggregation) and enable them to join together (coalesce) despite their protective layers of surfactant [68]. As droplets grow larger, they settle and further coalesce to form a separate layer of water at the tank bottom that may be drained away. The alkoxylated-APFCs have been shown to be most effective as coalescers of small water droplets, but ineffective at aggregation. High molecular weight cross-linked polyols, such as those cross-linked by toluene diisocyanate, are ineffective at coalescing but are highly effective at aggregating droplets, so are especially useful at collecting up finer water droplets that are not being brought together by settling. Consequently, these two types of demulsifier are synergistic in their activities [76].

Interfacial tensions of emulsions are at their lowest when the most effective alkoxylated-APFCs are used – those giving fastest water separation also gave the lowest interfacial tensions, which occurs for those compounds that contain about 40% EO/PO copolymer [76]. EO/PO ratios, such as 50/10 molecular units per APFC molecule (molecular weights 840–2,000), were used by one group to provide intermediate effective demulsifiers that had HLBs ranging from 8 to 12 [90,91]. Their modest final molecular weights (3,000–5,000) allows them sufficient mobility to migrate to the surface of emulsion droplets and their higher HLBs enable them to displace the low HLB surfactants stabilising the emulsion. Having displaced the oil soluble, low HLB stabilising surfactants on the contingent surfaces of two water droplets, they enable the fuel to drain away from between them, thinning the film that separates the two droplets till it collapses and they coalesce [90]. Higher HLB surfactants than these would tend to form stable oil-in-water emulsions in the separated layers, which would give further problems.

Low molecular weight surfactants (<3,000), such as the ammonium salts of tetracosyl[32] benzene sulphonate [67], can diffuse more rapidly to the droplet surfaces than the higher molecular weight alkoxylated-APF resins that produce stronger, multipoint adsorption [91]. Combinations of these two types are synergistic in their demulsifying activities.

Commercial demulsifiers maybe a grade of alkoxylated-APF resin alone for water-fuel haze problems. Alternatively, more potent fuel demulsifiers may be combinations of the components that address each of the three approaches to moving dispersed droplets of water into a separated bulk layer: a relatively high molecular weight poly-EO/PO glycol, that may be cross-linked by a diisocyanate or diepoxy resin, to act as a droplet agglomerants; an alkoxylated-APF resin of moderate molecular weight to act as the most effective droplet coalescer; and an alkylbenzene sulphonate, which maybe a less effective coalescer but can react more quickly, being of lower molecular weight. Such an approach is used in commercial gasoline demulsifier blends, such as those made up of a polyalkoxylated polyol, an alkoxylated-APF resin and an ammonium-(alkylbenzene or alkylphenol) sulphonate [67,92,93].

7.6 Anti-Icing

7.6.1 The Problem of Icing

Problems caused by bulk water collecting in low points of fuel lines and freezing in cold weather, blocking fuel flow, can arise in any fuel. Dry fuel leaves the refinery but may pick up water by adventitious, avoidable routes, such as rain and leaking tank covers, or by the tanks drawing in moist air, as temperatures fall, from which water condenses inside the tank. Such problems are mostly met by good housekeeping along the fuel supply lines, preventing simple leakage of water into fuel systems and regular draining of settled water from bulk fuel containers. Anti-icing additives developed for jet fuel and gasoline add further protection.

Jet fuel and gasoline have further problems caused by icing. While extra procedures are in place to minimise the amount of water in jet fuel, even the small amounts that are present can cause icing problems as a result of the very low temperatures, down to or below −40°C, which jet fuel experiences during flights. The gasoline problem has a different cause which is the extra cooling of moist air in a carburettor due to the evaporation of gasoline. While the same additives are used to treat both problems, they will be discussed separately because of their technical differences.

7.6.2 The Gasoline Icing Problem

At the time in which carburettor systems were important in the motor car industry, there was much patent activity, until the mid-1970s, concerned with mechanical systems and additives to obviate the icing problem. Since then, carburettors have been replaced by fuel injectors which have little risk of icing as the fuel is heated by the mechanical energy of the fuel injection pump; also, the area around the injector is rapidly heated by the combustion of fuel in the engine. However, anti-icing additives are still in use in fully formulated aviation gasoline [94] because some rotary gasoline engines, on light aircraft, use carburettors, such as the 200 series models made by Continental Motors[33], while the 300 and 500 series are available as either carburetted or with fuel injection [95]. Consequently, anti-icing is still included in aviation gasoline specifications as optional or by agreement.

Gasoline engines using carburettors suffer from the problem of ice formation in the carburettor when the relative humidity is above 65%–80% and the ambient temperature is in the region of −6°C to + 12°C [15,61,96]. Under these conditions, the extra cooling caused by the evaporation of gasoline in the carburettor causes ice crystals, from the water in the air, to form on the throttle plate and body. When the throttle plate is almost closed, as in idling or low load, the ice restricts air flows causing the engine to stall shortly after starting [97,98].

The first step on the path to effective anti-icing additives in gasoline was to apply the molar melting point depression effects of low boiling alcohols (the low flash point not being a problem in low flash gasoline). Alcohols used were methanol, ethanol, isopropanol or tetrahydrofurfuryl alcohol [96], all of which have volatilities similar to those of gasoline and burn readily in the gasoline spark ignition engine – fortunate, as they were needed at 1%–2% to be effective. Isopropanol is still permitted for use in aviation gasoline, as recommended by the manufacturer but not to exceed 1% [24] [99]; used at this level, such alcohols are at the borderline between fuel additives and blend-components (section 1.2.5).

It was found that surfactants that adsorb onto metal surfaces, such as corrosion inhibitors, also inhibit carburettor icing. When these additives adsorb onto metal surfaces in the carburettor, they render them hydrophobic, rather than hydrophilic. Hydrophobic surfaces repel the condensation of water or ice. Engine tests indicated the effectiveness of various additives as carburettor icing inhibitors [96]: in one test, a six-cylinder Chevrolet engine was run under icing-critical conditions of 95% relative humidity and at 4°C, thus causing cold, water-saturated air to be drawn into the carburettor. Running on cycles of 20 seconds at low engine speed then 10 seconds at high engine speed, with no anti-icing agent, the engine would stall after five cycles. By contrast, gasoline gave no stalling until 11 cycles if it were provided with the anti-icing activity of the corrosion inhibitor dimer-trimer acid (at 67 ppm) (compound described in section 5.7.1), or no stalling for 25 cycles with a 1/2 mix of dimer-trimer acid and the dodecyl monoether of heptapropylene glycol (67 ppm) [96]. By comparison, another patent [100], using the same test, reports that gasoline containing 2% of isopropanol was less effective, giving 7 or 8 stalls per 25 cycles (compared with 18 for the untreated gasoline).

Aviation gasoline may suffer both the carburettor icing problem and that of jet fuel, since it too may experience low, sub-zero temperatures that cause ice crystals to form from water that was dissolved in the fuel at ambient temperatures. Now that the use of DiEthylene Glycol Monomethyl Ether (DIEGME) has been approved for use in aircraft, through its use in jet fuel, this is also approved for use in aviation gasoline. The specifications allow the use of DIEGME at 0.10%–0.15%, or isopropanol at an amount specified by the manufacturer but at less than 1.0% [24]. Glycol ethers have some oleophilic character, from the alkyl groups that they carry, so can have the double activity of solubilising any water and preventing it from freezing, as does isopropanol, and of absorbing onto metal surfaces making them hydrophobic.

7.6.3 The Jet Fuel Icing Problem

Icing in jet fuels is always a concern because of the extremely low temperatures to which jet fuel, in aircraft wing tanks, is exposed. Civilian aircraft are routinely exposed to air temperatures of −50°C, though air frictional heating keeps the fuel temperatures closer to −40°C [17]. In addition, civilian aircraft usually rely upon fuel heaters and the heat from the heat exchangers, in which the fuel is used to cool the aircraft's lubricating system, to minimise ice formation [101]. Use of a Fuel System Icing Inhibitor (FSII as applied to aircraft) is allowed by agreement for jet A and A-1 [24], though it is not compulsory. Military aircraft, however, generally do not have fuel heaters and may fly at higher, colder altitudes, where air temperatures may be −60°C or lower. After a B-52 crash in 1959, that was attributed to icing, FSIIs became mandatory in most military jet fuel specifications in 1961 [102].

In January 2008, a Boeing 777 from China crash landed at Heathrow just short of the runway; it had flown over Siberia and Scandinavia at air temperatures of −65 to −74°C but the fuel temperature, apparently, never fell below −34°C [103]. The investigation concluded that ice built up in the fuel system during the long period of low fuel flow and this became dislodged by the high fuel flow and warmer conditions when the plane came in to land, blocking the fuel lines to both engines [104]. Recommendations to avoid a repeat of this problem were, long term, redesign of the fuel system and, short term, use of FSII, as in military aircraft, and special attention to the draining of water from the fuel tank sumps during maintenance. The general recommendation is: 'water has settled to the bottom of the tank. As a safety measure, fuel sumps should be drained (of water) before every flight during the pre-flight inspection' [105].

Bulk water contaminating the fuel is mostly removed by fibre filters that coalesce droplets or emulsions of water and separate it as the fuel is loaded onto a refuelling tanker at the airport [106,107]. However, some of the smaller droplets of water, down to 5 micron, may escape the water separator and the smallest droplets produced are unlikely to settle out in the tanks of an aircraft in use as their rates of separation are very slow, such as 1 cm per hour for 10 micron droplets in static conditions [103]. Even if jet fuel contains no bulk water as it is loaded, at ambient temperatures it is usually saturated with dissolved water[34]. The solubility of water in jet fuel is 0.005%–0.01 % at 20°C and, effectively, zero at −35°C [103]. The long-haul Boeing 777 holds 181,000 litres of fuel at take-off which would contain at least 9 litre of water in solution. When the temperature of the fuel drops by 55°C, this water comes out of solution as ice crystals that can build up on cold metal surfaces and block fuel filters and narrow fuel lines on the aircraft [108]. AAIB[35] investigations suggest that ice crystals form in fuel and adhere to the metal surfaces in the aircraft fuel systems at temperatures between −5°C and −20°C [103].

During the temperature cycling experienced by the fuel in aircraft, significant amounts of water accumulates in fuel tanks between services [103]. While it is recommended that water should be drained from the fuel tank sumps before take-off [105], turnaround times at airports can be as little as 30 minutes, so draining the sumps may have to wait until routine maintenance [103]. There are a variety of short-term maintenance procedures so this could come under the weekly maintenance, allowing quite a lot of water to accumulate in the fuel tanks[36] [103]. Accumulated bulk water will be dispersed as fine droplets into the fuel by the agitation of refuelling, take-off and landing, and turbulence; then, the aqueous dispersion can add to ice from water coming out of solution.

Water in fuel also evaporates and condenses on cooler parts of the wing tanks. This water is readily frozen as the wing tanks cool during flight. In this way, frost and ice, which is not tightly bound, can build up on the metal surfaces – especially in the empty space above the fuel (the ullage[37]) – and any accreted ice is easily dislodged to contribute to the ice suspended in the fuel, all of which may block fuel lines and filters.

7.6.4 Jet Fuel Anti-Icing Additives

The most widely used anti-icing additive for Jet fuel is Di-Ethylene Glycol Monomethyl Ether (DIEGME) (Figure 7.16) which replaced ethylene glycol -monomethyl and -monoethyl ethers, also known as methyl and ethyl cellosolves, because of their volatility and toxicity [102] – though the cellosolves, blended with methanol, may still be in use in former eastern bloc countries [24,109].

The anti-icing effectiveness of both DIEGME and ethylene glycol monomethyl ether (EGME), containing 10% glycerol, was demonstrated by in-flight tests using a B-52 aeroplane, which has eight engines that may be fed independently from different fuel tanks [110]. The fuel used (JP-4) was saturated with water by adding excess bulk water ('2 millilitres per gallon'). Three engines were protected with fuel system heaters, so gave no problems; one engine, with no heater and no additive, suffered repeated fuel flow interruptions and flame outs during the test flights; the two engines fed with fuel containing 0.1% of EGME/glycerol (9/1) and the two engines fed with fuel containing 0.1% of DIEGME/glycerol (9/1) gave no in-flight problems. After the tests, the fuel system that was unheated, and had no additive, had its filter screen blocked with ice while none of the other fuel screens had any ice on them. In addition, the water drain-valves on the fuel tanks that contained no anti-icing additive were frozen, while those containing the anti-icing additive were not, indicating that such glycol ethers do separate with the water and ice.

Since the mid-1980s, when required, DIEGME is added to jet fuel as an FSII at 0.10%–0.15% [24,101]. The use of DIEGME – mostly in military aircraft – is not without problems; it was found that evaporation of DIEGME out of warm (on the ground ambient) fuel and condensation in the ullage[37] was causing pealing of the topcoat epoxy resin from the metal surfaces. Pieces of separated

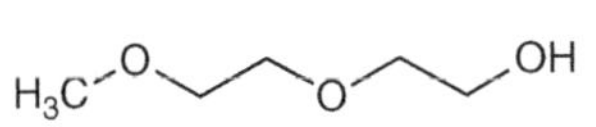

Figure 7.16 DIEGME Di-Ethylene Glycol Monomethyl Ether.

resin in the fuel could cause blockages in the fuel system. As a result, it was recommended that the less volatile triethylene glycol monomethyl ether should be used instead or that the upper limit for DIEGME should be 0.07% in JP-8 and more frequent removal of water from the fuel tank sumps [102]. This problem had been recognised earlier with the use of EGME prompting the use of EGME/polyol blends or other compounds to overcome the problem [108,110].

DIEGME and related compounds are believed to be effective because they are hygroscopic and strongly attract any water in a fuel. If the draining of sump water from the fuel tanks is carried out as part of the pre-flight inspections, the maximum amount of water dissolved in the fuel that might separate is 0.005%–0.01%. DIEGME at 0.10%–0.15% will then be in excess in the water/ DIEGME mix that will form. The proportions of DIEGME in bulk water in contact with jet fuel are greater than 50% and the freeze points of a mixtures of DIEGME/water are about −30°C, −60°C and −85°C for 50%, 60% and 100% DIEGME (respectively), so DIEGME can inhibit icing at most low temperatures that fuels in civilian aeroplanes are likely to experience [111].

Diesel fuel does not generally suffer from icing problems, though bulk water contamination could cause fuel pipeline blockages. Searching diesel additive suppliers, there is little or no mention of anti-icing, excepting the misleading use of the term when the products referred to are wax modifying cold flow improvers. However, the presence of surfactants – such as the detergent, corrosion inhibitor and lubricity improver – also help to provide anti-icing protection. In the development and application of diesel additive packages, care is taken to ensure that the treated diesel fuel sheds water with, if necessary, the aid of demulsifiers.

7.7 Biocides

7.7.1 Problems

The unavoidable contamination of fuel by fungi and bacteria (microbial organisms) leads to the growth of these organisms at water/fuel interfaces [70,112]. Despite widespread procedures to prevent it, water is always present in fuel storage systems as a result of ingress of liquid water, from various sources, and as water condenses from warmer, humid air in the space above the fuel, drawn in as the tank and fuel cool – day/night and seasonal changes in temperature. The water settles to the bottom in quiescent conditions and is remixed into the fuel during movement, as in vehicle tanks and delivery tankers, and during pumping into and out of the tank. These sources of water are the same as those providing icing problems (section 7.6). Various organisms can grow in the water as they feed off the fuel at the water/fuel interface – some of these organisms are aerobic and some are anaerobic so can live in air-free conditions. At risk are such systems as bulk-fuel storage tanks (any fuel), tankers, locomotive tanks, generator systems and diesel fuelled trucks, boats, and ships [113]. In marine fuels, water contamination is common, and ships have centrifugal separators to remove bulk water from the bunker fuel that they carry [70]; water is the major contaminant that supports microbial growth.

Over time, large microbe colonies form plaques on the walls and on the bottom of tanks containing gasoline, jet fuel, heating oil, diesel fuel or residual fuel – all liquid petroleum fuels. Over long periods, in poorly maintained tankage, sludge may form at the bottom of the tank. When these deposits become suspended into the fuel on agitation, they can be drawn into a vehicle or generator fuel system where they may block fuel lines and filters. Aerobic bacteria may produce carboxylic acids; anaerobic sulphate reducing bacteria may produce sulphur containing acids, acids that can lead to pitting and perforation of storage tanks [114]. Microbial growth has also been related to the corrosion of aluminium alloys in aircraft wing tanks, as well as causing jet fuel filter blocking [115].

Problems caused by the growth of microbial material have plagued water systems such as industrial water supplies, reservoirs, heat exchangers and cooling towers, as well as moist or wet biological products, wood pulp and animal hides or leather. It is for these problems that biocides to prevent bacterial and fungal growth have been developed and some of these have been selected for use in the petroleum industry.

7.7.2 Chemistries of Biocides Used in Fuels

In a formulation patent [116], concerned with packages to enable the storage stability of gasoline and diesel, there is a list of possible components in which various biocides are identified for use in these fuels, including: thiazoles, thiocyanates, isothiaxolines, mixed dioxaborinanes (BIOBOR™), bromo-compounds (halogen compounds are now not generally permitted for use as fuel additives), cyanobutane, dithiocarbamates, KATHON 886™ (aqueous solution of magnesium nitrate and magnesium chloride), ONDEO-NALCO 303MC™ and other commercial products.

There are a number of products promoted for use in petroleum derived fuels, such as Racor™ Diesel Biocide [113]. From Racor's registration, it contains two well established biocides: MBT, a bactericide, and TCMBT, a fungicide. MBT is methylene bis(thiocyanate), for which the simplest synthesis is the reaction of methylene dichloride with sodium or potassium thiocyanate in ethanol solution [117] (Figure 7.17). In a patent concerned with the delivery of MBT to No.2 fuel oil (heating oil or diesel fuel), a treat rate of 2 ppm in the fuel was recommended [118].

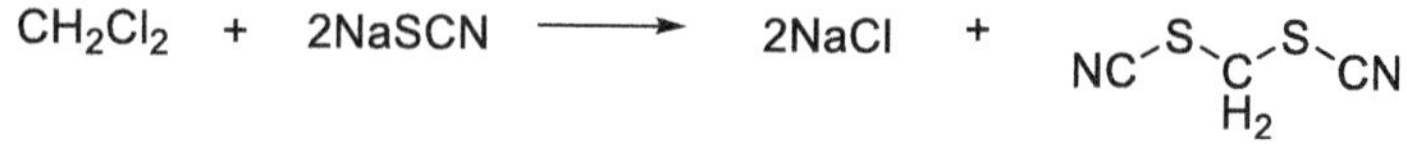

Figure 7.17 The Preparation of MBT.

TCMBT is 2-(thiocyanomethylthio) benzothiazole, which was first registered as a pesticide in 1949, one preparation is described in a 1969 patent which also claims the advantage of using it in combination with MBT [119]. TCMBT was tested for its ability to inhibit the growth of a problem bacterium in wood pulp, for which about 1 ppm [120] was sufficient to kill off this bacterium at pH 5.5–6.5, and 3 ppm prevented the growth of three varieties of problem fungi [119]. TCMBT is prepared by the reaction of the sodium salt of 2-mercaptobenzothiazole with chloromethylthiocyanate in alcohol solution (Figure 7.18). It was suggested that the activity of TCMBT as a biocide results from its slow hydrolysis to produce hydrogen cyanide, supported by an experiment in which a fifth of the TCMBT in a 0.025 molar aqueous solution, at 70°C, was hydrolysed to produce hydrogen cyanide in three hours [119].

The use of borates has a long history. Pre-1960, 8 tons of sodium tetraborate, dissolved in about 100 tons of heated water would need to be added to fuel storage tanks of volumes 10,000 m^3 to prevent the formation of a brown, gummy sludge [121]. The problems of making such large volumes of the borate solution in the vicinity of remote storage tanks promoted the Borax Company to investigate oil soluble borate esters that could be added to fuel. Heating sodium tetraborate in contact with excess ethylene glycol and removing the water of reaction, gave a range of water-soluble sodium glycolborate liquids that the patent recommended should be added to the fuel storage tank to form a layer on the bottom, which prevented formation of bacterial sludge [121]. However, it was preferred that fuel from a refinery should already be resistant to sludge formation and water-soluble borates were rapidly lost from a treated fuel into any adventitious water occurring along the fuel supply line.

Borate esters that are fuel soluble became the preferred solution. Cyclic esters, made by dehydrating boric acid combined with alkyl diols, such as 2-methyl-2,4-pentanediol, to give 2,2′-oxybis-1,3,2-dioxaborinanes (Figure 7.19), were found to have, low water solubility and are effective fuel-soluble biocides [122]. Dioxaborinane biocides, using the Biobor name, are the subject

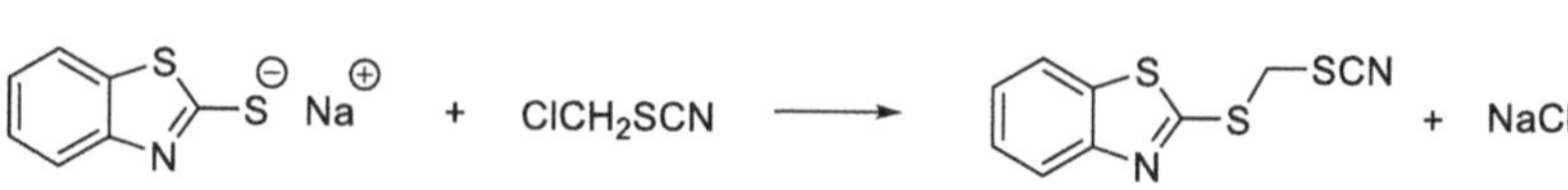

Figure 7.18 The Preparation of TCMBT.

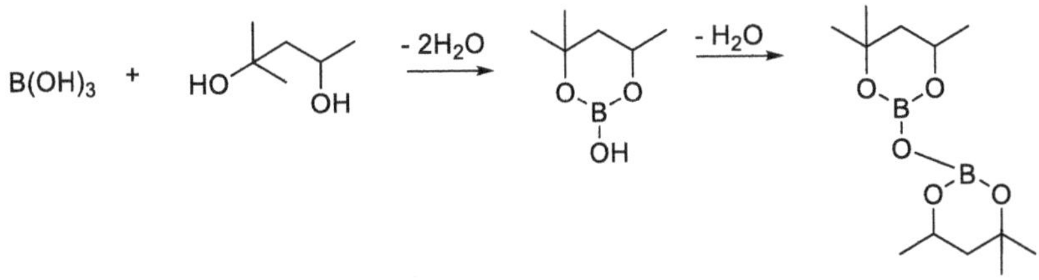

Figure 7.19 The Preparation of a Dioxaborinane Biocide.

of a Standard Oil of Indiana (Amoco) patent [115]: Biobor JF is described as a mixture of 2,2′-oxybis-(4,4,6-trimethyl-1,3,2-dioxaborinane) and 2,2′-(1-methyltrimethylenedioxy) bis(4-methyl-1,3,2-dioxa-borinane).

Kathon was a Dupont product but now trades under the Dow name after the DowDuPont merger in 2017. The KATHON™ trademark now belongs to Dow® Chemical[38], a company that produces a range of biocides [123], such as KATHON FP 1.5 which contains 1.5% of a mix of 5-chloro-2-methyl-4-isothazolin-3-one and 2-methyl-4-isothazolin-3-one – a combination known as KATHON 886F[12] – recommending 100–400 ppm of this product, which delivers 1.5–6 ppm of biocide, containing only 0.3–1.4 ppm of sulphur, to a fuel. This product was invented by Rohm and Haas chemists in the 1960s [124] and is one of many that they patented [125]; it has also been used in combination with other biocides, such as TCMBT [126]. The preparation [124] is an interesting piece of heterocyclic chemistry, involving the cyclization of 3,3′-dithiobis(N-methyl-propanamide) with a halogenating agent, such as sulphuryl chloride, which produces a mixture of the chlorinated and non-chlorinated isothiazolinones (Figure 7.20). The 3,3′-dithiobis(N-methyl-propanamide) is made by a series of standard chemical reactions: methyl acrylate with thioacetic acid gives the thioacetate by Michael addition, hydrolysis gives 3-thiopropionic acid and oxidation then gives the disulphide; amide formation can then be the simple process of making the acid chloride and reacting with methylamine (Figure 7.20).

Dow takes into account the USA requirement that fuel additives must have a maximum sulphur content of 15 ppm, the same as the fuels that they treat [123], while other countries only specify the ultra-low sulphur levels in the treated fuel. The isothiazoline biocides contain substantial proportions of sulphur, so Dow also provides non-sulphur products, such as FUELSAVER™, that rely upon the morpholine chemistries 4-(2-nitrobutyl)-morpholine and its diethyl derivative [123]. ONDEO-NALCO also provides non-sulphur biocides such as ONDEO-NALCO 303MC, whose MSDS describes it as containing an amine-substituted resin (probably a Mannich-type detergent, Chapter 3) and an imidazoline

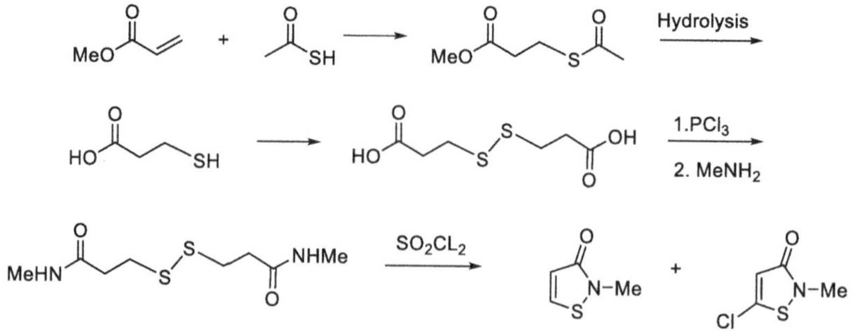

Figure 7.20 Preparation of KATHON Isothiazolinones.

derivative [127]. However, NALCO is also the applicant for several patents involving the use of MBT, which supports the view that such biocides are preferable when there is not a stringent sulphur limitation on the fuel additive.

There is another type of biocide in use, one that releases formaldehyde on hydrolysis in water [128]. These are oxazolidines and triazines. Triazines have already been mentioned with respect to inhibition of sulphur corrosion and the trapping hydrogen sulphide (section 5.7.2). The MSDSs of a group of commercial fuel biocides, advertised for use in diesel and biodiesel, show them to contain 3,3′-methylenebis(5-methyl-oxazolidine) [129].

Notes

1 See Chapter 3, for its preparation.

2 Different fuels are sent down the same pipeline in batches – a batch of gasoline may be pumped for several hours followed by a batch of jet fuel than a batch of diesel fuel. The separate batches are followed by a scheduling process and tracked at both ends of the pipeline.

3 **Needed to keep the particles separate and not coagulating to reform large masses.

4 The company that uses this type of process, Conoco, has several patents in the area. This is one example.

5 The terms 'static dissipater additive' and 'fuel conductivity improver' are used by Innospec, supplier of Stadis 450, the most widely used static dissipater [23]. Anti-static additive and conductivity improver are also used interchangeably [17], but 'static dissipater' is the term used in jet fuel specifications [24].

6 The Dutch company that made this patent application was a subsidiary of Royal Dutch Shell.

7 The calcium salt of 'Aerosol OT'. Aerosol OT is the sodium salt of the dioctyl ester of sulphosuccinic acid a widely used surfactant whose synthesis from

dioctyl maleate or fumarate and sodium bisulphite was first described in 1933 [130].

8 Estimating the molecular weight at 1135, as additive concentrations are given as gram atom Cr per litre.

9 Derived from Shell's wax cracking technology.

10 ASA-3 and Stadis 450 were both 'considered' by the US military after 1978 [30].

11 The patent also uses and claims a 1-decene/sulphur dioxide polymer that also contains 4.5 mole% maleic anhydride; polysulphone copolymers containing small proportions of an acid-bearing monomer appear in patents from time to time, so may also have been in use.

12 Sulphones of branched n-alkanes and the vinyl derivatives, such as alkyl methacrylates, have much lower ceiling temperatures.

13 Coco, from coconut oil, is mostly a mix of dodecyl and tetradecyl alkyls.

14 DuPont is the originator of the relevant patents but divested production to Octel in 1994. Innospec, the reformed Octel Corporation, is currently the manufacturer of Stadis 450. Baker Hughes, also provides this as a diesel additive under the trade name Tolad 3512.

15 The patent data used was for detergent A in jet fuel and detergent B in ULSD.

16 Additive package containing PIBSA-PAM detergent, lubricity improver, demulsifier and cold flow improver.

17 'Bonding' is the term used for electrically connecting different parts of a fuel handling system to enable charge to be quickly neutralised, for example, the bodies might be a fuel tanker that is bonded to ground, to the metal nozzle of the hose and to the receiving tank.

18 Taken from patents referred to in this chapter.

19 This was the lowest value of a treated gasoline [12].

20 Possibly from the oxidative step of the Merox process used to remove mercaptans from jet fuel. It involves oxidation with caustic soda, air and a cobalt phthalocyanine disulphonate catalyst [46,131]. Alkyl benzene-sulphonates are by-products.

21 Even the minimum of water defined by specifications, 0.02% in EN590, would provide one molecule of water per sulphone groups in 0.2% of the decene-polysulphone (repeat unit molecular weight is 11× that of water) and it is not known how much water the static dissipater contains.

22 Based on Crawford et al.'s chapter 6 of 'Chemistry and Technology of Lubricants' [49] and Callaghan's chapter 2 of 'Defoaming' [58].

23 In fuel, for surfactant read other fuel additives.

24 Large amounts of water are pumped into oil bearing rock for tertiary oil recovery processes and fracking.

25 Steam is injected into petroleum liquids in various parts of the distillation pipestills to aid the initial boiling and to remove lower boiling fractions from the side-streams, improving fractionation efficiency.

26 Such as oxidation products as alcohols and carboxylic acids and naturally occurring nitrogen heterocycles.
27 Molecular weight of polyalkylene glycols for such compounds appears to be 3,500 [77].
28 Credited as also being OndeoNalco Energy Services.
29 Patent description suggests the C24 branched chain because such an alkylbenzene-sulphonic acid is available, the Ca or Mg salt of which is used as a lubricating oil additive [132,133].
30 Details of multiple repetitions and proton transfers have been taken as read.
31 Hydrophile Lipophile Balance values are from 0, fully water soluble, to 20, water insoluble but hydrocarbon soluble.
32 This would be the highly branched octamer of propylene.
33 'Recognized world leader – – – for aircraft piston engines' [95].
34 General reading around this subject suggests that there are many routes to accidental water contamination while refuelling.
35 Air Accident Investigation Branch.
36 Fuel saturated with water at ambient temperature in Singapore could shed all of this water by the time it reaches Heathrow, where the plane would be refuelled with fresh fuel, saturated with water at ambient temperature, and so water would accumulate until the sump is drained.
37 Ullage is the free space in the wing tanks above the fuel.
38 Dow and DuPont merged to form DowDuPont in 2017 and then split into three separate companies in 2019: Dow, DuPont and Corteva [134]; Dow retained the Biocide part of the business.

References

1 Cuenca, F.G., Marin, M.G., and Diaz, M.B.F. (2008). Energy-Savings modelling of oil pipelines that use drag-reducing additives. *Energy & Fuels* 22: 3293–3298.
2 Culter, J.D. and McClaffin, G.G. (1972). Method of friction fluid loss reduction in oleaginous fluids flowing through conduits. US Patent 3,692,676. Granted 16 September.
3 Mohitpour, M., Yoon, M.S., and Russell, J.H. (2012). Hydrocarbon transmission pipeline and storage systems. Design and Operation. American Society of Mechanical Engineers, ISBN 9780791860007.
4 Smith, K.W., Haynes, L.V., and Massouda, D.F. (1994). Solvent free oil soluble drag reducing polymer. US Patent 5,449,732. Filed 6 May.
5 Guan, Z., Liang, J., Liu, C. et al. (2008). *Research and Application of Drag Reducer for Oil Products.* Calgary: American Society of Mechanical Engineers. Proceedings of IPC 2008-64230. 7th International Pipeline Conference. ISBN 978079184859.

6 Nesyn, G.V., Konovalov, K.B., Vetrova, O.V. et al. (2015). Laboratory evaluation of drag reduction additives effectiveness. Elsevier B.V. *Procedia Chemistry* 15: 371–377.

7 Nesyn, G.V., Manzhai, V.N., Suleimanova, Y.V. et al. (2012). *Polymer Drag Reducing Agents for Transportation of Hydrocarbon Liquids*. Pleiades Publishing Inc. Polymer Science, Ser.A, Vol. 54, pp. 6167.

8 Mack, M.P. (1983). Polymerization process for drag reducing substances. US Patent 4,493,903., Filed 7 November.

9 Johnston, R.L. and Lee, Y.N. (1997). Nonaqueous drag reducing suspensions. US Patent 6,172,151. Filed 11 September.

10 Fink, J. (2011). Drag reducers. Petroleum Engineers' Guide to Oil Field Chemicals and Fluids. Elsevier Inc., Ch 12, 379–389.

11 Johnson, D.D. (1972). Antistatic additive compositions. US Patent 3,811,848. Filed 30 June.

12 Skei, T. and Douglas, J.E. (1959). Liquid hydrocarbon compositions. US Patent 3,013,868. Filed 27 May.

13 Bergazli, C. and Henry, C.P. (2006). Synergistic additive composition for petroleum fuels. WO 2008,054,368. Filed USA, 12 September.

14 Schield, J.A. (1996). Anti-static additives for hydrocarbons. US Patent 5,672,183. Filed 1 June.

15 Marshall, E.L. and Owen, K. (1995). *Motor Gasoline*. Society of Chemical Industry. ISBN 0854044094.

16 ATC. Fuel (2004). Fuel additives and the environment. Technical Committee of Petroleum Additive Manufacturers in Europe. 1 January. http://www.atc-europe.org/public/doc52.pdf (Cited: 1 October 2013).

17 Fuel Additives: Use and Benefits, ATC Doc. 113 (2013). Technical committee of petroleum additive manufacturers in Europe. https://www.atc-europe.org/public/Doc1132013-11-20.pdf (Cited: 12 July 2020).

18 Van-der-minne, J.L. and Jan-Hermanie, P.H. (1956). Volatile organic liquids of increased electrical conductivity. US Patent 3,012,969. Filed 15 November.

19 Van-der-minne, J.L., Douwes, C., and Jan-Hermanie, P.H. (1961). Liquid hydrocarbons with increased electrical conductivity. GB Patent 1,001,942. Filed UK, 22 November.

20 Flett, L.H. (1939). Dry cleaning composition. US Patent 2,290,870. Filed 14 January.

21 Owen, K., Coley, T.R., and Weaver, C.S. (1995). *Automotive Fuels Reference Book*, 2nd edn. SAE. ISBN 1560915897.

22 Westbrook, S.R. and Shah, R.J. (2003). *Fuels and Lubricants Handbook: Technology, Properties, Performance and Testing* (ed. G.E. Totten), ASTM International. ASTM manual series: MNL37WCD. ISBN 0803120966.

23 Innospec Inc (2017). Refinery specialties – Static dissipator additives. http://www.innospec.com/fuel-additives/refinery-pipeline/static-dissipators (Cited: 28 November 2017).

24 ExxonMobil Aviation International Ltd (2005). *World Jet Fuel Specifications with Avgas Supplement*. ExxonMobil Aviation. ExxonMobil Aviation International Ltd.

25 Home, Wiki, Switch Loading. Petroleum Equipment Institute. https://www.pei.org/wiki/switch-loading (Cited 31 January 2022)

26 Kulinowsky, A.M. and Brennan, T.J. (2007). Conductivity of middle distillate fuels with a combination of detergent and cold flow improver. US Patent 2008,256,849. Filed 19 April.

27 Civil Aviation Authority of New Zealand (2015). Aircraft fuelling – Fire prevention and safety measures 14 October https://www.caa.govt.nz/Advisory_Circulars/AC091_22.pdf (Cited: 14 January 2018).

28 American Petroleum Institute Publishing Services. (1998). API 2003: Protecting against ignitions arising out of static, lightening, and stray currents. 49 CFR 195.405(a). September. https://law.resource.org/pub/us/cfr/ibr/002/api.2003.1998.pdf (Cited: 12 January 2018).

29 Dacre, B. and AbiAoun, W.G. (1997). Effects of fuel components on the performance of conductivity improvers in hydrocarbons. Elsevier. *Journal of Electrostatics* 39: 89–110.

30 Dacre, B. and Hetherington, J.I. (1998). The effects of contaminants on the behaviour of conductivity improvers in hydrocarbons. Elsevier Science B.V. *Journal of Electrostatics* 45: 53–68.

31 Cook, R.E., Dainton, F.S., and Ivin, K.J. (1957). Effect of olefin structure on the ceiling temperature for olefin polysulfone formation. John Wiley & Sons, *Journal of Polymer Science A: Polymer Chemistry* 26: 351–364.

32 Spassky, N., Sepulchre, M., and Sigwalt, P. (1992). Polysulfones. In: *Handbook of Polymer Synthesis: Part B* (ed. H.R. Kricheldorf), 1038–1046. Marcel Dekker Inc. Ch 16.

33 Sutkowski, A.C. (2005). Conductivity improving additive for fuel oil compositions. EP 1,568,756. Filed Europe, 22 February.

34 Fuel Additive Technology/Conductivity Improvers https://www.bakerhughes.com/products-and-services/downstream-chemicals/fuel-additive-technology/conductivity-improvers.(Cited: 2 January 2018).

35 Schield, J.A. (2001). Anti-static additive compositions for hydrocarbon fuels. US Patent 2001,048,099. Filed 16 April.

36 Henry, C.P. (1974). Compositions of olefin-sulphur dioxide copolymers and polyamines as antistatic additives for hydrocarbon fuels. US Patent 3,917,466. Filed 29 October.

37 Henry, C.P. (2002). Anti-static additive compositions for hydrocarbon fuels. US Patent 2004,107,635. Filed 5 December.

38 Oppenlaender, K., Wegner, B., Barthold, K. et al. (1990). Reaction products of aminoalkylene polycarboxylic acids with secondary amines and middle distillate compositions containing the aforesaid. US Patent 5,071,445. Filed 18 May.

39 Nottes, G. (1971). Chromium acyl-aminobenzoates – As antistatics for hydrocarbons active in very low concentration. DE Patent 2,116,556. Filed Germany, 5 April.

40 Udoetok, E.S. and Nguyen, A.H. (2011). Grounding resistance for control of static electricity ignition hazards. Elsevier. *Journal of Electrostatics* 69: 23–29.

41 Roberts, R. (2011). Using physical and chemical properties to manage flammable liquid hazards. February http://www.roberts-roberts.com/documents/UsingPropertiestoManageFlammableLiquidHazards.pdf (Cited: 15 January 2018).

42 Process Safety Beacon, American Institute of Chemical Engineers (2007). Oil tanker fire caused by static discharge. December http://sache.org/beacon/files/2007/12/en/read/2007-12-Beacon-s.pdf (Cited: 27 July 2020).

43 Accident Investigation Summary (2011). osha.gov, Report ID 100232107, Accident 200232551. https://www.osha.gov/pls/imis/establishment.inspection_detail?id=315027870 (Cited: 15 January 2018).

44 Xu, W. and Angell, C.A. (2003). Solvent free electrolytes with aqueous-solution-like conductivities. October. *Science* 302 (5644): 422–425.

45 Smith, G. and Watts, D. (2015). Oil demulsification – Electrical methods. Society of Petroleum Engineers International. 6 July http://petrowiki.org/Oil_demulsification#Electrical_methods (Cited: 27 July 2020). Page ID: 9418.

46 Fahim, M.A., Al-Sahhaf, T.A., and Elkilani, A.S. (2010). Fundamentals of petroleum refining. Elsevier B.V.

47 Grabowski, W. and Haubrichs, R. (1999). Diesel fuel and lubricating oil anti foams and method of use. US Patent 6,221,815. Filed 8 June.

48 Pillon, L.Z. (2008). Interfacial properties of petroleum products. Taylor&Francis ISBN 9781420051001.

49 Crawford, J., Psaila, A., and Orszulik, S.T. (1997). Miscellaneous additives and vegetable oils. In: *Chemistry and Technology of Lubricants*, 2nd edn. (ed. R.M. Mortier and S.T. Orszulik), n. Springer Science+Business Media,Dortrecht. Ch 6, 183-202. ISBN 978940171-237.

50 Trautman, C.E. and Ambrose, H.A. (1944). Improvements in preventing of foaming of hydrocarbon oils. GB Patent 591,767. Filed UK, 15 March.

51 Grabowski, W. (1994). Efficient diesel fuel antifoams of low silicon content. US Patent 5,542,960. Filed 7 July.

52 Grabowski, W. and Haubrichs, R. (1998). Diesel fuel antifoam composition. US Patent 6,093,222. Filed 15 December.

53 Jeffrey, G.C. (1997). Foam reducing fuel additive. GB Patent 2,322,138. Filed UK, 17 February.

54 Morehouse, E.L. (1963). Siloxane-alcohol ethers. US Patent 3,381,019. Filed 27 August.

55 Adams, G. and Jones, M.A. (1986). Foam control. US Patent 4,690,688. Filed 4 February.

56 Brook, M.A. (2000). *Silicon in Organic, Organometallic and Polymer Chemistries*. John Wiley and Sons Inc. ISBN 0471196584.

57 Colas, A. (2018). Silicone chemistry overview – silicone tutorial. https://www.dow.com/content/dam/dcc/documents/en-us/mark-prod-info/95/95-726-01-xiameter-silicone-chemistry-overview.pdf?iframe=true (Cited: 31 July 2020).

58 Callaghan, I.C. (1993). Antifoams for non-aqueous systems in the oil industry. In: *Defoaming: Theory and Industrial Applications* (ed. P.R. Garrett), Marcel Dekker, New York. Ch 2. ISBN 0824787706.

59 Fuel Oil No. 2 Environment Canada, emergencies, science and technology division. http://www.etc-cte.ec.gc.ca/databases/Oilproperties/pdf/WEB_Fuel_Oil_No._2.pdf (Cited: 31 July 2020).

60 Hilberer, A. and Chao, S.H. (2014). Anti-foaming agents. [book auth.] Mark H.F. In: *Encylopedia of Polymer Science and Technology*, 4th edn. (ed., A. Seidel and M. Peterca), Vol. 1. 662–686. Wiley.

61 Richards, P. (2014). *Automotive Fuels Reference Book*, Third Edition. SAE International, Pennsylvania.

62 Gattuso, M.J. (1974). Additives for petroleum distillates. GB Patent 1,511,748. Filed UK, 31 July.

63 Richardson, W.L. (1970). Improved distillate fuel compositions. GB Patent 1,296,585. Filed UK, 3 November.

64 Biswell, C.B. (1955). Fuel oil compositions. US Patent 2,805,925. Filed 28 November.

65 Armour and company (1956). Improvements in or relating to fuel oil compositions. GB Patent 817,262. Filed UK, 26 January.

66 Zimmerman, A.A. and Vardi, J. (1987). Improved fuel composition. EP 234,753. Filed 28 January.

67 Jackisch, P.F. and Worrel, C.J. (1977). Gasoline compositions. US Patent 4,116,644. Filed 21 January.

68 Kelland, M.A. (2014). Demulsifiers. [book auth.] In: *Production Chemicals for the Oil and Gas Industry* (M.A. Kelland), Taylor and Francis group. Ch 11, 287-305. ISBN 9781439873793

69 Breen, P.J., Wasar, D.T., Kim, Y.H. et al. (1996). Emulsions and emulsion stability. In: *Emulsions and Emulsion Stability* (ed. J. Sjoblom), Marcel Dekker Inc. Ch 4, 237-286.

70 Vermeire, M.B. (2012). Everything you need to know about marine fuels. PDF4PRO.com. June (Cited: 12 June 2020).

71 EAZY Combustion Improver (2015). September https://www.eazychem.co.uk/eazy-combustion-improver (Cited: 13 July 2020).

72 Yang, M., Stewart, A.C. and Davies, G.A. (1996). Interactions between chemical additives and their effects on emulsion separation. Paper presented at the s, SPE Annual Technical Conference and Exhibition. 6-9 October, Denver, Colorado, 1996. Paper number SPE-36617-MS. https://doi.org/10.2118/36617-MS.

73 Smith, G. and Watts, D. (2015). Oil demulsification – Chemical methods. Petrowiki. 6 July https://petrowiki.org/index.php?title=Oil_demulsification&action=info (Cited: 1 August 2020). Page ID: 9418.

74 Sharples-Specialty-Company. (1921). Processes for resolving emulsions. GB Patent 180,447. Filed UK, 3 March.

75 Barnickel, W.S. (1923). Process for treating petroleum emulsions. GB Patent 225,617. Filed UK, 4 September.

76 Pena, A.A. (2004). *Dynamic Aspects of Emulsion Stability*. Ph.D.Thesis. Department of Chemical Engineering, Rice University. 207–238.

77 Oppenlaender, K., Liebold, G., and Buettner, E. (1971). Crude oil demulsifiers. US Patent 3,786,081. Filed 3 December.

78 Lordo, S.A. (2013). *Oil Production Facilities and Chemicals* (ed. H.N. Giles), Seattle: COQA. Proceedings Crude Oil Quality Association Meeting.

79 Fink, J. (2012). Chapter 23. Demulsifiers. In: *Petroleum Engineers' Guide to Oil Field Chemicals and Fluids* 705–725. Elsevier Inc. ISBN: 9870123838445

80 Maddox, J., Zimmermann, M.J., and Tuggle, R. (1974). Demulsification compositions. AU Patent 492,350. Filed Australia, 29 July.

81 Miller, C.O. (1969). Ester-containing composition. GB Patent 1,306,529. Filed UK, 12 May.

82 Moeller, A. (1937). Improvements to the breaking of emulsions. DE Patent 702,012. Filed Germany, 9 June.

83 DeGroote, M. and Keiser, B. (1948). Chemical manufacture. US Patent 2,499,368. Filed 16 February.

84 DeGroote, M. and Keiser, B. (1949). Process for breaking petroleum emulsions. US Patent 2,557,081. Filed 28 November.

85 Vardi, J. and Zimmerman, A.A. (1987). Improved fuel composition. EP 234,753. Filed Europe, 28 January.

86 Clayton, C.W. and Miller, D. (2007). Formulating fuel compositions. US Patent 2007,240,361. Filed 18 April.

87 DeGroote, M. and Shen, K.T. (1956). Certain polyepoxide-modified oxyalkylated phenol-aldehyde resins and method of making same. US Patent 2,854,428. Filed 15 June.

88 DEMTROL™ High-Performance Crude Oil Demulsifier Bases (2014). APRIL http://msdssearch.dow.com/PublishedLiteratureDOWCOM/dh_0904/0901b80380904d18.pdf?filepath=oilandgas/&fromPage=GetDoc (Cited: 3 August 2020).

89 Riff, I., Alonso, T., and vanVoorst, R. (2011). Process using bisphenol a aminated and alkoxylated derivative as demulsifier. US Patent 2013,237,641. Filed 15 November.

90 Al-Sabagh, A.M., Nasser, N.M., and El-Hamid, M.A. (2013). Investigation of kinetic and rheological properties for the demulsification process. Elsevier B.V. *Egyptian Journal of Petroleum* 22: 117–127.
91 Al-Sabagh, A.M., Hassan, M.E., Desouky, S.E.M. et al. (2016). Demulsification of W/O emulsion at petroleum field and reservoir conditions using some demulsifiers based on polyethylene and propylene oxides. Elsevier B.V. *Egyptian Journal of Petroleum* 25: 585–595.
92 Honnen, L.R. (1976). Fuel additive for distillate fuels. US Patent 4,128,403. Filed 29 June.
93 Colucci, W.J., Smith, I.L., and Aradi, A.A. (1997). Gasoline compositions containing ignition improvers. US Patent 5,782,937. Filed 19 May.
94 Bazzani, R.V., Bennett, P.J., Butler, G. et al. (2007). Fuel composition. US Patent 2008,289,998. Filed 31 October.
95 Continental Motors Group (2017). Excellence in piston engines. http://www.continentalmotors.aero/engines/gasoline-engines.aspx (Cited: 25 November 2017).
96 Garth, B.H. (1973). Process for improving anti-icing characteristic of gasoline. GB Patent 1,418,526. Filed UK, 12 December.
97 Polss, P. (1973). Gasoline. US Patent 3,873,278. Filed 29 November.
98 Andress, H.J. and Gee, P.Y.C. (1959). Glycine alkenyl succinamic acids in distillate fuels. US Patent 3,039,861. Filed 1 December.
99 Gasoline Aviation: Grades UL91, 100/130 and 100/130LL. (2015). Joint service designation AVGAS 100 and AVGAS 100LL. Ministry of Defence, Defence Standard 91-090 Issue 4. UK Defence Standardization.
100 Lovett, W.E., Walker, R.W., and Jordan, J.M. (1958). Anti-icing gasoline. US Patent 2,981,614. Filed 27 February.
101 Innospec (2017). Fuel system icing inhibitors. innospecinc.com. http://www.innospecinc.com/our-markets/fuel-specialties/refinery-specialties/fuel-system-icing-inhibitors (Cited: 3 August 2020).
102 Zabarnick, S., Adams, R., West, Z. et al. (2010). Compatilibility of DiEGME and TriEGME fuel system icing inhibitor additives with BMS 10-39 aircraft tank topcoat material. *American Chemical Society* Energy & Fuels. 24: 2614–2627.
103 Baena, S., Repetto, S.L., Lawson, C.P. et al. (2013). Behaviour of water in jet fuel. A Literature Review. Elsevier, July. *Progress in Aerospace Sciences* 60: 35–44.
104 Sleight, P.A. and Carter, R.D.G. (2010). Report on the accident to boeing 777-236ER, GYMMM, at London Heathrow Airport on 17 January 2008. Air Accidents Investigation Branch, Department of Transport. Crown Copyright Aircraft Accident Report 1/2010, EW/C2008/01/01.
105 U.S.Department of Transportation Federal Aviation Administration (2009). Pilot's handbook of aeronautical knowledge, 2009. ISBN: 9780160823671.

106 The Effect of Stadis 450 on MSEP Rating and Coalescence (1999). *Technical Basis of Re-Doping Turbine Fuels with Stadis 450*. Air Force Research Laboratory. Coordinating Research Council, . AFRL-PR-WP-TR-1999-2118. CRC report 601.

107 Filter Water Separators Hydro-carbon filtration and separation. http://www.hydro-carbon.nl/separation-of-water-from-hydrocarbons/filter-waterseparators-coalescers-separators (Cited: 3 August 2020).

108 Jaffer, J.H. (1961). Jet engine fuel. US Patent 3,103,101. Filed 3 February.

109 Klopov, B.N. and Plakhova, R.N. (1979). Rapid determination of content of anti-icing additives in jet fuels under airfield conditions. *Plenum Publishing Corporation* Chemistry & Technology of Fuels & Oils. 15 (1): 68–70.

110 Shotton, J.A. (1960). Additives and fuels containing the same. GB Patent 906,891. Filed UK, 30 December.

111 Taylor, K.B. (2010). *Comparative Study of Alternative Fuel Icing Inhibitor Additive Properties and Chemical Analysis of Metal Speciation in Aviation Fuels*. The School of Engineering, University of Dayton. Thesis.

112 Biocides for Fuels and Lubricants – Application and Regulations (2016). *Lubrizol*. Stuttgart: s.n. UNITI Mineral Oil Technology Congress.

113 Racor Additives (2017). Racor diesel fuel biocide, product bulletin 7642. 11 May http://www.parker.com/parkerimages/Racor/7642(DieselBiocide112005).pdf (Cited: 4 August 2020).

114 O'Reilly, K.T., Moir, M.E., and O'Rear, D.J. (2002). Fuel blend having biocide activity. GB Patent 2,383,798. Filed UK, 7 October.

115 Amick, J.W. (1974). Oil-soluble biocide combination for distillate fuels. US Patent 4,166,725. Filed 15 October.

116 Todd, T.A. and Webber, R.F. (2003). Fuel additive systems. US Patent 2004,060,229. Filed 17 September.

117 Matt, J., Hunter, E.W., and Goretta, L.A. (1966). Method for preparing methylene bis (thiocyanate). US Patent 3,507,901. Filed 23 December.

118 McDermott, A.R. (1977). Method for preventing microorganism induced corrosion of hydrocarbon liquid storage tanks. US Patent 4,086,066. Filed 22 February.

119 Buckman, S.J., Buckman, J.D., Pera, J.D. et al. (1968). Compositions containing an organic thiocyanate and 2-hydroxyalkyl esters of organic thiosulfonic acids and processes of utilizing the same. US Patent 3,639,604. Filed 8 March.

120 Donofrio, D.K. and Whitekettle, W.K. (1989). Biocidal compositions and use thereof containing a synergistic mixture of 2-(Decylthio)ethanamine hydrochloride and a mixture of 5-Chloro-2-Methyl-4-Isothiazolin-3-one and 2-Methyl-4-Isothiazolin-3-one. CA Patent 1,325,950. Filed Canada, 6 September.

121 Willcockson, G.W. (1962). Method for inhibiting bacterial activity in petroleum. CA Patent 642,919. Published Canada, June.

122 Killian, L.N. and Degray, R.J. (1962). Preservation of kerosene during storage. CA Patent 648,972. Published Canada, September.

123 Browne, B.A. and Hermsen, N.Biocide selection guide for fuel applications. www.dowmicrobialcontrol.com (Cited: 25 February 2018). Changed to Dupont Microbial Control, combined with iff through microbialcontrolsolutions.com/our-markets/oil-and-gas/fuel. (Cited 23 April 2022).

124 Lewis, S.L. and Miller, G.A. (1969). Process for the preparation of 3-Isothiazolones and 3-Hydroxyisothiazoles. US Patent 3,849,430. Filed 14 June.

125 Lewis, S.N.A., Miller, G.A., and Law, A.B. (1969). 3-Isothiazolones. US patent 3,761,488. Filed 25 June.

126 LaMarre, T.M., Martin, C.H., and Wilharm, M.T. (1985). Synergistic Biocide of 2-(Thiocyanomethylthio) Benzothiazole with a Mixture of 5-Chloro-2-Methyl-4-Isothiazolin-3-one and 2-Methyl-4-Isothiazolin-3-one. US Patent 4,595,691. Filed 22 July.

127 NALCO (2010). *Safety Data Sheet NALCO® 303MC*. Nalco Company.

128 Appiah-Amponsah, E., Dyer, D., and Zegalia, K. (2016). Method for measuring formaldehyde in solutions containing certain formaldehyde-releasing biocides. US Patent 2017,052,155. Filed 11 August.

129 Wynn's (2020). Fuel biocide disinfects small fuel tanks of trucks, boats, vehicles and stationary. wynns.eu. https://www.wynns.eu/product/fuel-biocide-3/#:~:text=Disinfectssmallfueltanksof,Preventsmicrobia lre-growth (Cited: 5 August 2020).

130 Jaeger, A.O. (1933). Esters of sulphodicarboxylic acids. US Patent 2,028,091. Filed 28 July.

131 Ganguly, S.K., Das, G., Kumar, G. et al. (2013). Catalytic oxidation of mercaptans in light oil sweetening: kinetics and reactor design. *Chemical Engineering Transactions* 32: 661–666.

132 Marsh, J.F. and Cleverley, J.A. (1979). A process for the production of basic magnesium sulphonates and the products obtained by this process. EP 015,341. Filed Europe, 7 December.

133 Jao, T.C. and Passut, C.A. (2006). Detergent formulation in lubricants. In: *Handbook of Detergents, Part D: Formulation* (ed. M.S. Showell), 437–471. Taylor & Francis group. Ch 15.

134 Root, A. (2019). *DowDuPont Is Splitting into 3 Companies*. Here's Everything You Need to Know. BARRON'S. 30 April.

Index

Note: Page numbers in **bold** refers to tables.

Fuel Additives: Chemistry and Technology, First Edition. Robert D. Tack.

o

p

Printed and bound by CPI Group (UK) Ltd, Croydon, CR0 4YY
16/08/2022
03141868-0001